C0-ALU-557

Early Life History of Fish

Early Life History of Fish

An energetics approach

Ewa Kamler

Institute of Ecology, Polish Academy of Sciences
Dziekanów Leśny, 05–092 Łomianki, Poland

CHAPMAN & HALL

London · New York · Tokyo · Melbourne · Madras

Published by Chapman & Hall, 2–6 Boundary Row, London SE1 8HN

Chapman & Hall, 2–6 Boundary Row, London SE1 8HN, UK

Van Nostrand Reinhold Inc., 115 5th Avenue, New York NY10003, USA

Chapman & Hall Japan, Thomson Publishing Japan, Hirakawacho Nemoto Building, 7F, 1–7–11 Hirakawa-cho, Chiyoda-ku, Tokyo 102, Japan.

Chapman & Hall Australia, Thomas Nelson Australia, 102 Dodds Street, South Melbourne, Victoria 3205, Australia

Chapman & Hall India, R. Seshadri, 32 Second Main Road, CIT East, Madras 600 035, India

First edition 1992

© 1992 Chapman & Hall

Typeset in 10/12pt Photina by Interprint Limited, Malta
Printed in Great Britain by T.J. Press (Padstow) Ltd., Padstow, Cornwall

ISBN 0 412 33710 X   0 442 31428 0 (USA)

Apart from any fair dealing for the purposes of research or private study, or criticism or review, as permitted under the UK Copyright Designs and Patents Act, 1988, this publication may not be reproduced, stored, or transmitted, in any form or by any means, without the prior permission in writing of the publishers, with the exception of Polish language rights, or in the case of reprographic reproduction only in accordance with the terms of the licences issued by the Copyright Licensing Agency in the UK, or in accordance with the terms of licences issued by the appropriate Reproduction Rights Organization outside the UK. Enquiries concerning reproduction outside the terms stated here should be sent to the publishers at the London address printed on this page.
    The publisher makes no representation, express or implied, with regard to the accuracy of the information contained in this book and cannot accept any legal responsibility or liability for any errors or omissions that may be made.

A catalogue record for this book is available from the British Library

Library of Congress Cataloging-in-Publication data

Kamler, Ewa, 1937–
Early life history of fish: an energetics approach/Ewa Kamler.
– 1st ed.
p. cm. – (Chapman & Hall fish and fisheries series; 4) Includes bibliographical references and index.
ISBN 0–442–31428–0
1. Fishes–Development. 2. Fishes–Larvae. 3. Fishes–Food. 4. Bioenergetics.
I. Title. II. Series: Fish and fisheries series; 4.
QL639.25.K36 1991   91–15371
597′.03–dc20   CIP

QL
639
.25
.K36
1992

This work is dedicated to Jacek who tolerated my devotion to baby fish and helped me to rear our sons

Contents

Acknowledgements

Dr A. Duncan and Professor R. Z. Klekowski taught me bioenergetics and encouraged me to write this book. I am enormously indebted to them for their support.

I am grateful to Dr O. Matlak, Dr H. Żuromska and Dr T. Kato who shared their experience during our common work on carp, vendace and rainbow trout, respectively. Over the years it has been my good fortune to have many stimulating discussions with Professor T. Backiel, Professor Z. Fischer, Dr L. Horoszewicz, Dr M. Szlamińska and Dr E. Urban-Jezierska.

My work at the Nikko Branch, National Research Institute of Aquaculture was supported by the bilateral exchange programme between the Japan Society for the Promotion of Science and the Polish Academy of Sciences. The guidance of Professor T. Miura is heartily appreciated.

During the past twenty years I have performed most of my work as a guest of the Experimental Fish Culture Farm, Polish Academy of Sciences at Gołysz. I wish to thank Director J. Broda and Dr J. Szumiec. We have worked together with some members of the Gołysz staff: Dr M. Lewkowicz, Dr S. Lewkowicz, Mrs B. Barska, Mrs M. Jakubas, Mr M. Kuczyński and Mr L. A. Stanny: cooperation with them was a real pleasure. The remaining colleagues, especially Mrs M. Janik, Mrs W. Olech, Mrs W. Pilarczyk, Mr J. Adamek, Mr H. Białowas, Mr T. Olech, Dr A. Pilarczyk and Mr S. Zarzycki, extended to me their experience, space and equipment. They helped me to overcome difficulties and shared my pleasure at any success. I thank them very much.

Thanks are due to Dr A. Dowgiałło for his advice on chemical techniques, to Professor M. I. Shatunovskij for corrections of fish scientific names, and to Dr T. Wierzbowska and Dr J. Uchmański for checking my mathematics; they helped me to avoid the blunders one is liable to make in unfamiliar areas of research. Dr M. C. Swift and Dr D. Cacamise corrected my English; the major improvements were done with an admirable competence and accuracy by Dr C. Hollingworth, to whom I am greatly indebted. For assistance in the hatchery and laboratory I thank Mrs H. Irie, Miss H. Kamińska, Mrs E. Karczewska, Mrs M. Tezuka and Mr Y. Fukuda. In the preparation of the manuscript I was greatly aided by Mrs E. Krajczyńska and Mr R. Dąbrowski who contributed heroic efforts in word processing the reference list and in proof-reading.

My sincere appreciation is directed to all the friends who generously provided me with literature during the last stages of the writing in Algeria.

I express my gratitude to the authors and publishers who have graciously allowed me to reuse published figures: Dr K. Raciborski, Dr M. Zalewski, Polish Scientific Publishers (PWN) and *Acta Universitatis Lodziensis*.

I thank three anonymous reviewers who read the manuscript and offered many constructive comments.

This work was in part financially supported by the Polish Academy of Sciences, within Projects MR II/15 and C.P.B.P.04.09.

Ewa Kamler

Series foreword

Among the fishes, a remarkably wide range of biological adaptations to diverse habitats has evolved. As well as living in the conventional habitats of lakes, ponds, rivers, rock pools and the open sea, fish have solved the problems of life in deserts, in the deep sea, in the cold Antarctic, and in warm waters of high alkalinity or of low oxygen. Along with these adaptations, we find the most impressive specializations of morphology, physiology and behaviour. For example we can marvel at the high-speed swimming of the marlins, sailfish and warm-blooded tunas, air breathing in catfish and lungfish, parental care in the mouth-brooding cichlids and viviparity in many sharks and toothcarps.

Moreover, fish are of considerable importance to the survival of the human species in the form of nutritious and delicious food of numerous kinds. Rational exploitation and management of our global stocks of fishes must rely upon a detailed and precise insight of their biology.

The Chapman and Hall *Fish and Fisheries Series* aims to present timely volumes reviewing important aspects of fish biology. Most volumes will be of interest to research workers in biology, zoology, ecology and physiology, but an additional aim is for the books to be accessible to a wide spectrum of non-specialist readers ranging from undergraduates and postgraduates to those with an interest in industrial and commercial aspects of fish and fisheries.

This book by Ewa Kamler is the fourth volume in the *Fish and Fisheries Series* and reviews the early life of fishes from an energetics perspective. Periods during early life are often crucial in determining cohort strength and hence recruitment to fish populations exploited by man. These critical periods entail complex and controversial interactions between the ecology, behaviour and physiology of the baby fishes to which Dr Kamler is devoted. The natural mechanisms which generate the volatility of the world's major commercial fisheries cannot be fully understood without considering the budget of energy within and between organisms.

The theme of Dr Kamler's book follows the life-history sequence of gonad maturation, factors affecting egg quality and hatching success, early free-swimming life while relying on nutrition from the yolk sac, the switch from this endogenous nutrition to searching for and eating food from the environment, and problems of the early free-living stage. The book finishes with a section on the feeding of fish larvae in aquaculture.

The author is not a native English speaker and we have to thank our copy-editor extraordinaire Charles Hollingworth for bringing the final touches of eloquence to the manuscript. With a large number of references to the extensive world literature in this field, I hope that this book will become a valuable reference volume as well as providing a stimulating perspective on the critical early life of fishes.

Dr Tony J. Pitcher
Editor, Chapman and Hall Fish and Fisheries Series
Special Research Fellow, Imperial College, London

CHAPMAN & HALL FISH AND FISHERIES SERIES

1. Ecology of Teleost Fishes
   R. J. Wootton
2. Cichlid Fishes: Behavior, ecology and evolution
   Edited by M. H. A. Keenleyside
3. Cyprinid Fishes: Systematics, biology and exploitation
   Edited by I. J. Winfield and J. S. Nelson
4. Early Life History of Fish: An energetics approach
   E. Kamler
5. Fisheries Acoustics
   D. N. MacLennan and E. J. Simmonds
6. Fish Chemoreception
   Edited by T. Hara

Forthcoming Titles

Behaviour of Teleost Fishes, 2nd edn
Edited by T. J. Pitcher
Fish Swimming
J. Videler
Sea Bass
G. Pickett and M. Pawson
Fisheries Ecology, 2nd edn
Edited by T. J. Pitcher and P. Hart

Symbols
and abbreviations

A	assimilation (in terms of energy, $A = P + R$; in terms of protein, $A = P + U$)
c	as a subscript, denotes parameters of a cumulative budget (e.g. P_c, cumulative production)
C	consumption of food
C_Y	yolk absorbed
$C.e.$	caloric equivalent (energy content in an organism) ($J\,indiv^{-1}$)
$C.e._0$	initial energy content of an egg at fertilization ($J\,egg^{-1}$)
CF	condition factor ($CF = 100 \times W / L^3$)
CINRE	cumulative index of net reproductive effort ($CINRE = P_{rc} \times 100 / P_c$) (%)
CV	coefficient of variation ($SD \times 100 / mean$)
d	day
D	digestible energy
D_L	development (complexity) of a shoreline
$D°$	day-degrees ($D° = \tau t$)
$D°_{eff}$	effective day-degrees ($D°_{eff} = \tau(t - t_0)$)
E	eyeing
$e.c.$	energy in the egg capsule (chorion)
$E.n.$	absolute fecundity (number of eggs laid per female)
F	faeces excreted
Fe	fertilization (or other triggering of cell division, i.e. activation of development)
G	specific growth rate ($G = (\ln W_2 - \ln W_1) / (\tau_2 - \tau_1))(d^{-1})$
$G.d.$	energy in the germinal disc
GSI	gonadosomatic index ($GSI = gonad\,wt \times 100 / total\,body\,wt$)
h	caloric value of matter (J per unit wt)
H	hatching

i	as a subscript, denotes parameters of an instantaneous budget (e.g. P_i, instantaneous production)
IINRE	instantaneous index of net reproductive effort (IINRE $= P_{ri} \times 100 / P_i$) (%)
K_1	gross conversion efficiency, the efficiency of consumed energy (or matter) utilization for growth ($K_1 = P \times 100 / C$) (%)
K_2	net conversion efficiency, the efficiency of assimilated energy (or matter) utilization for growth ($K_2 = P \times 100 / A$) (%)
K_A	assimilation efficiency ($K_A = A \times 100 / C$) (%)
K_e	efficiency of egg energy (or matter) transformation into body tissue ($K_e = P_c \times 100 / C.e._0$) (%)
L	length
L_f	fork length
L_s	standard length (body length)
L_t	total length
L_∞	length at infinite age (parameter of the Bertalanffy growth equation)
P	production ($P = P_g + P_r$)
P_g	somatic growth
P_r	reproductive growth
PNR	point of no return (see page 177)
Q_1	van't Hoff temperature coefficient for a temperature difference of 1 °C ($Q_1 = Q_{10}^{0.1}$)
Q_{10}	as above, for a difference of 10 °C
R	metabolism
R_a	activity metabolism (energy used for locomotory activity)
R_f	feeding metabolism
R_r	resting metabolism (recorded in unfed animals at rest) (also called standard metabolism)
R_t	metabolic rate at temperature t (see Equation 4.31)
RCF	relative condition factor ($RCF = W \times 100 / L^b$)
Re	completion of yolk sac resorption
s	swimming speed
S	free-swimming, initiation of external feeding
Sa	surface area of a water body
SCP	single-cell protein
SD	standard deviation
SDA	specific dynamic action (costs associated with food conversion)
t	temperature (°C)
t_{eff}	effective temperature ($t_{eff} = t - t_0$) (°C)
t_0	threshold temperature, at which development is theoretically arrested (°C)
U	nonfaecal excretion (urinary and other excreted wastes)

V	developmental rate (rate at which the subsequent developmental events appear) (τ^{-1})
W	weight
W_d	dry weight
W_w	wet weight
Y	yolk
Y_0	initial yolk (at fertilization)
$Y.r.$	remaining yolk
λ	growth coefficient (see Equation 4.25)
τ	developmental time (age)
τ_a	time from fertilization to the establishment of embryonic axis
τ_m	duration of a single mitotic cycle in early cleavage
τ_0	age zero, when the life cycle begins
τ_{opt}	optimal age, at which females produce the largest eggs

Chapter one

Introduction

It is a well-known phenomenon that all changes in a population depend on reproduction, growth and mortality. The last two processes are most intensely manifested in the early developmental stages (Hjort, 1914; Allen, 1951; Marr, 1956; Bernatowicz et al., 1975; Mahon et al., 1979; Eldridge et al., 1981a, b; and many others). Planktivorous fish larvae are among the most efficient predators at transferring biomass from one particle size to a larger size (Borgmann and Ralph, 1985). Therefore the early life history of fish was recognized as one of the key issues in fishery science (Azeta, 1981; Kernehan et al., 1981; Sherman and Lasker, 1981; Nielsen et al., 1986; Viljanen, 1988). Reproductive strategies and adaptations for early development define the ecological guilds of fishes proposed by Balon (1975a). Evaluating studies on early life history of fishes, Alderdice (1985) found that they often receive less recognition and support than they merit, and drew attention to the multiple advantages of such studies when they are associated with practical goals. He produced a list of ten activities by which these goals could be achieved; they included study of the bioenergetics of growth and development, and of the physiological ecology of early life stages.

In the last decade a number of papers have appeared pertaining to ecology energetics and the transformation of matter in fishes (reviews: Elliott, 1979, 1982; Brett, 1979; Brett and Groves, 1979; Shatunovskij, 1980; Fischer, 1983; Tytler and Calow, 1985; Gershanovich et al., 1987). They were concerned with postlarval developmental periods: earlier development has been less attractive for such studies because the processes occur rapidly, are complicated, and the fish are small. Information on energy and matter transformations during the early development of fish is therefore far from complete.

In studies aimed at optimization of egg incubation in hatcheries and rearing of fish larvae in aquaculture, survival and growth are most often evaluated. Morphometric, histological, physiological or bioenergetical investigations, or studies of the causes of size variability, are rarely employed. These methods can provide information on mechanisms that produce the observed effects. Such an approach can be more fruitful, and less expensive, than blindly manipulating numerous variables in search of optimal combinations (e.g. by composing artificial diets for fish larvae).

The topics discussed in this book are energy and matter transformation during fish gonad formation, the endogenous feeding period, and the larval period, as well as energetical aspects of the feeding of fish larvae in aquaculture. Consideration is given to the specificity of these transformations, as compared with analogous processes taking place during fish postlarval development on the one hand and during reproduction and early development of other animals, mostly aquatic invertebrates, on the other hand. Attention is paid to the effects of endogenous and external factors. Methods are evaluated when necessary. An attempt has been made to cover both Western and Eastern literature.

Chapter two

Gonad formation

Four stages of oocyte growth were recognized by Wallace and Selman (1981): first, primary growth; second, formation of yolk vesicles from which cortical alveoli subsequently develop; third, vitellogenesis, during which hepatically derived vitellogenin is transferred via maternal blood into yolk spheres or granules; fourth, maturation, which is accompanied by water uptake and followed by ovulation.

2.1 SEASONAL CHANGES OF MAIN CONSTITUENTS

Seasonal changes in body growth and ovary formation are shown in Fig. 2.1, using a polycyclic, single-spawning, non-migratory planktivorous coregonid, *Coregonus albula* (vendace) as an example. In spring, when temperature and zooplankton biomass in the lake increase, the body size of this species also increases (Fig. 2.1(A)). The increase begins slowly as the recovery after winter depletion of body reserves occurs. Later, body size increases rapidly (Fig. 2.1(A)); according to Marciak (1962) about 70% of annual growth in length and weight of vendace occurs in June and July. In an analogous period, Dabrowski (1982a) reported a rapid growth of fillet mass and a considerable increase (from about 3% to about 20% of dry matter) in lipid storage in muscles of a related species, *C. pollan*. However, growth of ovaries in *C. albula* (Fig. 2.1(B)) is negligible until June. Vendace body size stops increasing in July/August (Fig. 2.1(A)).

When ovarian growth rate increases (July/August, Fig. 2.1(B)), somatic growth decreases (Fig. 2.1(A)); meanwhile a rapid accumulation of lipids occurs (Fig. 2.1(C)), reaching the very high value of 41% of dry matter. Similarly, during early maturation stages of *Anguilla anguilla* (change from yellow phase, GSI < 0.7, to silver, GSI > 1.3), Lewander *et al.* (1974) found an increase of lipids from 11.3% to as much as 25.7% of wet matter. During this period the percentage of proteins in vendace ovaries does not change (Fig. 2.1(D)).

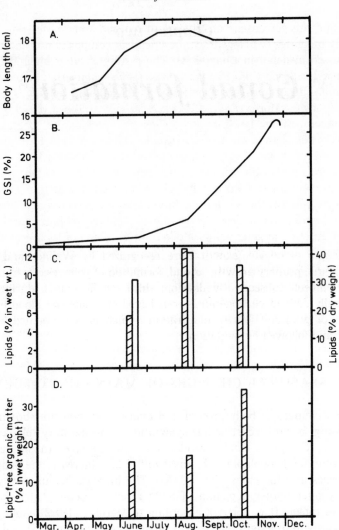

Fig. 2.1 Seasonal changes in body growth and ovary formation in *Coregonus albula*. (A) Body length (age group 2, data from Ciepielewski, 1971); (B) gonadosomatic index (data from Backiel, 1952); (C) lipids (%) in ovarian wet matter (hatched bars) and in ovarian dry matter (open bars; note different vertical scale) (data from Lizenko *et al.*, 1973); (D) lipid-free dry matter (mostly protein) expressed as % ovarian wet matter (data from Lizenko *et al.*, 1973).

The clear acceleration of relative growth in vendace ovarian weight (Fig. 2.1(B)) does not begin until body growth is complete (Fig. 2.1(A)); ovarian weight increases greatly up to the spawning period (in Poland, November/December). At this time the lipid fraction of ovarian dry matter (Fig. 2.1(C)) decreases to about 28%, typical for mature eggs of salmonid fishes, and the percentage of proteins in ovarian wet matter (Fig. 2.1(D)) abruptly increases (vitellogenesis; the chemical composition of mature eggs is discussed in detail in Chapter 3). The results of other chemical studies (Suyama, 1958a on *Oncorhynchus mykiss* (*Salmo gairdneri*)*; Stepanova and Tyutyunik, 1973 on *Ctenopharyngodon idella*; Chechenkov, 1973 on *Coregonus albula*; Moroz et al., 1973 on *Acipenser*; Kulikova, 1973 on *Gobius melanostomus*; Craig, 1977 on *Perca fluviatilis*; Boëtius and Boëtius, 1980 and Epler et al., 1981b on *Anguilla anguilla*; Dąbrowski, 1982a on *Coregonus pollan*; review: Love, 1970), calorimetric studies (Nishiyama, 1970 on *Oncorhynchus nerka*) and histological studies (Backiel, 1952 on *Coregonus albula*) also indicate that during the period of intensive growth of fish ovaries and oocyte volume, there is a rapid increase in the percentage of proteins in ovarian matter and a decrease in the percentage of lipids (and thus a decrease in the energetic value of dry matter). Some of these relationships are given in Table 2.1. Decreased fertilizability of *Ctenopharyngodon idella* eggs containing an augmented percentage of lipids is attributed by Stepanova and Tyutyunik (1973) to incomplete maturation of the eggs. These alterations of chemical composition, if expressed in terms of concentration, mean only that during vitellogenesis the rate of lipid accumulation in ovaries is slower than that of protein (Bagirova, 1976; Ridelman et al., 1984). The total content of those compounds in ovaries (Craig, 1977; Dąbrowski, 1982a) and the energy content per ovary (Dąbrowski, 1983) increase very rapidly during this period.

Changes in ovarian lipid content correspond to changes in somatic body lipid content. Shulman (1972) reported on data by Khromov for a poorly feeding population of *Sardinella aurita* during the period of gonad formation.

Table 2.1 Relationships between fish ovarian dry matter properties (P, % protein; L, % lipids; H, caloric value, $J\,mg^{-1}$) and gonadosomatic index (GSI, gonad weight $\times$ 100 / total body weight) or egg diameter (ED, mm)

Equation*	n	Correlation coefficient†
1. $P = 12.65 \times GSI^{0.296}$	29	0.617***
2. $L = 103.2 \times GSI^{-0.218}$	29	−0.565**
3. $L = 87.75 \times GSI^{-0.354}$	5	−0.998***
4. $H = 29.23 - 0.161\ GSI$	28	−0.774***
5. $H = 30.39 - 0.559\ ED$	25	−0.751***

*1,2: *Anguilla anguilla*, GSI = 12.5–46.8%, source Epler et al., 1981b; 3: *Coregonus pollan*, GSI c. 5–20%, recalculated from figure 1 and table 3 in Dąbrowski, 1982a; 4,5: *Oncorhynchus nerka*, source Nishiyama, 1970.
†Significance levels: **, $P < 0.01$; ***, $P < 0.001$.

---

* The American Fisheries Society's Committee on Names of Fishes has recently adopted *Oncorhynchus mykiss* as the scientific name of *Salmo gairdneri* (Kendall, 1988).

An especially large decrease in lipid share in wet body weight of this fish occurred when its gonads were at initial maturity stage II (Mejen's scale). This decrease slowed at passage from stage II to stage III and stopped during the period of intense growth of oocytes – stages III and IV. The main source of energy for coregonid gonad formation is muscle; this is illustrated in very informative diagrams by Dąbrowski (1982a, 1983).

Thus in the reproductive cycle of some fish species one can distinguish two alternate (although sometimes overlapping) phases, in which priority is given either to body growth or to ovarian growth; at the beginning of the second phase, lipids are accumulated more rapidly, and at its end, proteins.

In some species such a cycle occurs only once before death; these are **monocyclic** or **semelparous** species (*sensu* Cole, 1954) such as *Anguilla* or *Oncorhynchus*. In other species (**polycyclic** or **iteroparous**), this cycle occurs repeatedly every few years or each year. However, in fish with continuous oocyte growth (**serial spawners** e.g. *Seriphus politus*, DeMartini and Fountain, 1981 and *Tinca tinca*, Horoszewicz, 1981a, b) the picture is not so clear. Females of *T. tinca* cultured under the same conditions can exhibit different reproductive rhythms. In addition, they can resorb mature and maturing oocytes, directing the recovered energy to other needs, with or without inhibiting vitellogenesis of the developing oocytes (Epler *et al.*, 1981a). The above three temporal patterns of reproduction are related to the patterns of ovarian maturation (Wallace and Selman, 1981): synchronous, group-synchronous and asynchronous.

The main trends of changes in concentration of important chemical components during fish ovary formation are therefore well known. However, until recently we could only study the dynamics of changes occurring in single oocytes during their development indirectly, because of difficulties in isolating immature oocytes from the remaining components of an ovary (e.g. connective tissue). An oocyte's weight could be estimated from its volume, but information about its chemical composition could only be inferred from qualitative histochemical investigations or from quantitative chemical determinations carried out on the whole ovary, of which 77–96% is eggs (Opuszyński, 1983).

Ozernyuk (1970) devised a method of mass isolation of oocytes from *Misgurnus fossilis* ovaries; fish oocytes at different maturation stages can now be separately examined. At the stage of slow ovarian growth, oocyte size (wet and dry weight and protein content per oocyte) increased about two-fold, and oxygen consumption per oocyte and per hour increased more than four fold. The subsequent stage of rapid ovarian growth comprises previtellogenesis and vitellogenesis; during that stage an increase of oocyte size by a few dozen times was accompanied by a tenfold increase of oxygen consumption. The major factor regulating the increase of oxygen consumption during ovarian growth was an increase in mitochondrial mass, and consequently an increase in cytochrome oxidase activity. During the final stage of ovarian maturation, neither the oocyte dry weight nor the protein content per oocyte exhibited any substantial changes; the increase in oocyte wet weight resulted from the accumulation of water. At that time, oxygen consumption decreased. A decrease in oxygen consumption during final maturation of eggs has been observed by several authors in other fish species and in invertebrates (review: Ozernyuk, 1985).

2.2 REPRODUCTIVE EFFORT

The concept of **reproductive effort** was defined by Fisher (1930) as an organism's investment in any act of reproduction. This concept is central to contemporary discussion on life history. Calow (1979) considered two hypotheses to explain the negative relationship between reproductive output and life span and/or subsequent fecundity which is known for many organisms. The first hypothesis suggests evolutionary (selective) or ecological (density-dependent) influences. The second hypothesis suggests more direct ethological or physiological influences. Calow (1979) reported a great deal of evidence in favour of the physiological component of the second hypothesis. The physiological basis of the relationship between reproductive output and life span and/or fecundity is 'competition' for a limited supply of energy or other resources. This trade-off between present reproduction and other organismic processes, decisive for future expected reproduction, is supposed to occur within a parental organism (Wootton, 1985). It is therefore worthwhile to review the foundation of the theory of energy transformations.

Papers by Ivlev (1939a, b, c) and Winberg (1956) underlie the theory of energy flow through the biological system (e.g. through a population or an individual). Using symbols applied by Petrusewicz (1967), Ricker (1968), Klekowski (1970), Klekowski and Duncan (1975a), and by other authors, an energy budget for a growing, reproducing organism can be presented as follows:

$$C = P + R + U + F \qquad (2.1)$$

$$\text{and} \quad P = P_g + P_r \qquad (2.2)$$

where C is consumption, P is production (P_g is growth of the somatic body, P_r is reproductive growth), R is metabolism, U is nonfaecal excretion, and F is faeces.* An instantaneous energy budget can be calculated for a moment in time during an animal's life. Its parameters are C_i, P_i, R_i, U_i and F_i, and they are expressed in energy units (J or cal) per unit time (e.g. hour, day or year). In a cumulative energy budget (first proposed by Klekowski *et al.*, 1967, developed by Klekowski, 1970, and Klekowski and Duncan, 1975a,

---

*A complete list of symbols and abbreviations used in this book may be found on pages xiii–xv.

and proved to be useful in fish early developmental stages by Kamler and Kato, 1983), all the parameters are cumulated from the beginning of the life cycle (time τ_0) to the end of an identifiable period of the life cycle (time τ_n). Its parameters are C_c, P_{gc}, P_{rc}, R_c, U_c, and F_c and they are expressed in energy units per life period. In practice, the value of P_{gc} is accepted as the final energy content of an individual, because in many species energy content at the beginning of embryonic development in the germinal disc is negligible.

Analogous budgets of parameters other than energy (e.g. protein, lipids, nitrogen, phosphorus, carbon) – Penczak *et al.* (1982), Fischer (1983), Urban (1984) – can be calculated, although there is less information on this subject.

Review of approaches

This Chapter will consider the energy requirements associated with ovarian maturation. The energy cost for testicular growth is usually smaller (review: Wootton, 1985).

There are numerous ways in which to express reproductive effort (Calow, 1979). The simplest are (ı) egg number per female or (ıı) egg biomass per female. Both provide only a crude approximation of reproductive effort. Egg number per female (ı) ignores both the biomass of eggs produced (production of a larger number of smaller eggs can be energetically less expensive than production of a smaller number of larger eggs) and the female body biomass. Egg biomass per female (ıı) ignores only female body biomass, but this leads to over-estimation of reproductive effort in larger females in comparison to the smaller ones.

Gonad wet weight as a percentage of body weight (ııı) has been widely applied for many years in fishery practice (sometimes gonad weight has been related to length cubed). Initially this index was called the relative weight of the ovaries; later, the name **gonadosomatic index** or gonosomatic index (GSI) was adopted. This method avoids the errors mentioned above and is easy; a great number of GSI data have been accumulated to quantify reproductive preparedness. The index is based on the assumptions that the regression of egg biomass on parental biomass is linear, has a zero intercept, and does not change with gonadal development. However, these criteria are often violated in fishes (DeVlaming *et al.*, 1982; Erickson *et al.*, 1985 and literature herein). In serial spawners, GSI greatly under-estimates the reproductive effort (Wootton, 1979; DeMartini and Fountain, 1981; Mills and Eloranta, 1985). In addition, the comparison of wet weights of fish ovaries and fish body tissues can be misleading, because in mature female fish, energy is more 'dilute' in body wet matter than in ovarian wet matter. Both the lower degree of hydration and the higher caloric value of ovarian matter (Chapter 3) contribute to this difference. Hence, GSI is not a reliable estimate of reproductive effort. Pianka and Parker (1975) used index ııı to estimate reproductive effort in reptiles; both offspring size and parental body size were alternatively given in terms of weight, volume or energy.

Another measure of reproductive effort is the ratio of the energy (or matter) output in reproduction to that taken in. This relationship (IV) can be described using symbols given in Equations 2.1 and 2.2 as $P_r \times 100 / C$. Thus it is a 'gross reproductive effort'. It was used by Hirshfield (1980) for *Oryzias latipes* and by Wootton *et al.* (1980) for *Gasterosteus aculeatus.* Wootton (1979, 1982, 1985) reviewed the results of measurements of gross reproductive effort; the annual estimations were of the order of 10%. Calow (1979) discusses the advantages of this way of expressing the reproductive effort, but he also draws attention to its disadvantages. Both the rate of food consumption and the reproductive output can vary (see also Hirshfield, 1980 and review in Wootton, 1982). According to Calow (1979), $C = P_r + rest$, where rest denotes 'other metabolic purposes'. He contends that index IV is not a good measure of reproductive effort. For example, when food is not limited and when an increase of P_r is supported by an increase in consumption, then despite an increase in index IV, the parental organism has not been more exploited to produce the increase in reproductive output.

For this reason Calow (1978) has proposed index V:

$$1 - [(C - P_r) / rest^*]$$

where rest* is determined in pre-reproductive adults. Index V has been used by Wootton *et al.* (1980) for *Gasterosteus aculeatus.* According to Calow (1979) it is a good, although not an ideal, measure of the cost of reproduction.

The extensive use of indices IV and V should be restricted for several reasons. First, consumption (C) is very difficult to measure in ecophysiological studies on poikilotherms (e.g. Fischer, 1970a; Duncan and Klekowski, 1975; Klekowski and Fischer, 1975; Striganova, 1980). This is especially true in studies on natural populations (for fish, see review in Wootton, 1986). Second, food consumption decreases or ceases in some fishes during gonad maturation. Third, in fish the energy used for gonad formation originates not only directly from food, but also from energy stored earlier in the somatic body, mainly as fat deposits or in muscle (Moroz *et al.*, 1973; Craig, 1977; Hirshfield, 1980; Wootton *et al.*, 1980; Dąbrowski 1982a, 1983). Dąbrowski (1985) shows that the seasonal course of reproductive effort cannot be illustrated by index IV in species which form their gonads at the expense of body constituents. Moreover, the precise determination of reproductive cost using indices IV and V requires long-lasting, complex studies. Hirshfield (1980) reviewed theoretical studies of **reproductive effort**, defined as the relation between energy in eggs and energy in food, index IV, and he concluded that their empirical base was weak, partly because of the lack of information on food intake.

Thus, reproductive effort will be further illustrated by the relationships between the energy released in gametes (P_r) and the energy increment in somatic body growth (P_g). These relationships have been examined in aquatic invertebrates and fish by Streit (1975), Kamler and Mandecki (1978), Kamler and Żuromska (1979), Kamler *et al.* (1982), Dąbrowski (1983), Khmeleva and Golubev (1984), Khmeleva and Baichorov (1987) and Dąbrowski *et al.* (1987); the earlier Western studies are reviewed by Wootton (1979).

An index (VI) of '**net reproductive effort**' can be defined as:

$$P_r \times 100/(P_g + P_r) \ (\%) = P_r \times 100/P(\%)$$

This index appears to apply well to fish because many species continue to increase their body mass after reaching sexual maturity.

However, this definition of reproductive effort is less than ideal, since it takes into account the energy incorporated in matured gametes, but neglects other energy expenditures associated indirectly with reproduction, e.g. secretions (pheromones, mucus) and increased metabolism associated with maturation processes, pregnancy, spawning migrations, courtship, and spawning activity. For the last activity fish can use 10–40% of the fat stored earlier in the somatic body (Shulman, 1972). In aquatic Gastropoda, an increase in metabolic rate associated with reproduction has been reported by Berg *et al.* (1958) and Kamler and Mandecki (1978); this phenomenon is also well known in mammals (Kleiber, 1961). For various species of fish, Stroganov (1962) provided data which indicated both an increase and a decrease in respiratory rate in connection with reproduction. Wootton (1985) pointed out that fish have the ability to control energy partitioning between reproduction and other life processes. An example is given by Koch and Wieser (1983): swimming activity of *Rutilus rutilus* was reduced during the period of gonad formation by about 40% and 8% at 18 and 6 °C, respectively. In general, then, the index of net reproductive effort (VI) neglects many aspects of the problem.

In keeping with Klekowski *et al.*'s (1967) definition of instantaneous and cumulative energy budgets, reproductive effort will be further defined in two ways:

(a) Instantaneous index of net reproductive effort

$$IINRE = P_{ri} \times 100 / P_i \ (\%) \tag{2.3a}$$

In this case both P_{ri} and P_i are expressed in joules or in mg of matter per individual and per unit time (day or reproductive period).

(b) Cumulative index of net reproductive effort

$$CINRE = P_{rc} \times 100 / P_c \ (\%) \tag{2.3b}$$

where both P_{rc} and P_c are expressed in joules or in mg of matter per individual; P_{rc} and P_c are cumulated for the whole life span.

These two ways of expressing the reproductive effort (IINRE and CINRE) are compared in Fig. 2.2. It is not advisable, however, to express reproductive effort using P_r in its instantaneous form (P_{ri}) and P_g in its cumulative form (P_{gc}), as was done by Dąbrowski (1983).

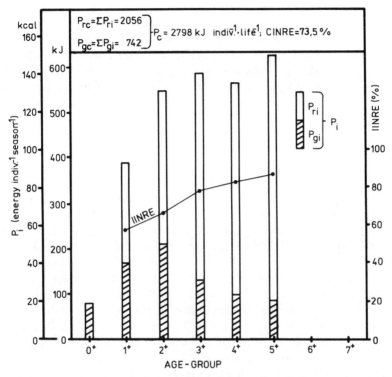

Fig. 2.2 Examples of computation of instantaneous index of net reproductive effort (IINRE) and cumulative index of net reproductive effort (CINRE), using data for females of *Coregonus albula* (a polycyclic species, which hatches in spring, starts to reproduce at age 1^+, and spawns once a year in late autumn) from Lake Pluszne (Poland), 1976, $n=26$. P_{gi} calculation: the body length (L) of each spawner measured just prior to spawning and its body length at the beginning of the last growing season (back-calculated from scales) were both converted to body minus gonad wet weight (W_w) using a W_w on L regression equation computed for this population. Next, ΔW_w was converted to ΔW_d assuming 21.75% of W_d in W_w and to P_{gi} using 20.63 J mg^{-1} dry weight. Mean values of P_{gi} for each age class were then obtained. P_{ri} calculation: fecundity was estimated from the regression of fecundity on body length (Zawisza and Backiel, 1970) and then converted to P_{ri} using the caloric equivalent of an egg, which had been measured separately for each fish; the mean values of P_{ri} for each age class were then obtained (unpublished data by Kamler *et al.*, 1982).

Reproductive effort in fishes and invertebrates

In Tables 2.2 and 2.3 the instantaneous and cumulative indices of net reproductive effort are given for representative fish species and are compared with those of invertebrate species. Only spawning individuals were included; energy that had accumulated in the somatic body prior to the beginning of intense gonadal growth but was transferred to gonads and/or used for metabolic purposes during gonad formation has been disregarded in determining P_g. These two subcomponents are especially high (Boëtius and Boëtius, 1980) in silver eel, *Anguilla anguilla*, which does not feed during its long period of migration or during breeding. Hence Dąbrowski (1983) was right in saying that presentations that neglect changes of P_r and P_g during the period of gonad formation do not reveal their maximum values, which can occur at different periods but well before spawning. The problem of temporal changes in the transfer of matter (energy) from the somatic body to gonads (Shulman, 1972; Moroz *et al.*, 1973; Craig, 1977; Hirshfield, 1980; Wootton *et al.*, 1980; Dąbrowski, 1982a, 1983) is very interesting from the physiological point of view, but for an ecologist a more important question is what fraction of energy or matter is invested in the parental organism and what part leaves it irreversibly and contributes to the following generation. This latter approach is represented in the present Chapter.

Both instantaneous and cumulative indices of the net reproductive effort very often exceed 50% (Tables 2.2 and 2.3; Figs 2.3, 2.4). Similarly, out of 34 CINRE values as many as 23 values exceeded 50%, calculated from data for 30 crustacean species (table 25 in Khmeleva and Golubev, 1984, of which only one is presented in Table 2.3). This means that reproductive growth in poikilotherms is very often higher than somatic growth. Sometimes P_{ri} can be higher than P_{gi} by ten times or more; examples of IINRE of 90–100% can be found in Table 2.2 and in Figs 2.3 and 2.4. In extreme cases, and most probably for only short time intervals (as compared with the whole life span), body reserves as well as food are used for reproduction; values of P_{gi} are then negative and IINRE values are greater than 100% (Table 2.2 – Hirshfield's (1980) data for *Oryzias latipes* at high temperatures). There are also some cases of low reproductive effort, illustrated by low indices (less than 20%) in Table 2.2. In these cases less than one-quarter as much energy is allocated to reproduction as to somatic growth; such cases are rather rare. Similarly, only two of 34 CINRE values compiled by Khmeleva and Golubev (1984) were lower than 20%. These conclusions are valid for both ways of expressing reproductive effort, IINRE and CINRE, although values of the latter, as based on P_{rc} and P_c, are somewhat more 'smoothed' (compare e.g. CINRE values for *Coregonus albula* listed in Table 2.3 with IINRE values for the same species shown in Fig. 2.3). In general, values of CINRE do not much exceed 90%.

Shulman (1972) reviewed papers dealing with changes in body lipid content of fish. Females of some salmonids spend obout 90% of body fat

Table 2.2 Comparison of IINRE (instantaneous index of net reproductive effort) in fish and invertebrate females; P_{ri}, P_{gi} and P_i are expressed in J per individual and per reproductive period (RP) or per day (D)

Species		Age	P_{ri}	P_{gi}	P_i	IINRE (%)
Fishes						
Oncorhynchus nerka[1]	(RP)	–	–	–	–	20.0
Anguilla anguilla[2]	(RP) a	–	17.1×10^5	61.9×10^5	79.0×10^5	21.6
	(RP) b	–	17.4×10^5	43.9×10^5	61.3×10^5	28.3
Esox lucius[3]	(RP)	1 y	7.2×10^5	10.6×10^5	17.8×10^5	40.5
	(RP)	2 y	11.0×10^5	12.2×10^5	23.2×10^5	47.4
	(RP)	3 y	18.0×10^5	19.9×10^5	37.9×10^5	47.5
	(RP)	4 y	18.5×10^5	19.5×10^5	38.0×10^5	48.7
Tinca tinca[4]	(RP) a, C	2^+–3^+ y	57.3×10^4	9.6×10^4	66.9×10^4	85.6
	(RP) a, H	2^+–3^+ y	108.8×10^4	23.0×10^4	131.9×10^4	82.4
	(RP) b, C	2^+–3^+ y	61.5×10^4	0.4×10^4	62.0×10^4	99.2
	(RP) b, H	2^+–3^+ y	85.4×10^4	17.2×10^4	102.6×10^4	83.2
Coregonus muksun[5]	(RP)	2^+–3^+ y	2.5×10^4	40.0×10^4	42.5×10^4	5.9
	(RP)	9^+–10^+ y	58.2×10^4	100.6×10^4	158.8×10^4	36.6
C. lavaretus pitschian[6]	(RP)	9^+–10^+ y	386.8×10^3	334.9×10^3	721.7×10^3	53.6
C. albula[7]	(RP) min.	1^+ y	11.7×10^3	5.9×10^3	32.2×10^3	33.4
	(RP) max.	7^+ y	617.0×10^3	328.6×10^3	714.1×10^3	97.3
Perca fluviatilis[8]	(RP)	8–9 y	167.4×10^3	66.1×10^3	233.6×10^3	71.7
Gasterosteus aculeatus[9]	(RP)	0^+ y	15.7×10^2	13.4×10^2	29.1×10^2	53.9
Tilapia mossambica[10]	(D) a, 25 °C	6–7 mo.	1.8×10^1	95.2×10^1	97.1×10^1	1.9
	(D) a, 28 °C	6–7 mo.	13.4×10^1	61.3×10^1	74.7×10^1	17.9
	(D) a, 31 °C	6–7 mo.	15.5×10^1	22.8×10^1	38.3×10^1	40.4

cont'd overleaf

Table 2.2 *cont'd*

Species	Age	P_{ri}	P_{gl}	P_i	IINRE (%)
(D) a, 25 °C	7–8 mo.	4.8×10^1	100.0×10^1	104.8×10^1	4.6
(D) a, 28 °C	7–8 mo.	30.3×10^1	97.9×10^1	128.2×10^1	23.6
(D) a, 31 °C	7–8 mo.	46.6×10^1	65.1×10^1	111.7×10^1	41.7
(D) b, 25 °C	7–8 mo.	3.9×10^1	99.6×10^1	103.5×10^1	3.8
(D) b, 31 °C	7–8 mo.	24.2×10^1	45.6×10^1	69.8×10^1	34.7
Oryzias latipes[11]					
(D) a	–	13.5	16.2	29.7	45.5
(D) b	–	33.9	–3.9	30.0	113.0
(D) c	–	35.6	–10.9	24.7	144.1
Poecilia reticulata[12]					
(D)	60–120 d	5.0	21.9	26.9	18.6
Invertebrates					
Physa acuta[13]					
(D) a	60th d	0.4	1.4	1.8	22.2
(D) b	170th d	23.3	1.4	24.7	94.3
Idotea baltica basteri[14]					
(D) a	40–70 d	16.9×10	16.7×10	33.6×10	50.2
(D) b	280–298 d	67.2×10^{-1}	2.3×10^{-1}	69.5×10^{-1}	96.6
Daphnia pulex[15]					
(RP)	8–48 d	–	–	–	>70.0
Simocephalus vetulus[16]					
(D) a	9th d	123.9×10^{-3}	80.4×10^{-3}	204.3×10^{-3}	60.6
(D) b	20th d	212.6×10^{-3}	10.0×10^{-3}	222.6×10^{-3}	95.5
Plectus palustris[17]					
(D) a	26th d	179.6×10^{-5}	10.0×10^{-5}	189.6×10^{-5}	94.7
(D) b	17th d	510.3×10^{-5}	56.5×10^{-5}	566.8×10^{-5}	90.0

[1] A monocyclic fish, sockeye salmon (a predator); wild population on spawning grounds (source: Brett and Groves, 1979).
[2] A monocyclic fish. European eel (a predator); hormone-matured, maturation probably incomplete: (a) GSI 32–47% (mean value for 8 most mature fish selected from table 6 in Boëtius and Boëtius, 1980); (b) source: Epler *et al.* (1981b). Original results given in chemical composition have been converted to energy units.

[3] A polycyclic fish (spawns once a year in late April or early May), northern pike (a predator); wild population in Lac Ste Anne, Canada (latitude 53°43′ N). (Recomputed from table 3 in Diana and Mackay 1979.)

[4] A highly polycyclic fish (batch-spawning between May and August), tench; field experiments: (a) 1973–74 in (C) an unheated (cold) pond and in (H) a pond heated by 2.0–2.5 °C in winter and by 3.5–4.7 °C in spring–summer; (b) 1974–75 in (C) a cold pond and (H) one heated by 2.1–4.3 °C in summer. Calculations based on: first, raw numerical data by Horoszewicz (1981a) on somatic growth (wet wt) between the end of the previous reproductive season and the end of a given reproductive season; second, total number of eggs deposited by the average female tench over a given reproductive season (F_c^{-s} in Morawska's (1984) table vi); third, average caloric value of tench somatic body, 4.1 J per mg wet wt and average energy content 3.98 J per egg (Stachowiak and Kamler, unpublished data).

[5,6] Polycyclic fishes; based on Ryzhkov's (1981) compilations.

[7] A polycyclic fish (spawns once a year in late autumn), vendace (planktivorous); wild populations from Polish and Finnish lakes. The minimum and maximum values of P_{ri} were selected from 46 values given in Kamler et al. (1982); the P_{gi}, P_i and IINRE values were similarly selected, therefore they do not correspond on rows. For further details see Figs 2.2, 2.3 and 2.7.

[8] A polycyclic fish (spawns once a year in spring), perch (a predator); wild populations in Windermere, England. A rough estimation for a standard female (26 cm length, Craig, 1977). The theoretical age (8.66 y) at which the fish attains 26 cm was obtained from table i in Craig and Kipling (1983); the mean values of the growth statistics were used. Lengths at age 8 and 9 y were computed and converted to wet wt using the length–weight relationships from table 2 in Craig. 1980 (data for spent females). Caloric value of the wet somatic tissue of the standard female (Craig, 1977) was used and P was computed. P_r was estimated from figures 5 and 6 in Craig (1977).

[9] A polycyclic fish (3–4 batches in late spring and summer), threespined stickleback (a predator); laboratory and field studies of the population inhabiting backwaters of the River Rheidol (Mid-Wales). Recalculated: mean value for total egg production (1565) was taken from table iv in Wootton et al. (1980) and somatic growth in mg dry wt between September (the onset of gonad formation) and July from their table i (the missing growth from July to August was assumed to be equal to the growth from June to July and was added). A caloric value of 18.84 J per mg dry somatic wt was assumed.

[10] A highly polycyclic fish (2–23 spawnings at ages from 4.5 to 9 months), tilapia; females incubate eggs in their mouths; food: Tubificidae in (a) limited and (b) surplus food supply; laboratory experiments. Selected results from Mironova (1977).

[11] A highly polycyclic fish (breeds every day), Japanese medaka (27.1 mm initial length); food: Tetramin staple, 193 J per individual and per day; laboratory experiments at (a) 25 °C, (b) 27 °C and (c) 29 °C. Recomputed from Hirshfield (1980).

[12] A highly polycyclic, viviparous fish, guppy (a benthophage); food: Tubifex in restricted supply, 4 h d^{-1}, in laboratory experiments at 25 ± 1 °C. The original data, given by Krishnamurthy et al. (1984) in mg dry wt, were conveted into energy units using the caloric value 20.93 J per mg dry wt.

[13] A warm-water snail; detritus feeder; food: powdered formula + live Oscillatoria; laboratory experiments at 26 °C, on (a) the first and (b) the last day of the reproductive period (source: Kamler and Mandecki, 1978; full data are presented in Figs 2.4 and 2.6).

[14] An isopod; laboratory experiments at 20 °C at (a) the beginning and (b) the end of the reproductive period (full data in Figs 2.4 and 2.7; values recomputed from tables 23–25 in Khmeleva and Golubev, 1984).

[15] A cladoceran; filter-feeder; food: Oocystis in laboratory experiments at 18–20 °C (source, Pechen and Kuznetsova, 1966).

[16] A cladoceran; filter-feeder; food: Chlorella in laboratory experiments at 22 °C (a) the beginning and (b) the end of the reproductive period (data of Klekowski and Ivanova, cited in Klekowski et al., 1972; full data in Fig. 2.4).

[17] A benthic nematode; food: Acinetobacter, (a) 6–9 × 10^8 cells cm^{-3} and (b) 6–9 × 10^9 cells cm^{-3} in laboratory experiments at 20 °C; data for the last day of the reproductive period (source: Schiemer et al., 1980).

Table 2.3 Comparison of CINRE (cumulative index of net reproductive effort) in fish and invertebrate females: P_{rc}, P_{gc} and P_c are expressed in J per individual and per life span

Species		Life span	P_{rc}	P_{gc}	P_c	CINRE (%)
Fishes						
Oncorhynchus nerka[1]		–	–	–	–	20.0
Anguilla anguilla[2]	a	–	17.1×10^5	61.9×10^5	79.0×10^5	21.6
	b	–	17.4×10^5	43.9×10^5	61.3×10^5	28.3
Coregonus albula[3]	H, 77	$0–7^+$ y	31.7×10^5	9.5×10^5	41.2×10^5	76.9
	P, 77	$0–5^+$ y	22.2×10^5	8.2×10^5	30.4×10^5	73.0
	P, 76	$0–5^+$ y	20.6×10^5	7.4×10^5	28.0×10^5	73.6
	M, 76	$0–6^+$ y	15.4×10^5	5.6×10^5	21.0×10^5	73.3
	S, 77	$0–3^+$ y	13.1×10^5	8.6×10^5	21.7×10^5	60.4
	N, 77	$0–4^+$ y	8.2×10^5	5.1×10^5	13.3×10^5	61.7
	N, 76	$0–4^+$ y	5.9×10^5	4.1×10^5	10.0×10^5	59.0
	Ou, 76	$0–7^+$ y	4.6×10^5	3.2×10^5	7.9×10^5	58.2
	Kg, 77	$0–5^+$ y	1.7×10^5	1.4×10^5	3.1×10^5	54.8
Coregonus pollan[4]		$0–5^+$ y	14.7×10^5	8.7×10^5	23.4×10^5	62.8

Invertebrates

Idotea baltica basteri[5]		0–298 d	125.8×10^1	45.6×10^1	171.4×10^1	73.4
Physa acuta[6]	a	0–260 d	146.7×10^1	17.0×10^1	163.7×10^1	89.6
	b	0–203 d	126.2×10^1	27.2×10^1	153.3×10^1	82.3
Tribolium castaneum[7]		0–106 d	250.3	26.0	276.3	90.5
Tubifex tubifex[8]		0–90 d	10.5	8.5	19.0	55.3
Simocephalus vetulus[9]		0–20 d	238.6×10^{-2}	146.1×10^{-2}	384.7×10^{-2}	62.0
Rhizoglyphus echinopus[10]		0–70 d	213.5×10^{-2}	37.7×10^{-2}	251.2×10^{-2}	85.0
Plectus palustris[11]	a	0–26 d	119.7×10^{-4}	48.6×10^{-4}	168.3×10^{-4}	71.1
	b	0–17 d	169.1×10^{-4}	80.0×10^{-4}	249.1×10^{-4}	67.9

[1] A monocyclic, predatory fish; see also 1 in Table 2.2 (source: Brett and Groves, 1979).

[2] A monocyclic, predatory fish; see also 2 in Table 2.2 (based on: (a) Boëtius and Boëtius, 1980: (b) Epler et al., 1981b).

[3] A polycyclic, planktivorous fish; wild populations from Polish (H, P, M, S, N) and Finnish (Ou, Kg) lakes (abbreviations are explained in Fig. 2.3); see also 7 in Table 2.2 (based on Kamler et al., 1982).

[4] A polycyclic, planktivorous fish; wild population from Lough Neagh (Northern Ireland). Based on Dąbrowski (1982a). The wet weights of eggs and somatic body were estimated graphically from figure 5 in Dąbrowski (1982a). Egg wet weight was converted to dry weight (28.41%, table 3 in Dąbrowski, 1982a) and to joules using caloric value 30.1 J per mg dry wt (Kamler et al., 1982 for C. albula eggs having 29.8% lipids). Wet grams of somatic body were converted to joules to joules using caloric value 4.2 J per mg wet wt (Kamler and Żuromska, 1979).

[5] An isopod; laboratory experiments at 20 °C; see also 14 in Table 2.2 (based on Khmeleva and Golubev, 1984).

[6] A gastropod; laboratory experiments at (a) 22 °C and (b) 26 °C; see also 13 in Table 2.2 (based on Kamler and Mandecki, 1978).

[7] A beetle, stored-product deposit feeder; food: wheat flour; laboratory experiments at 29 °C (based on Klekowski et al., 1967).

[8] A worm, detritus feeder; food: decomposing lettuce; laboratory experiments at 24 °C (based on Kosiorek, 1979).

[9] A cladoceran; see also 16 in Table 2.2 (based on Klekowski and Ivanova, cited in Klekowski et al., 1972).

[10] A mite, stored-product deposit feeder; food: rye germ; laboratory experiments at 25 °C (approximate data from Klekowski and Duncan, 1975a).

[11] A nematode; see also 17 in Table 2.2 (based on Schiemer et al., 1980).

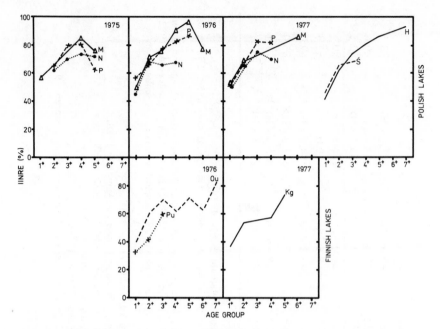

Fig. 2.3 Age course of the instantaneous index of net reproductive effort (IINRE) during the reproductive period of *Coregonus albula*, using field data collected in 1975–77 from five Polish lakes (M. Lake Maróz, 53°32′N, 20°24′E; P, Lake Pluszne, 53°36′N, 20°24′E; N, Lake Narie, 53°55′N, 20°01′E; H, Lake Hańcza, 54°16′N, 22°49′E; Ś, Lake Śrem, 52°06′N, 17°00′E) and from three Finnish lakes (Ou, Oulujärvi Lake, 64°08′–64°33′N, 26°45′–28°00′E; Pu, Puruvesi Lake, 61°48′–62°05′N, 29°25′–29°50′E; Kg, Kangosjärvi Lake, 67°45′N, 23°50′E). (Reproduced with permission from Kamler *et al.*, 1982.)

reserves, and Acipenseridae about 80%, for egg maturation and spawning; they expend a lot of energy for migration to spawning grounds. In other species this percentage is as follows: *Osmerus eperlanus*, 60%; *Stizostedion lucioperca* and *Clupea harengus membras*, 50%; *Clupea harengus*, 15%.

To summarize, two conclusions can be drawn. First, there is a considerable drain on energy reserves due to egg production, therefore evaluation of the role of a species in energy and matter transformation in an ecosystem will not be complete if reproductive growth is overlooked. Second, effort devoted to reproduction may vary over a broad range, so one should consider factors that modify the reproductive effort.

Factors affecting reproductive effort

Taxonomic position

It is tempting to assume that organisms differing in taxonomic position should differ in the effort put into reproduction. However, the data reported

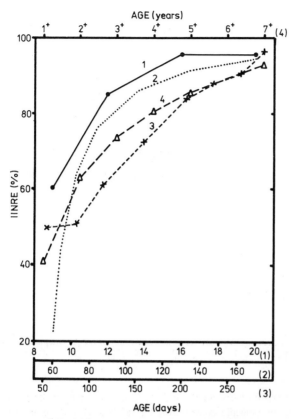

Fig. 2.4 Age course of IINRE during the reproductive period. (1) to (3) Invertebrates (laboratory experiments – details in Table 2.2): (1) *Simocephalus vetulus* (data of Klekowski and Ivanova in Klekowski *et al.*, 1972); (2) *Physa acuta* (data from Kamler and Mandecki, 1978); (3) *Idotea baltica* (data from Khmeleva and Golubev, 1984); (4) fish, *Coregonus albula* (field data, Hańcza Lake, 1977, from Kamler *et al.*, 1982).

by Khmeleva and Golubev (1984) do not suggest the existence of differences among different orders of Crustacea. Similarly, data gathered in Table 2.2 and Figs 2.3 and 2.4 seem to indicate that fish do not differ substantially from invertebrates in terms of IINRE, which can change considerably even within one species (e.g. data for *Coregonus albula* and two invertebrates, *Physa acuta* and *Idotea baltica basteri*, in Table 2.2 and Figs 2.3 and 2.4). However, values of cumulative index of net reproductive effort (CINRE) for fish are rather low (*c.* 20–75%) as compared to those for invertebrates (*c.* 55–90%; Table 2.3, Fig. 2.5). An association of long life span and low reproductive effort on one side, and a combination of short life span and high reproductive effort on the other (e.g. Stearns, 1976; 1977) is not surprising.

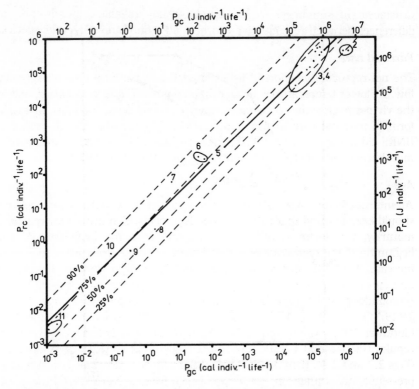

Fig. 2.5 Relationship between P_{rc} and P_{gc} (both in energy units per individual and per life span). Interspecific comparison for fish (2–4) and for invertebrate (5–11) species (2–11 as in Table 2.3). Solid line, mean regression line: $P_{rc} = 3.260\ P_{gc}^{0.946}$ joules base; 95% confidence limits for the intercept, 1.945–5.464, and for the slope, 0.892–1.000; the correlation coefficient $r = 0.993$ is highly significant ($P < 0.001$) for d.f. = 19. Broken lines, CINRE percentages for slope = 1. Log–log scale.

Mono- and polycyclic behaviour

The net reproductive efforts in two monocyclic fishes, *Oncorhynchus nerka* and *Anguilla anguilla*, 20–28%, are clearly low compared with values for polycyclic species; they are in the lower range of values listed in Tables 2.2 and 2.3. Although Calow (1979) presented examples of negative correlations at an interspecific level between high reproductive effort and breeding frequency in plants and animals, his own data (table 3 in Calow, 1979) indicate that there is no such dependence in fish. That is probably owing to the high plasticity of fish growth rate, egg size and fecundity (Svärdson, 1949; Stearns, 1977). Nor did Khmeleva and Golubev (1984) find differences between reproductive efforts of monocyclic and polycyclic species of

Crustacea. In summary, we do not have any reliable generalization of differences between reproductive efforts in all mono- and polycyclic species.

Parental care

The net reproductive effort of fishes engaging in parental care is especially low. At lower temperatures, in an oral incubator, *Tilapia mossambica*, and in the viviparous *Poecilia reticulata* (viviparity is considered the most advanced form of parental care in fish: Balon, 1977; Opuszyński, 1983), values of IINRE did not exceed 20% (Table 2.2). The gonadosomatic index of mature mouthbrooding females is exceptionally low (Wootton, 1979).

Age

An internal factor – age – strongly affects the net reproductive effort, but one should keep in mind that in such animals as fish which grow after reaching maturity, this factor is difficult to separate from the body size. At the beginning of the reproductive period, values of IINRE are lowest, then they increase. This is illustrated in Figs 2.3 and 2.4 and by data for *Esox lucius* and *Coregonus muksun* in Table 2.2. An increase in the energy demands for reproduction with age is a well-known phenomenon in fish (e.g. Orton, 1929; Craig, 1977, 1978; Wootton, 1979; Ryzhkov, 1981; Mills and Eloranta, 1985; Mann and Mills, 1985; Garcia and Brana, 1988) and in reptiles (Pianka and Parker, 1975). At first, IINRE increases very abruptly (Figs 2.3 and 2.4), then it slows down. Sometimes, a decrease is observed in pre-senile individuals (cf. *C. albula*, Lakes Maróz, Narie and Pluszne in 1975 and L. Maróz in 1976 – Fig. 2.3). Patterns of change with age in values of IINRE do not differ in fish as compared with those in invertebrates (cf. Figs 2.3 and 2.4); if the age axes are adjusted to the length of reproductive periods, these patterns become similar in poikilotherms of very different taxonomic position, biomass or life span (Fig. 2.4). This pattern can be regulated environmentally. A maximum GSI at an early age was found by Cowx (1990) to be favourable for *Leuciscus leuciscus* in an unstable environment.

The changes in IINRE with age discussed above can result from changes in P_{ri} while P_{gi} remains constant during the whole reproductive period; such a situation occurs in a gastropod, *Physa acuta* (Fig. 2.6). Both P_{gi} and P_{ri} can also vary over time. Wootton (1985) has reported that in female *Gasterosteus aculeatus* (a serial-spawning fish) from a backwater of the River Rheidol, the amount of energy stored in ovaries from mid-September (the onset of gonad formation) to mid-May (the start of the breeding season) was 0.51 kJ, and in the soma 0.66 kJ, per female; hence the IINRE was 43.6%. My calculation of P_{ri} and P_{gi} (Table 2.2) of the same population of sticklebacks for the whole reproductive period (September–August), based on data in Wootton *et al.* (1980), shows that during the breeding season (May–August) P_{ri} increased about threefold, and P_{gi} increased only about twofold. Thus, IINRE cal-

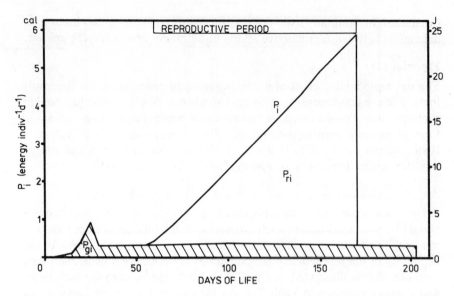

Fig. 2.6 The course of instantaneous somatic growth (P_{gi}) and reproductive growth (P_{ri}) in a gastropod (*Physa acuta* at 26 °C. Explanations in Table 2.2 (data from Kamler and Mandecki, 1978).

culated for the whole reproductive period was higher (53.9%). From data presented in Figs 2.2 and 2.7 for a fish (*Coregonus albula*) and an invertebrate (*Idotea baltica*), it is clear that in younger reproductive females somatic growth can still be high, but in older individuals P_{gi} decreases and reproductive growth attains higher priority. Similarly Craig (1977) has observed an increased requirement of energy for reproduction at the expense of somatic growth in *Perca fluviatilis* of increasing age; the same applies to *Melanogrammus aeglefinus* (table 9 in Shatunovskij, 1980). So, P_{gi} and P_{ri} can compete for the total energy used for production (P_i). Two examples of a negative dependence between P_{ri} and P_{gi} within the same population of *Coregonus albula* in the same year are presented in Fig. 2.8 – it is evident that this relationship is controlled in this case by the age of spawning females.

In summary, reproductive effort (like ageing, Craig, 1985) is determined by a biological clock. Reproductive effort is also controlled by external factors; the high degree of plasticity in the response of fish to their environment makes them an inviting subject for such studies. Two environmental factors, temperature (abiotic) and food (biotic), are particularly important factors affecting reproductive effort.

Temperature

As temperature increased, an increase in net reproductive effort ($P_r \times 100/P$; P_r and P based on carbon measurements) was observed in a freshwater gastropod, *Ancylus fluviatilis* (7 °C, 0%; 10 °C, 15%; 13–25 °C, 81.2% on

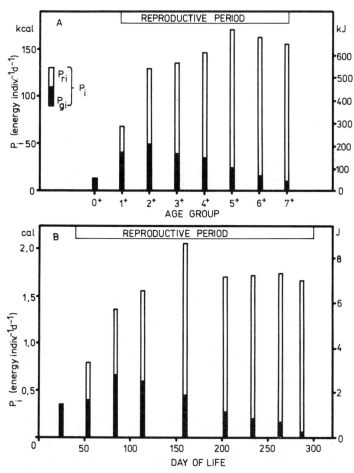

Fig. 2.7 The course of P_{gi} and P_{ri} in: (A) a field population of *Coregonus albula* in Lake Hańcza, 1977, mean values for age groups (modified from Kamler *et al.*, 1982); (B) a laboratory 'population' of a crustacean, *Idotea baltica basteri* (recomputed from Khmeleva and Golubev, 1984). Explanations in Table 2.2.

average: Streit, 1975). Similarly, in the fish *Tilapia mossambica* (irrespective of age and food supply) and *Oryzias latipes*, an increase in temperature resulted in an increase of IINRE (Fig. 2.9). On the other hand, *Tinca tinca* (Table 2.2) reared in a cold pond (C) exhibited a higher IINRE than in a heated one (H): 85.6% (C) and 82.4% (H) in 1973–74; 99.2% (C) and 83.2% (H) in 1974–75. It should be stressed that in all these three species the increase in temperature caused an increase in energy directed to reproduction (see P_{ri} in Fig. 2.9 and Table 2.2). At the same time P_{gi} responded to increased temperature in *T. mossambica* and *O. latipes* in a different way to that in *Tinca tinca*. In *T. mossambica* and *O. latipes*, the

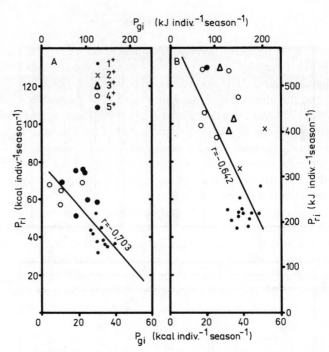

Fig. 2.8 Relationship between instantaneous reproductive growth (P_{ri}) and instantaneous somatic growth (P_{gi}) in individual *Coregonus albula* females from (A) Lake Maróz, 1975, $n=21$, and (B) Lake Pluszne, 1976, $n=26$. Least-squares regression lines are shown. Both correlations are highly significant ($P<0.001$) (partially adapted from Kamler *et al.*, 1982).

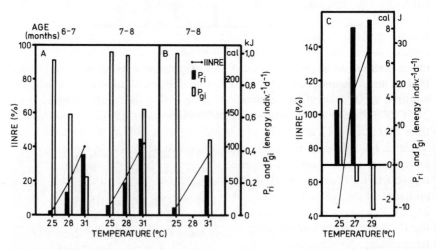

Fig. 2.9 Influence of temperature on the instantaneous net reproductive effort (IINRE), reproductive growth (P_{ri}) and somatic growth (P_{gi}). (A) and (B) *Tilapia mossambica*: (A), limited and (B), surplus food supply (data from Mironova, 1977); (C) *Oryzias latipes* (data from Hirshfield, 1980). Explanations in Table 2.2.

increase in P_{ri} with temperature was associated with a decrease in P_{gi} (Fig. 2.9; in *O. latipes*, P_{gi} at higher temperatures was even negative). In *T. tinca*, on the other hand (Table 2.2), P_{gi} in the heated pond was higher than P_{gi} in the cold pond, and the increase in P_{gi} (2.4 times in 1973–74 and as high as 43 times in 1974–75) was even more intense than the increase in P_{ri} (1.9 and 1.4 times, respectively). The two subcomponents of total production, energy directed to somatic growth and to reproduction, can therefore react differently to temperature, depending upon the range of temperature. Temperature optima and deviations from them are probably more important than simple increases or decreases in temperature.

The effect of temperature on IIRE can be indirect. In poikilotherms, high temperature accelerates development (this problem will be discussed in detail in Sections 4.1 and 6.1) and the ageing process. Thus animals reared at higher temperature will be physiologically older than animals reared at lower temperature, despite having the same calendar age. Age per se strongly affects IINRE (Figs 2.3 and 2.4) by changing both its components differently (see P_{gi} and P_{ri} in Figs 2.2, 2.6, 2.7 and 2.8). Therefore there is no unequivocal answer to the question: how does instantaneous net reproductive effort vary as temperature increases?

Difficulties in interpretation of IINRE resulting from the effects of temperature on physiological age should not occur with CINRE values, because P_g and P_r are cumulated until the end of the life span (Table 2.3). However, neither are the effects of temperature on CINRE unequivocal, simple relationships. In *Coregonus albula* from more northerly Finnish lakes (latitude 64°08′–67°45′ N; 3 Ou and 3 Kg in Table 2.3) the average CINRE amounted to 56.5%, and was lower than in more southerly Polish lakes (68.3%) (52°06′–54°16′ N; 3 H–3 N in Table 2.3). Although in the Polish lakes the somatic growth of fish (P_{gc}) was higher (2.9 times on average) than in the Finnish lakes, the difference between the reproductive growth (P_{rc}) was much greater: values were 5.2 times, on average, higher in the Polish than in the Finnish lakes. A freshwater gastropod, *Physa acuta*, on the other hand, cultured at 22 °C showed higher CINRE (89.6%) than when cultured at 26 °C (82.3%). It is interesting that in *Physa acuta* at 22 °C, P_{rc} was higher than at 26 °C (1.47 and 1.26 KJ per individual per life span, respectively – Table 2.3), although from the data discussed above it appears that an increase of temperature increases the amount of energy directed to reproduction. It turned out that *Physa acuta* is not an exception, since at the midpoint of the reproductive period (137th day at 22 °C and 115th day at 26 °C), P_{ri} amounted to 8.3 and 11.2 J per individual per day, respectively; the P_{ri} at the higher temperature was 1.3 times as high as that at the lower temperature. Simultaneously at 26 °C a greater reduction (by a factor of 1.7) of the length of the reproductive period was observed (from 183 days at 22 °C to 110 days at 26 °C), hence the P_{rc} value was lower at 26 °C than at 22 °C.

Considering the ways in which temperature affects CINRE one can assume
that (as in the case of IINRE) the following factors have an important effect
on CINRE: different temperature optima for P_r and P_g and indirect effects of
temperature. Indirect effects of temperature on CINRE manifest themselves
through changes in length of the reproductive period.

There is a general tendency for an increase in the amount of energy
allocated to reproduction at higher temperature, although the ways in
which temperature affects ultimate reproduction are various. An increase in
the share of energy allocated to reproduction are various. An increase in the
share of energy allocated to reproduction at higher temperature can lead to
increased mortality during the post-spawning period (Hirshfield, 1980 for
Oryzias latipes) and/or to a decrease in the amount of energy invested in
somatic growth. This will have an adverse effect on future reproduction
because fecundity is positively related to body size (this problem will be
discussed later). Moreover, high temperature accelerates the developmental
rate. Hence an increase in temperature shifts the position of organisms along
the r–K continuum (Pianka, 1974) in the direction of r.

Food availability

In an invertebrate, the nematode *Plectus palustris*, the cumulative net
reproductive effort decreased as food availability increased (Table 2.3);
although P_{rc} at the higher food level was higher than at the lower one, its
increase (by a factor of 1.4) was less than that of somatic growth P_{gc} (by a
factor of 1.6). Similarly, the instantaneous net reproductive effort in *Tilapia
mossambica* (Mironova, 1977) fed a surplus ration was always lower than
the IINRE of fish fed a limited food supply (Fig. 2.9). This difference was
maintained irrespective of fish age and habitat temperature. The gross
reproductive effort was inversely related to food level in *Oryzias latipes* at
27 °C and 29 °C despite increased fecundity at higher food levels (Hirshfield,
1980). Hirshfield calculated reproductive effort as the proportion of total
energy intake (index IV, page 9). Iwata *et al.* (1983) showed with a crab,
Scopimera globosa, that females having a higher lipid level in their bodies (mg
lipids per g wet weight) directed a lower percentage of lipids to their eggs.
Wootton (1985) considered energy partitioning in two short-lived fish
(*Gasterostus aculeatus* and *Oryzias latipes*) and in a relatively long-lived
species (*Pseudopleuronectes americanus*). The short-lived species fed restricted
rations sacrificed growth rather than reproduction. The long-lived species,
which breeds several times, supported growth at the expense of reproduction
under low food conditions.

Thus, at a high food level the allocation of energy to growth or reproduction becomes less critical and the reproductive effort decreases. Generally,
fish do not seem to deviate much in this respect from other poikilotherms.
One can, however, expect differences in the strategy of energy partitioning
between species with different life histories; this problem needs more
empirical studies.

Food type

The effect of foot type on reproductive effort is not very clear. A typical predator, *Esox lucius*, showed low IINRE (below 50%, even in 4-year-old fish); *Gasterosteus aculeatus* also had a low IINRE, but another predator, *Perca fluviatilis*, had average IINRE values (Table 2.2). Probably the effect of food type, if any, can be masked by other factors and no generalization can be made at present.

Relationship between reproductive and somatic growth

The relationship between reproductive and somatic growth can be either negative or positive. Negative relationships between P_{ri} and P_{gi} can be produced at the individual level during gonad formation when energy from somatic tissue is transferred to the gonads (Craig, 1977; Wootton, 1979; Hirshfield, 1980; Wootton *et al.*, 1980; Dąbrowski, 1982a, 1983). Negative relationships between P_{ri} and P_{gi} determined for spawning specimens were observed also at the population level (field or experimental); these relationships can be age- (Fig. 2.8) or temperature-controlled (Fig. 2.9). In contrast, large-scale comparisons (Fig. 2.5) between populations inhabiting different biotopes and interspecific comparisons show an increase in P_{gc} accompanied by an increase in P_{rc}. It is obvious that organisms with larger biomass can direct more energy both to growth and to reproduction. Khmeleva and Golubev (1984) found a highly significant positive correlation between the log of daily P_{ri} and the log of mean biomass for 22 species of Crustacea.

Ways in which reproductive growth can be modified

In monocyclic species, reproductive growth can be expressed as:

$$P_r = E.n. \times W \times h \qquad (2.4)$$

and

$$C.e._0 = W \times h \qquad (2.5)$$

where *E.n.* is the number of eggs laid per female (**absolute** or **total fecundity**), W is the weight of individual egg, h is the caloric value per unit weight of egg matter and $C.e._0$ is the **caloric equivalent** of an egg. It is worth mentioning that the caloric value ($J g^{-1}$) and caloric equivalent ($J egg^{-1}$) have often been confused (Winberg, 1971). For polycyclic species Equation (2.4) becomes:

$$P_r = \sum_{i=n}^{n} E.n._i \times W_i \times h_i \qquad (2.6)$$

Therefore, any changes in P_r result from changes in fecundity, egg weight, or caloric value of egg matter; change can pertain to one, two or three elements in various combinations.

A strong negative correlation $(r = -0.93)$ between log relative fecundity and egg diameter was found at the interspecific level by Albaret (1982) for 44 Ivory Coast freshwater fish species. For example *Barbus sublineatus*, *Mormyrus hasselquistii* and *Papyrocranus afer* produced on average 677 000, 24 300 and 531 eggs per kg, respectively, while the mean egg diameters were 0.85, 1.85 and 3.60 mm, respectively. An intraspecific trade-off between egg size and number was also found (Hempel and Blaxter, 1967; Blaxter, 1969; Bagenal, 1969; Constantz, 1979; Wootton, 1984; Mann and Mills, 1985), but no significant relationships between egg size and egg number were found as well (Bartel, 1971; Townshend and Wootton, 1984) or larger fish had both larger and more numerous eggs (Thorpe *et al.*, 1984). Nishino (1980) summarizes consequences of different combinations of fecundity and egg size as a function of female size for invertebrate and fish species.

Absolute fecundity has been examined and reviewed in many papers, therefore it will be only briefly summarized here. It is known that absolute fecundity $(E.n.)$ increases with female body length (L):

$$E.n. = a \times L^b \tag{2.7}$$

This relationship has been found in oviparous teleost fish (e.g. Morawska, 1964, 1967; Zawisza and Backiel, 1970; Craig, 1980, 1982; DeMartini and Fountain, 1981; Penczak, 1985; Fadeev, 1987 and many others: reviews in Hoar, 1957; Blaxter, 1969 and Wootton, 1979). A semi-logarithmic transformation was applied by Mills and Eloranta (1985). Many studies (McFadden *et al.*, 1965; Bagenal, 1973; Wootton, 1979, 1985; Craig, 1980; Nishino, 1980; Kamler *et al.*, 1982; Craig and Kipling, 1983; Mann and Mills, 1985; Backiel and Zawisza, 1988; Cowx, 1990) have shown that the regression of $E.n.$ on L as calculated for a given species is not necessarily valid for all its popultions. The constant a and/or the exponent b vary in different populations of the same species (Fig. 2.10) and even for different cohorts of the same population. Consequently, the absolute fecundities of fish of the same length may be different.

Fecundity of young and old invertebrate females is low; females of average age are most fecund (Crustacea: Richman, 1958; Khmeleva and Golubev, 1984). In fish, age and body size are strongly correlated, therefore the relation of fecundity to age can only be detected after elimination of the effect of size by a statistical technique. Wootton (1979) reviewed the age effect on fecundity in fishes; he found it to be strong or weak or non-existent. Recently Schneider (1985) showed in *Perca fluviatilis* that fecundity is more strongly related to length than to age. In *Gasterosteus aculeatus*, Wootton (1985) found a much stronger relationship between fecundity and body weight than between fecundity and daily food ration.

Under poor food conditions fecundity decreases, as a rule, both in invertebrates (Schiemer *et al.*, 1980 for *Plectus palustris*; Richman, 1958 and

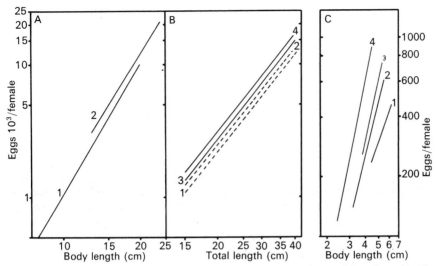

Fig. 2.10 Examples of different relationships between fecundity and length in (A, B) different populations of two fish and (C) an invertebrate species. (A) *Coregonus albula*: 1, Finnish populations (based on the formula given by Kamler *et al.*, 1982); 2, Polish populations (formula by Zawisza and Backiel, 1970); (B) *Salmo trutta*, Pennsylvania, USA: 1 and 2, streams of low basic productivity, Kettle Creek and Six Mile Run, respectively; 3 and 4, streams of high productivity, Spring Creek and Tionesta Creek, respectively (formulae by McFadden *et al.*, 1965); (C) *Palaemon paucidens* (a freshwater shrimp), Japan: 1, Lake Akan (latitude 43°12′ N); 2, Lake Ono (34°05′ N); 3, River Tomoe (34°55′ N); 4, Lake Biwa (35°15′ N) (formulae by Nishino, 1980). Log–log scale.

Redfield, 1981 for *Daphnia*) and in fishes (Scott, 1962 and Springate *et al.*, 1985 for *Oncorhynchus mykiss*; McFadden *et al.*, 1965 and Bagenal, 1969 for *Salmo trutta*; Wooton, 1977, 1985 for *Gasterosteus aculeatus*; Hirshfield, 1980 for *Oryzias latipes*; Townshend and Wootton, 1984 for *Cichlasoma nigrofasciatum*; reviews: Wootton, 1979, 1982; Opuszyński, 1983). However, fish are able to react to a food deficiency with an increase in fecundity (Mironova, 1977 for *Tilapia mossambica*), or food restrictions during gonad formation may produce no significant effect on fecundity of females of equivalent size (Kato, 1975 and Springate *et al.*, 1985 for *Oncorhynchus mykiss*). Food restriction is known to decrease fecundity directly, and an indirect effect of food on fecundity mediated through body size can be observed (Springate *et al.*, 1985). Dahlgren (1985) observed decreased fecundity of *Poecilia reticulata* exposed to high population densities.

Evidence of increased fecundity associated with increased temperature was given for fish by Mironova (1977) for *Tilapia mossambica*, Hirshfield (1980) for *Oryzias latipes*, Morawska (1984) for *Tinca tinca*, and Penczak (1985) for *Fundulus heteroclitus* (reviews: Wootton, 1979; Opuszyński, 1983).

In general, values of the exponent b in Equation 2.7 cluster around 3.0. Extensive interspecific comparisons (124 observations from 62 species, Wootton, 1979) revealed a broad range of b, 1.0–7.0, with modal class 3.250–3.794.

Changes in the two remaining components of reproductive growth – weight of an individual egg (W) and caloric value of egg matter (h) – will be considered in Chapter 3.

Concluding remarks

The amount of energy (or matter) transferred to progeny and/or invested in the parent has been considered in this Chapter. That does not exhaust the list of parameters that affect reproductive costs. Specific problems omitted from this discussion include the effect of reproductive effort on future fecundity and/or longevity of a parent (Pianka and Parker, 1975; Mann and Mills, 1985). Calow (1985) analysed the conflicting advantages and disadvantages of the growth–reproduction equilibrium shifted to the growth side. A large body results in high fecundity, but requires prolonged development, which is associated with increased risk of parental mortality. Ware (1982) discussed fish egg production (believed to be proportional to parent stock size) and density-dependent larval mortality. The resulting excess of reproductive effort would be high at high stock levels; possible compensatory mechanisms involved in the minimization of over-reproduction were considered.

Reznick (1985) grouped studies of reproductive costs into four categories. (A) Phenotypic correlations: these investigations include both field and laboratory measurements pertaining to the relationship between an index of reproductive effort and the costs borne by a parent, e.g. growth, subsequent survival, etc. (B) Experimental manipulations: these papers examine how reproductive effort responds to experimental manipulations of various factors. (C) Genetic correlations and (D) correlated responses to selection. Reznick (1985) stressed the usefulness of these last two approaches for theoretical studies of life history evolution. However, studies of types (A) and (B), which included the majority of papers he reviewed, were useful for creating a theoretical concept of 'costs'. Studies reviewed in this Chapter fall into categories (A) and (B).

Chapter three

Characteristics of fish reproductive products

The terms 'strategy' and 'tactics', both derived from military science, are useful in the analysis of fish reproductive processes (Wootton, 1984). Wootton considers **reproductive strategy** as a 'overall pattern of reproduction typically shown by individuals in a species'; it consists of a complex of traits, e.g. age at first reproduction, fecundity and many others, including size and nature of gametes (Wootton, 1982, 1984). Although the original meaning of the terms strategy and tactics implied rational planning, this implication has no relevance to biological processes. The latter occur in variable environments; both the overall reproductive strategy and is tactical variations are adaptive. One approach to the study of such adaptive mechanisms is by analysing the relationships between traits of reproductive strategy and environmental variables.

In this Chapter, attention will be focused on the size and composition of fish eggs (Sections 3.1 to 3.5) and on the mechanisms determining the properties of fish offspring (Sections 3.6 to 3.10).

ENERGY CONTENT AND COMPOSITION OF MATURE EGGS

The main component of eggs is yolk – the source of energy and materials for developing embryos. Along with external factors, the amount and quality of yolk are decisive for the successful embryonic and post-embryonic development of many fish species and the consequent recruitment of new fish into the population (Nikolskij, 1974). Fish egg size and composition have therefore been extensively studied.

3.1 EGG SIZE

Egg size can be expressed in many ways. A single diameter is commonly used (reviews: Blaxter, 1969; Ware, 1975; Wootton, 1979; Coburn,

1986), but two diameters are sometimes quoted: the longest diameter, egg length, and a second one perpendicular to it, egg breadth (e.g. Bartel, 1971; Ciechomski, 1973; Kamler and Malczewski, 1982; Kato and Kamler, 1983; Townshend and Wootton, 1984). Other measures of egg size include egg volume (review: Bagenal, 1971), wet weight and dry weight. From an energetic point of view the best measure of egg size is the caloric equivalent of an egg (energy content per egg, J egg^{-1}), since it indicates the amount of energy available to a developing embryo. Although a majority of teleost eggs have a regular spherical shape, there are numerous exceptions, e.g. the eggs of *Rhodeus amarus* are ellipsoidal and those of *Glossogobius brunneus* are rod-shaped. The shape of elasmobranch eggs deviates considerably from the spherical (review: Opuszyński, 1983). It should also be remembered that the hydration of eggs varies considerably (Section 3.3). Thus relationships between the energy content per egg and the other measures of egg size listed above will become stronger in the sequence given above. The weakest correlations can be expected for egg diameter, and the strongest for dry weight (mg egg^{-1}).

At the present state of development, bioenergetics deals mostly with the 'average individual'; individual variability is neglected. However, there is no average individual in nature. Hence, in the future bioenergetics will have to acknowledge that properties of individual organisms are variable. Therefore here I present, when possible, frequency distributions for the properties of eggs. The comparisons of fish egg size listed below are based mainly on data collected for ripe, unfertilized (unswollen) eggs. Their variations will be considered at three levels: interindividual variations will be discussed at (1) the interspecific and (2) the intraspecific level, and (3) differentiation of the size of individual eggs derived from the same female and the same batch will be discussed at the intra-individual level.

Interspecific comparisons

Among the smallest of fish eggs are those from *Cymatogaster aggregata*, with a diameter below 0.3 mm (data from Eigenmann in Wallace and Selman, 1981), *Notropis buchanani* – 0.64 mm (Coburn, 1986), *Acanthurus triostegus* – 0.7 mm (Blaxter, 1969) and *Arnoglossus spp.* – *c.*0.60–0.70 mm (recomputed from Bagenal, 1971). Among Osteichthyes the eggs of the coelacanth, *Latimeria chalumnae*, are exceptionally large, 85–90 mm in diameter (Anthony and Millot, 1972). Similarly large are eggs of Elasmobranchii: *Chlamydoselachus anguineus* – 90–97 mm, *Raja batis* – 70 mm (Opuszyński, 1983) and *Scyliorhinus caniculus* – 65 mm in diameter (Blaxter, 1969). Thus egg size variation among fish species is enormous, with the wet weight of the largest eggs about 34 million times that of the smallest ones.

Fish egg diameter, volume and wet weight distribution considered at the interspecific level are presented in Fig. 3.1. In Fig. 3.1(A)–(F), data compiled

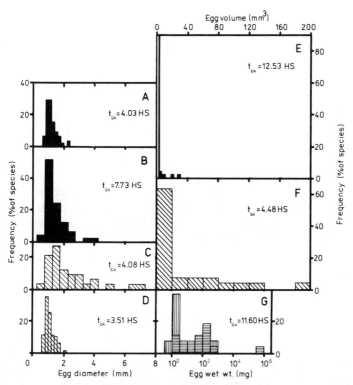

Fig. 3.1 Frequency distributions of fish egg size. ((A) to (D), egg diameter; (E) and (F), egg volume; (G), egg wet weight); interspecific comparisons. Each species is represented by a single value. Skewness, calculated according to Snedecor (1956), was tested with the t-test: $t_{sk} = sk/\sqrt{[6n(n-1)/(n-2)(n+1)(n+3)]}$ with d.f. $= \infty$; HS, skewness highly significant. (A) 65 marine pelagic fish species compiled by Ware (1975) (note: Ware excluded four species that produce eggs greater than 2.3 mm in diameter); (B) 101 marine European fish species, compiled by Wootton (1979); (C) 33 freshwater fish species, compiled by Wooton (1979); (D) 71 freshwater fish species (eastern North American cyprinids), compiled by Coburn (1986) (note: Coburn averaged the mean values from literature data and his own measurements for each species, regardless of the number of females in each sample; when only a range was available, he used the average of the minimum and maximum values); (E) 46 marine species of fish with planktonic eggs, based on Bagenal's (1971) table I, (using the averages of his minimum and maximum values); (F) 27 freshwater fish species, based on Bagenal's (1971) table VI (averages were computed as in E and only one average value was computed for species shown more than once in the table); (G) 27 marine and freshwater fish species: horizontal shading, Latimeridae, Acipenseridae and Salmonidae; vertical shading, Cyprinidae, Cobitidae, Gadidae, Percidae and Notothenidae; compiled from: Ogino and Yasuda (1962), Martyshev *et al.* (1967), Smirnov *et al.* (1968), Nishiyama (1970), Anthony and Millot (1972), Rakusa-Suszczewski (1972), Nikolskij (1974), Semenov *et al.* (1974), Kato (1975, 1979), Kamler (1976, 1987), Potapova (1978), Lapin and Matsuk (1979), Kamler and Malczewski (1982), Heming (1982), Kato and Kamler (1983), Ridelman *et al.* (1984), Ozernyuk (1985), Shatunovskij (1985), Springate *et al.* (1985), Trzebiatowski and Domagała (1986). Note the log scale for weight.

by other authors were used, whereas the distribution of egg weight (Fig. 3.1(G) is based on the original data from the literature. One should keep in mind that in this type of compilation it is not possible to include intraspecific variability of egg size. In addition, the average values reported in original works are based on different sample sizes, and authors often give only ranges for egg size. However it seems that for such a general comparison the procedure given by Coburn (1986) – see Fig. 3.1(D) – is, despite some shortcomings, acceptable for interspecific comparisons in which only one value has to be used for each species.

Fishes that scatter their eggs, such as pelagophils or phytophils, produce small eggs, the eggs of brood hiders (e.g. Salmonidae) are larger, and the largest eggs are derived from fishes that had more geological time to develop parental care: e.g. internal bearers such as *Latimeria* and *Scyliorhinus* (Balon, 1977).

Eggs of some freshwater fish are larger than 5 mm in diameter (Fig. 3.1(C)) or 65 mm^3 in volume (Fig. 3.1(F)); marine fishes have, in general, smaller eggs (cf. Figs 3.1(B), (C) or Fig. 3.1(E), (F); see also Blaxter, 1969). However, Ariidae, living in marine coastal waters, have perhaps the largest eggs of all oviparous teleosts, 10–25 mm in diameter (Fuiman, 1984). They are primarily freshwater fishes and are oral incubators. Parental care shifts the position of a species in the K direction along the r–K continuum. The distribution of egg size, especially egg volume and egg weight, is positively skewed, i.e. skewed to the right (Fig. 3.1). This means that the commonest fish egg sizes are distributed in the lower part of the size range whereas species producing large eggs are less frequent. For example, in 162 fresh-water fish species (Wootton 1984), egg diameters ranged from 0.75 to 6.55 mm; two-thirds of the species produced small eggs ($\leqslant 2$ mm in diameter), whereas large eggs (> 4 mm in diameter) were produced by only 14% of the species.

The positively skewed distribution of egg size may have some adaptative significance. It is known that the number of eggs from a given reproductive mass is inversely proportional to egg size. Ware (1975) has shown from extensive material that the mortality rate during early development is inversely related to the size of fish eggs or larvae. Consequently species with long incubation periods at low temperatures tend to produce large eggs. In other words, species employing a K strategy (i.e. producing few, large eggs) would be associated with waters having relatively low temperatures during spawning and early development, and species employing an r strategy (producing many, small eggs) with warmer waters. The positively skewed distribution (Fig. 3.1) could simply reflect the existing preponderance of species reproducing in warm waters in comparison to the number of species reproducing in cold waters and distributed in the temperate zone of the Northern Hemisphere. However, to explore the problem in a little more detail one should take into account its bioenergetic component. Our investigations on *Oncorhynchus mykiss* (Kamler and Kato, 1983) have shown that the yolk of small eggs was more effectively used in embryo tissue formation, and further by the larvae, than the yolk of large eggs. In other words, fish developing from small eggs are more economical in using their energetic reserves. Moreover, we have found that offspring of the same parental pair more effectively utilized the egg yolk at higher temperatures than at lower ones (this problem is discussed in detail in Chapter 4). In consequence the weight of tissue of larvae originating from a given biomass of small eggs and/or from eggs developing in warm water will be higher than that of larvae originating from an identical biomass of large eggs. This means a higher reproductive output from small eggs for the same reproductive effort – a speculation that needs to be proven. Hence a possible, but not necessarily true, explanation would arise for the advantage of a positively skewed distribution towards small eggs. 'Younger' species, having had less geological time to develop, produce smaller eggs than those of Acipenseridae, *Latimeria* and Elasmobranchii (Fig. 3.1(G)).

Intraspecific comparisons

Egg size variability among individuals of the same species was neglected in the above large-scale interspecific comparisons. However, it is well known that fish belonging to the same species can produce eggs of different size. Bagenal (1971) lists the volume of planktonic eggs from 46 species of marine fish and includes a minimum value for each species, a maximum value, and the percentage difference, i.e. the range expressed in % of the minimum value. The eggs of *Pegusa lascaris* were the least variable (min. 1.32, max. 1.38 mm^3, difference 4.5%), and the most variable were eggs of *Hippoglossoides platessoides* (min. 1.38, max. 22.45 mm^3, difference 1531.4%).

Comparisons between populations

Differences between weights of eggs produced by females taken from different populations of the same species are shown for example in lacustrine populations of *Coregonus albula* (Fig. 3.2), in both lacustrine and reared populations of *Oncorhynchus mykiss* (Fig. 3.3) and in *Salmo trutta* from different rivers (Fig. 3.4).

Comparison of dry weights of eggs collected from autumn-spawning populations of *Coregonus albula* (Fig. 3.2) indicates that differences between these relatively well isolated populations were maintained for three years. The largest eggs were produced by females from Lake Maróz in Poland. The eggs from Lake Narie were smaller, while the eggs collected from autumn-spawning populations in Finland were smallest; these differences were statistically significant (Kamler *et al.*, 1982, p. 87). Out of a total of 387 females, the lowest dry weight of eggs was found in a female collected from Lake Puruvesi in 1976 (0.362 mg average egg weight within the spawning

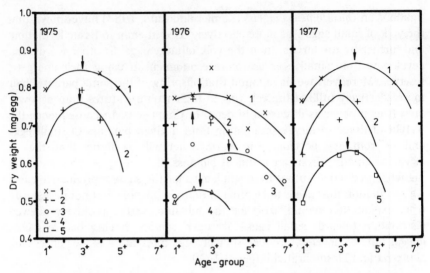

Fig. 3.2 Dry weight of *Coregonus albula* eggs from different Polish (1, Lake Maróz; 2, Lake Narie) and Finnish (3, Oulujärvi Lake; 4, Puruvesi Lake; 5, Kangosjärvi Lake) autumn-spawning populations in 1975–77. For the geographical position of the lakes see Fig. 2.3. Egg samples were taken from 21–30 females each year from each population. Points are mean values for age groups. Curves were computed from the equation $W_d = A + B_1 \tau + B_2 \tau^2$, where W_d is egg dry weight, τ is age group (age groups 1^+, 2^+ etc. were assumed to be 1, 2 etc.) and A, B_1 and B_2 are constants. The line fits were confirmed by the χ^2 test. Arrows, optimal age ($\tau_{opt} = B_1 / -2B_2$). Selected data from Kamler *et al.* (1982).

mass), and the maximum weight for the autumn-spawning populations was found in a female from Lake Pluszne in 1977 (0.965 mg, 2.7 times the minimum). Much larger eggs were produced by winter-spawning *C. albula* from Kajoonjärvi Lake, Finland (not shown in Fig. 3.2): the average weight of an egg in 1976 was 2.04 mg (95% confidence limits 1.96–2.12 mg), and exceptionally large eggs were found in a female from this population (2.42 mg, 6.7 times that of the female from Lake Puruvesi).

The wet weight of eggs produced by *Oncorhynchus mykiss* from different populations (Fig. 3.3) can differ by about 6.6 times, whereas comparisons within age groups differ by only 3.4–4.6 times.

For three years the dry weight of sea trout, *Salmo trutta*, eggs was investigated from four rivers on the Polish coast of the Baltic Sea (Fig. 3.4). These studies also showed a vast range in egg size, from 18.28 mg in a fish from the Rega in 1982 to 48.05 mg in a fish from the Vistula in 1984, 2.6 times the lower value. However, differences between these populations were not always significant and did not persist during the three-year study period. This could have been an artefact caused by different female age structure, but it might also have been the result of human activity, e.g. creation of barriers preventing natural migration of spawners to their native rivers or introduction of smolts into rivers.

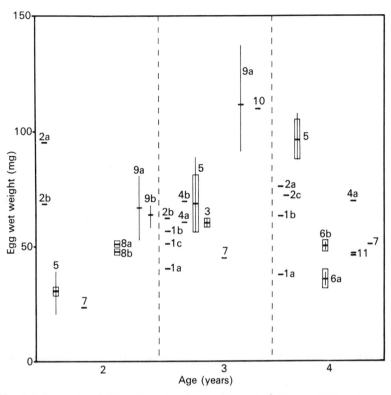

Fig. 3.3 Egg wet weight of *Oncorhynchus mykiss* natural (1a and 1b) and reared (1c to 11) populations. 1, Scott (1962)*, Canada, British Columbia: (1a) Paul Lake, (1b) Pennask Lake (highly significant differences in egg sizes were found between these two lakes, within age groups), and (1c) Smith's Fall Station; 2, Islam *et al*, (1973)*†: (2a) South Japan, Okayama Pref., Hiruzen Stn, (2b) Central Japan, Nagano Pref., Akashina Stn, and (2c) Central Japan, Tochigi Pref., Nikko Stn in 1969–72; 3, Kato (1975)†, Nikko Stn, in 1968–72; 4, Kato (1979): (4a) Nikko strain, and (4b) the large-sized egg strain; 5, Kato and Kamler (1983), Nikko Stn in 1980; 6, Pchelovodova (1976)†, USSR, a station in the Leningrad District: (6a) first spawn, and (6b) second spawn; 7, Savostyanova and Nikandrov (1976), USSR, a station in the Leningrad District; 8, Springate *et al*. (1985), Great Britain: females fed (8a) half ration or (8b) full ration; 9, Leitritz and Lewis (1976)†, USA, California stations: (9a) Mt Whitney Stn, domestic spawners, spring spawn, and (9b) Hot Creek Stn, select spawners, fall spawn; 10, Ridelman *et al*. (1984)*, USA, Seattle, University of Washington Experimental Stn; 11, Galkina (1970), USSR, Ropsha Stn. *Selected examples, †recalculated; modified and completed from Kato and Kamler (1983). Thick lines, means; thin lines, ranges; boxes, 95% confidence intervals.

Comparing populations of *Salmo salar* inhabiting the Neva and Narova Rivers, Kazakov *et al*. (1981) found significant differences ($t = 7.4$) in wet weight of swollen eggs, 151.4 and 127.0 mg, respectively. This is interesting because the population from the Narova is artificial, originating from stocked material that came mainly (>90%) from the Neva. In a northern

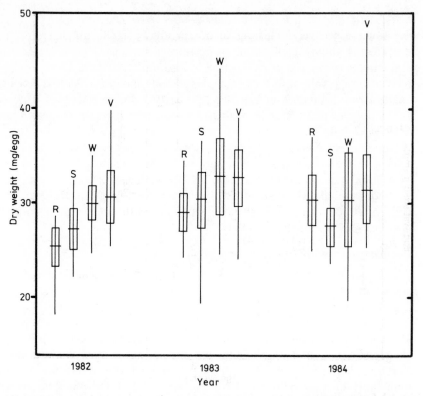

Fig. 3.4 Dry weight of sea trout, *Salmo trutta*, eggs from the Rega (R), Słupia (S), Wieprza (W) and Vistula (V) rivers in 1982–84. Spawn portions from 12 females were analysed each year from each popultion. Horizontal lines, mean values; vertical lines, ranges; boxes, 95% confidence intervals. (Reproduced with permission from Kamler, 1987.)

(Nova Scotia, Canada) population of *Fundulus heteroclitus* examined by Penczak (1985), egg size was smaller than in populations living farther south. Egg volume in *Noemacheilus barbatulus* from Lake Konnevesi (central Finland) was 2.5 times greater that from the River Frome (Southern England) (Mills and Eloranta, 1985). Smirnov *et al.* (1968) have reported size differences in eggs among local populations of *Oncorhynchus* and *Salmo.* Potapova (1978) reported diameters of ripe unfertilized eggs originating from 13 populations of *Coregonus albula* inhabiting the Karelian lakes. For example in Lake Pereslavskoe egg diameter ranged from 1.10 to 1.40 mm (mean 1.25 mm), and in Lake Chuchmozero from 1.40 to 1.90 mm (1.60 mm). The extreme values ranged from 0.50 mm (the smallest eggs in Lake Uros) to 2.36 mm (the largest eggs in Lake Munozero), a 4.7-fold difference. Shatunovskij (1985) indicates that *Oreoleuciscus potanini* from the Mongolian Lake Bon–Tsagan–Nur produced eggs whose wet weight diminished from 0.88 mg in 10-year-old females to 0.85 mg in 16-year-old females, but eggs from Lake Dayan–Nur were heavier, from 1.32 to 1.27 mg, respectively. Turning to invertebrates, significant differences in egg volume among populations of freshwater shrimps, *Palaemon paucidens* and *Paratya compressa*, were reported by Nishino (1980 and 1981, respectively); average volumes for extreme populations showed a 7-fold difference. Examples of differences in egg size (wet weight and dry weight) between populations of many species of Crustacea are presented by Khmeleva and Golubev (1984).

Summing up, differences in egg size among populations can occur within fish species as well as invertebrate species. These differences probably do not exceed one order of magnitude; they can be persistent or transient.

The above discussion dealt with the overall magnitude of egg size differences that can occur among populations. Egg size variations within populations will be presented next.

Comparisons within populations among years

Egg size within a population may vary among collections made in different years. Eggs produced by *Salmo trutta* in the Rega River, during 1983 were 1.15 times as heavy as those in 1982 (Fig. 3.4); this difference was significant ($t = 3.004$, d.f. $= 22$, $P < 0.01$). The egg weight differences between 1982 and 1983 were not significant in other rivers. Galkina (1970) has examined eggs from 100 spawners of Baltic Sea *Salmo salar* obtained at the Narva station (USSR) in 1953 and 1954, eggs of 104 White Sea *S. salar* from the Umb station in 1960–64 and eggs of 171 *Oncorhynchus mykiss* from the Ropsha station in 1962–65. The egg wet weight ranges were 92–188 mg, 62–174 mg and 22–102 mg, respectively, so the smallest eggs were respectively one-half, one-third and one-fifth the size of the largest. No corrections for possible variations in the age composition of spawners were made in the above comparisons. Within-population differences in egg size among years can be also found within age groups. For example, the average wet weight of eggs of *Coregonus albula* from the Karelian Vendyurskoe Lake (Fig. 3.5) was about twice as high in 1965 as in 1964; in 1967 it diminished again. These differences were observed for all age classes. Egg size can vary among years in a 'population' of domestic broodstock as a result of selective breeding (Leitritz and Lewis, 1976). For example in Nikko Station, Japan, where environmental conditions are very stable (Kato, 1979; Kato and Kamler, 1983), the mean wet weight of *Oncorhynchus mykiss* eggs produced by three-year-old females in 1968–72 was 59.9 mg (95% confidence limits 58.0–61.8 mg – Kato, 1975). Later (Kato, 1979) the weight of eggs produced by three-year-old females of the Nikko strain had not changed (59.7 mg), whereas a new, large egg strain was selected, which produced clearly heavier eggs (69.4 mg). The eggs of three-year-old females examined later in 1980–1 (Kato and Kamler, 1983) showed a weight similar to the latter (68.5 mg).

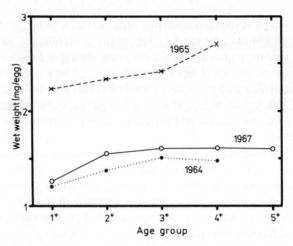

Fig. 3.5. Mean wet weight of eggs of *Coregonus albula* populations from Vendyurskoe Lake (Karelia) in 1964, 1965 and 1967 ($n = 610$, 520 and 2050 eggs, respectively). Based on Potapova's (1978) table 35.

Comparisons within populations within spawning seasons

Changes in size of eggs produced by fish from the same population over the course of a spawning season have been reported by several authors.

A continuous decrease in mean monthly egg volume as the spawning season progressed from winter to late spring was shown by Bagenal (1971) for ten marine species of fish with pelagic eggs. Similarly, Ciechomski (1973) has shown a decrease in egg volume over the course of a spawning season for *Engraulis anchoita* in the waters of the continental shelf off the Argentinian and Uruguayan coasts. In a related, multiple-spawning fish, *Engraulis encrasicolus*, the yolk reserve in eggs from the first batch was higher than that in eggs of later batches (Pavlovskaya, in Nikolskij, 1974). Also, in another serial spawner, *Seriphus politus*, egg diameter decreased by 39% as the spawning season progressed (De Martini and Fountain, 1981).

On the other hand, egg diameter may increase during the season. Egg diameter increased by a factor of *c.* 1.08 during the breeding season (October–April) in *Sardina pilchardus* (Daoulas and Economou, 1986). In a tropical serial spawner, *Cichlasoma nigrofasciatum*, eggs tended to be larger in successive spawnings (Townshend and Wootton, 1984).

In some other fish, egg size tends to be larger in the middle of the breeding season, and smaller at the beginning and end. This pattern was observed by Marsh (1984) in a multiple-spawning species, *Etheostoma spectabile*, whose breeding season stretched from December to March/April. The differences between the largest eggs collected in the middle of the season and the smallest ones was at most twofold. In another multiple-spawning fish, *Misgurnus anguillicaudatus*, in subsequent spawnings induced by monthly injections of hormones, the mean egg diameter increased steadily from the first to the 12th spawning (1.07-fold difference); later, between the 15th and 25th spawnings, the eggs were smaller (Suzuki, 1983). In *Tinca tinca* several batches of eggs are produced during late spring, summer and early autumn. In this species, egg size also initially increased and then it decreased (Stachowiak and Kamler, unpublished data). Mean dry weights per egg in the first five batches were 0.149, 0.151, 0.173, 0.146 and 0.144 mg, respectively.

Thus, for most species of fish, eggs derived from members of the same population can vary over the breeding season. However, the course of these changes differs among species. Three patterns can be identified: a decrease, an increase, and an increase followed by a decrease. In fish that spawn only once per season, the decease in egg size towards the end of the spawning season is explained by a change in age composition of spawners (Bagenal, 1971; Ciechomski, 1973). Older fish spawn first, and the young fish start spawning later on (Kato, 1975); the latter produce smaller eggs (Section 3.7). In multiple-spawning fish with decreasing egg size during the breeding season, the exhaustion of energy reserves can be a decisive factor. An increase in egg size during the spawning season can be determined physiologically. In *Oncorhynchus mykiss*, which spawns once a year, Pchelovodova (1976) has shown for four-year-old females that those spawning for the first time produced smaller eggs than those spawning for the second time. Ware (1975) considered relationships between egg survival, size and incubation time; his model predicts that egg size in summer spawners should decrease over the course of the spawning season, whereas it should increase in autumn spawners. The availability of food for larvae and adults can also explain increases or decreases in egg size through the spawning season (DeMartini and Fountain, 1981). These problems are discussed in detail in Sections 3.7 to 3.10.

Comparisons between females collected together

Variability of dry weight of eggs produced by females collected at the same time and place is shown in Table 3.1 for four species of freshwater fish. Eggs obtained from females collected from the same population in the same year are compared. Eggs from females producing the largest eggs were 1.5 to 2 times heavier than eggs from females producing the smallest eggs. The coefficients of variation calculated for these populations were scattered within the rather narrow range of 8.5 to 17%. It is noteworthy that these variabilities were similar for all species despite considerable interspecific differences in mean egg dry weight – from a fraction of a milligram (*Tinca tinca*) to several tens of milligrams (*Salmo trutta*). Eggs of a freshwater crustacean, *Paratya compressa*, from different water bodies in Japan (Nishino, 1981) were still smaller – about one-tenth as light as eggs of *T. tinca*. However, analogous coefficients of variation calculated from 20 samples of

Table 3.1 Variability and distribution of dry weight of eggs produced by fish collected at the same time and place: *Tinca tinca*, three-year-old females reared in a heated pond in Poland (Stachowiak and Kamler, unpublished data); *Coregonus albula*, lacustrine Finnish and Polish populations, females aged 1–7 years (Kamler et al., 1982); *Oncorhynchus mykiss*, two-year-old females reared in a cold pond in Japan (Kato and Kamler, 1983); *Salmo trutta*, riverine populations, Polish coast of the Baltic Sea, age not determined (Kamler, 1987)

Species	Population*	No. of females	Mean egg weight (mg)	Variability† W_{max} to W_{min} ratio	CV (%)	Skewness‡ sk	t_{sk}	Sign.	d.f.	Normality§ χ^2_{obs}		$\chi^2_{(0.01)}$
T. tinca	Siekierki Stm, 1978	19	0.15	1.7	17.0	−1.242	−4.527	**	1	8.523	>	6.635
C. albula	Puruvesi, 1976	24	0.53	2.0	15.9	−0.009	−0.042	NS	2	2.674	<	9.210
	Kangosjärvi, 1977	30	0.59	1.9	12.7	−0.083	−0.456	NS	2	1.906	<	9.210
	Oulujärvi, 1976	24	0.61	1.5	10.2	0.284	0.000	NS	1	0.810	<	6.635
	Hańcza, 1977	30	0.71	2.0	14.0	−0.759	−4.167	**	3	1.786	<	11.345
	Pluszne, 1976	26	0.72	1.6	15.1	−0.014	−0.070	NS	2	8.406	<	9.210
	Narie, 1976	23	0.72	1.5	11.5	0.090	0.386	NS	2	0.491	<	9.210
	Narie, 1977	30	0.73	1.8	16.9	0.632	3.465	**	2	3.014	<	9.210
	Narie, 1975	22	0.74	1.6	10.4	0.175	0.726	NS	1	4.228	<	6.635
	Pluszne, 1977	30	0.76	1.5	12.1	0.570	3.128	**	2	6.044	<	9.210
	Pluszne, 1975	22	0.76	1.4	9.4	0.070	0.290	NS	1	0.649	<	6.635
	Maróz, 1976	26	0.77	1.4	8.5	0.200	0.964	NS	1	0.699	<	6.635
O. mykiss	Nikko Stn, 1980	22	10.80	1.9	16.8	0.196	0.680	NS	1	3.739	<	6.635
S. trutta	Rega, 1982	12	25.25	1.6	12.0							
	Słupia, 1982	12	27.28	1.5	12.3							
	Wieprza, 1982	12	29.98	1.4	9.8							
	Vistula, 1982	12	30.64	1.5	14.2							

*Listed in order of mean egg weight.
†W_{max}, mean weight of eggs derived from the female producing the largest eggs; W_{min}, mean weight of eggs from the female producing the smallest eggs; CV, coefficient of variation (SD × 100 / mean).
‡Calculated as in Fig. 3.1: NS, not significant ($P > 0.05$); **, $P < 0.01$.
§Normality was tested following Snedecor and Cochran (1967) and verified by χ^2 goodness-of-fit test. If $\chi^2_{obs} > \chi^2_{(0.01)}$ the hypothesis of normality is rejected at the 1% level.

eggs ranged from 3.6 to 17%, so the range was similar to that presented for the fish in Table 3.1. Single batches of 50 eggs from each of 22 female *Oncorhynchus mykiss* were individually weighed (Table 3.1). Differences in egg dry weight among these females were highly significant (one-way analysis of variance, $F = 118.15$; d.f. $= 21$, 1078; $P < 0.001$, Sokal and Rohlf, 1969). Similarly in *Etheostoma spectabile* inhabiting rivers in central Texas, Marsh (1984) has shown statistically significant intra-locality among-female differences in dry weight of eggs. This has been observed in a majority of field collections and also among females reared in the same laboratory troughs (Marsh, 1984, 1986).

So differences have appeared in egg size between females collected at the same time and place, both under field conditions and in populations cultured in a pond, as well as under controlled laboratory conditions. This means not only that environmental factors can affect differences in egg size between females, but also that egg size is genetically determined.

The distribution of egg dry weight was normal and unskewed in the majority of populations presented in Table 3.1. Similarly Marsh (1984) reported that egg dry weights within a collection of *Etheostoma spectabile* generally did not deviate from the normal distribution. Thus, the most numerous group of females in each population produces medium-sized eggs. This indicates that selection occurs for optimal egg size in any given environment.

Intra-individual comparisons

In this section we shall consider the variability of eggs obtained from one female within one spawn portion, i.e. the variability derived from measurements of individual eggs. The ratio between the smallest and largest egg diameter was 1:1.5 in a spawn portion of *Clupea harengus* (Blaxter and Hempel, 1963). Egg diameter of *Engraulis anchoita* varied by a factor of 1.5–1.6 within the same female (Ciechomski, 1966). Values for these factors (S_{max} to S_{min} ratios) obtained during last two decades are given in Table 3.2. For linear properties (diameter, length, breadth) they range from 1.1 to 1.6, and for gravimetric parameters (wet weight, dry weight) they are, of course, higher and range from 1.2 to 3.6. Coefficients of variation for linear properties of eggs obtained from a single female amounted to c. 1–10%, and for gravimetric properties c. 1–30%.

The data compiled in Table 3.2 do not show any clear interspecific differences in variability of size of eggs produced by one female. On the other hand, intraspecific differences do seem to occur. Studies on *Oncorhynchus mykiss* (egg length, breadth, wet and dry weight) and on *Coregonus albula* (egg diameter) are used to illustrate the question. The increase in egg size with age was paralleled by a decrease in egg variability (Table 3.2); the variability of the least variable eggs was more uniform in all age groups than

Table 3.2 Intra-individual egg size variability

Species (age, years)	No. of females	Variable measured	Egg hydration*	Mean egg size	S_{max} to S_{min} ratio‡	CV(%)	Source§
		Linear Properties					
Misgurnus anguillicaudatus (?)	3	Diameter,	p.u.e.	0.8††	1.3–1.6††	–	1
Cyprinus carpio (?)	2	Diameter,	p.u.e.	1.2††	1.2††	–	2
C. carpio (8)	6	Diameter,	s.e.	1.9	1.1–1.3	2.4–7.1	3
Rutilus rutilus heckeli (?)	113	Diameter,	p.u.e.	–	–	mean 2.7	4
Abramis brama (?)	83	Diameter,	p.u.e.	–	–	mean 3.1	4
Coregonus albula (1⁺)	8	Diameter,	p.s.w.	1.8	–	1.6–5.4	5
C. albula (2⁺)	4	Diameter,	p.s.e.	1.9	–	1.5–2.7	5
Oncorhynchus mykiss (2)	22	Length,	s.e.	3.9	1.1–1.4	2.9–6.0	6
O. mykiss (3)	6	Length,	s.e.	5.1	1.1–1.3	2.5–6.0	6
O. mykiss (4)	5	Length,	s.e.	5.7	1.1–1.2	2.3–3.0	6
O. mykiss (2)	22	Breadth,	s.e.	3.7	1.1–1.3	2.8–9.4	6
O. mykiss (3)	6	Breadth,	s.e.	4.8	1.1–1.3	2.6–6.9	6
O. mykiss (4)	5	Breadth,	s.e.	5.5	1.1–1.2	2.1–3.7	6

Gravimetric properties

Oncorhynchus mykiss (2)	22	Dry wt,	s.e.	10.8	1.4–2.4	6.7–19.0	6
O. mykiss (3)	6	Dry wt,	s.e.	24.1	1.2–3.6	3.9–16.9	6
O. mykiss (4)	5	Dry wt,	s.e.	34.2	1.2–1.4	4.7–8.1	6
O. mykiss (3)	48	Wet wt,	p.s.e.	55.0††	–	7–27††	7
O. mykiss (3–7)	171	Wet wt,	s.e.	54.0	–	1–24	8
O. mykiss (2)	22	Wet wt,	s.e.	30.6	1.2–2.2	5.4–15.6	6
O. mykiss (3)	6	Wet wt,	s.e.	68.5	1.2–2.6	4.1–17.0	6
O. mykiss (4)	5	Wet wt,	s.e.	96.1	1.2–1.4	4.7–8.0	6
Oncorhynchus nerka (3)	47	Wet wt,	p.s.e.	65.0††	–	3–11††	9
Salmo salar (5–7)	204	Wet wt,	s.e.	130.7	–	2–10	8

*p.u.e., probably unswollen eggs; s.e., swollen eggs; p.s.e., probably swollen eggs.
†Mean size of eggs obtained from all females (mm for linear and mg for gravimetric properties).
‡S_{max} and S_{min}, individual size of the largest and smallest egg, respectively, derived from an individual female.
§Sources: 1, Suzuki (1976); 2, Tomita et al. (1980); 3, Kamler and Malczewski (1982); 4, Vladimirov (1974a); 5, Dąbrowski et al. (1987); 6, Kato and Kamler (1983); 7, Kato (1975); 8, Galkina (1970); 9, Kato (1978).
††Approximate value read from a graph.

that of the most variable eggs. In other words, some young females produced exceptionally variable eggs. Significant negative correlations were found between coefficients of variation and mean sizes of eggs obtained from individual *Oncorhynchus mykiss* females ($n = 33$): for egg length, $r = -0.404$, $P < 0.05$, and for egg dry weight, $r = -0.361$, $P < 0.05$. This result corroborates earlier data for wet weight of eggs laid by 48 three-year-old *Oncorhynchus mykiss* ($r = -0.63$, $P < 0.001$, Kato, 1975) as well as data for wet weight of eggs from 47 three-year-old *O. nerka* ($r = -0.61$, $P < 0.001$, Kato, 1978). It is interesting that these negative relationships are not limited only to egg weight. In *Misgurnus fossilis* the coefficient of variation of egg fertilizability was found to decrease significantly with the increased mean fertilizability of eggs derived from individual females; also the hatchability was less variable in spawn portions that exhibited a high percentage of hatched embryos ($r = -0.94$, $P < 0.001$ and $r = -0.95$, $P < 0.001$, respectively, Suzuki, 1983). Variability of egg weight of *Rutilus rutilus* was largest in young females producing small eggs (Lyagina, 1975), whereas females of average age produced large, less variable eggs (Nikolskij, 1974). In contrast, Galkina (1970) did not find any relationship between weight variability of eggs produced by females and their age. Potapova (1978) reported that in *Coregonus albula* the decrease in egg diameter was paralleled by an increase in egg variability in years of unfavourable trophic conditions. Fatty females of *Clupea harengus* produced eggs that were less variable than those derived from lean females (Anokhina, 1960).

Mechanisms of fish egg size regulation in ovaries, and the basis for variability, have been explained by Mejen (1940). Under favourable trophic conditions, all oocytes situated in the vicinity of either large or small blood vessels are supplied with sufficient nutrients. With decreased food availability, on the other hand, the flow of nutrients to oocytes situated near small blood vessels is restricted. This results in an increase in variability and a decrease in average size of spawned eggs. Ozernyuk (1985) reviewed yolk protein transport from the liver to the gonads and the pinocytic incorporation of these proteins by oocytes. Townshend and Wootton (1984) examined histologically the ovaries of *Cichlasoma nigrofasciatum* kept for 4 months at high, medium and low food rations. The proportion of oocytes that were vitellogenic was 38%, 12% and 3%, respectively.

A great diversity of oocyte recruitment strategies exists in teleosts with group-synchronous and asynchronous ovaries (review: Wallace and Selman, 1981). A few examples are given below. Size distribution of immature ovarian eggs is multimodal in many teleosts, i.e. the ovary contains oocytes of various size groups (Kato, 1975; Suzuki, 1976; Tomita *et al.*, 1980; Fenerich-Verani *et al.*, 1984; Garcia and Brana, 1988). Fenerich-Verani *et al.* (1984) analysed oocyte size composition in *Prochilodus scorfa* before hormone treatment; only eggs larger than 735.55 μm were considered. Females showing a multimodal ovarian egg distribution did not respond to

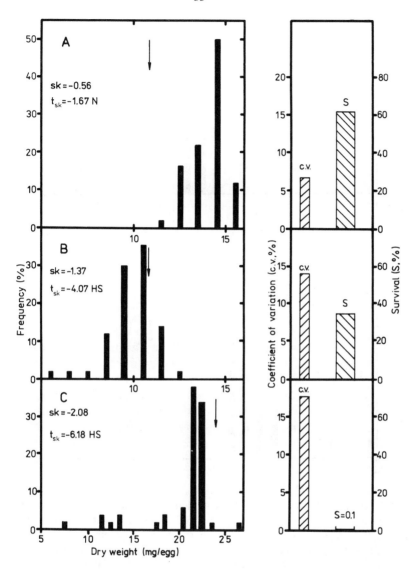

Fig. 3.6 Examples of intra-individual distribution of *Oncorhynchus mykiss* egg dry weight, as compared with embryo survival from fertilization to hatch. Dry weight of 50 eggs was measured individually in spawn portions obtained by artificial spawning from each of three females: (A) a two-year-old, (B) another two-year-old, and (C) a three-year-old. Skewness (see Fig. 3.1) and coefficient of variation of dry weight (%) are also shown; distribution (A) is unskewed, (B) and (C) are highly significantly skewed. Arrows, mean dry weight of eggs produced by all measured two-year-old females (22 females, 10.8 mg egg^{-1} in (A) and (B)) and three-year-old females (6 females, 24.1 mg egg^{-1} in (C)). Based on raw numerical data collected in the Nikko Branch by Kato and Kamler (1983).

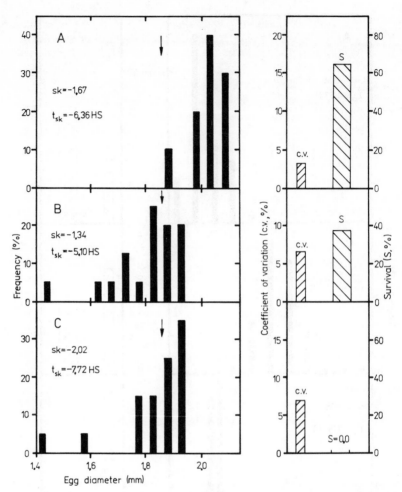

Fig. 3.7 Examples of intra-individual distribution of *Cyprinus carpio* egg diameter, as compared with embryo survival during the first 24 h after fertilization. The diameters of 20 eggs were measured individually in spawn portions obtained by an early (13–14 May), hormone-induced artificial spawning from each of three eight-year-old carp ((A), (B) and (C)). Arrows, mean diameter of eggs (1.857 mm). Based on raw numerical data collected in the Gołysz Fish Farm by Kamler and Malczewski (1982).

hormone treatment and those with a bimodal distribution produced nonfertilizable eggs. Successful fertilization was observed only in females in which the frequency distribution of ovarian eggs larger than $735.55\,\mu m$ was unimodal and symmetrical, and the mean egg diameters varied within a narrow range. Fenerich-Verani *et al.* (1984) therefore postulated that during the maturation process of species with total spawning, a synchronization of egg size occurs. Similar conclusions can be drawn from graphs presented by

Suzuki (1976) for *Misgurnus anguillicaudatus* and by Tomita *et al.* (1980) for *Cyprinus carpio*: small, immature ovarian eggs are multimodally distributed over a wide range of sizes, whereas the variability of large, spawned eggs is small and their size distribution is close to normal.

Kato and Kamler (1983) measured length, breadth, wet weight and dry weight of 50 eggs artificially spawned from each of 33 female *Oncorhynchus mykiss* (Table 3.2). They calculated the skewness of the distribution for a total of 132 samples. In the majority of results (78 out of 132) the distribution was unskewed – an example of an unskewed distribution is given in Fig. 3.6(A). Studies by Koyama and Kira (1956) on plants and by Wilbur and Collins (1973) on amphibians have shown that the weight frequency distribution is approximately normal at the initial stages of an organism's development. Turning to our example of *Oncorhynchus mykiss*, in 54 cases out of 132 a skewed distribution was observed. In all these cases but one, significant skewness was negative, i.e. the distribution was skewed to the left (Fig. 3.6(B) and (C)). This indicates the presence of a group of relatively small eggs in each of these spawn portions. Their presence can diminish the average size of eggs in a given spawn portion and increase variability of egg size (Fig. 3.6 and 3.7). Small eggs could have been undernourished and/or immature, and so could decrease the survival of embryos in the whole spawn portion (Fig. 3.6 and 3.7). The supposition of undernourishment in the case of rainbow trout in Nikko and carp in Gołysz might be invalid since spawners were fed very well. A more probable hypothesis is that some portion of small immature eggs were expelled as a result of handling during artificial spawning. An answer to this question could be from comparative studies of size frequency distributions in artificially and naturally spawned eggs.

3.2 CALORIC VALUE OF EGG DRY MATTER

The energy content of an organism is the total amount of energy per individual (J or cal indiv^{-1}), whereas the caloric value (h in Chapter 2) is the amount of energy per unit mass (J or cal mg^{-1}). Ranges of caloric values of dry matter for fish and aquatic invertebrates overlap (Table 3.3). It should be mentioned, however, that caloric values both of eggs and of body tissues are somewhat higher in fish than in invertebrates. Similarly, grand mean caloric values for aquatic invertebrates, as reported by Cummins and Wuycheck (1971), were lower than those of Osteichthyes. From Table 3.3 it can be seen that energy in eggs is more 'condensed' than in the body: the caloric values of egg dry matter, both in fish and in invertebrates, are higher by 20–25% than those of body tissues. Wootton (1979), using a large amount of material (60 observations for 50 species), has calculated the average caloric value of fish eggs as 23.48 J mg^{-1} dry weight (95%

Table 3.3 Caloric value ($J\,mg^{-1}$ dry weight) of eggs and soma of fish and aquatic invertebrates – a general comparison

Material		No. of species	Most frequent values		Median value
			Range	% spp.	
Eggs:	fish*	18	23.4–29.3	72	26.4
	invertebrate†	65	20.9–28.5	89	24.7
Soma:	fish‡	12	18.8–23.4	75	21.1
	invertebrate§	44	15.1–23.9	78	19.5

*Based on Table 3.4.

†Based on Khmeleva and Golubev (1984), 62 crustacean species; Pandian (1969) *Crepidula fornicata* (Mollusca); Pilarska (1977), *Brachionus rubens* (Rotatoria) resting eggs; Kosiorek (1979), *Tubifex tubifex* (Oligochaeta).

‡Compiled from Toetz (1966), *Lepomis macrochirus*; Fischer (1970b), *Ctenopharyngodon idella*; Cummins and Wuycheck (1971), *Cottus bairdi, Lepomis macrochirus, L. gibbosus*; Jezierska, (1974), Craig *et al.* (1978) and Dgebuadze and Kamler (unpubl.), *Perca fluviatilis*; Mironova (1977), *Tilapia mossambica*; Penczak *et al.* (1976, 1977), *Rutilus rutilus*; Penczak *et al.* (1978) and Moliński *et al.* (1978), *Leuciscus leuciscus*; Diana and Mackay (1979), *Esox lucius*; Kamler and Żuromska (1979), *Coregonus albula*; Staples and Nomura (1976) and Galicka (1984), *Oncorhynchus mykiss*; Stachowiak and Kamler (unpubl. data), *Tinca tinca*.

§Based on Prus (1970); for Mollusca only the data for bodies without shells were included.

confidence limits 22.75–24.21). So the caloric value of dry matter of fish eggs is especially high. This is the result of high concentrations of lipids; information on the chemical composition of fish eggs is presented in Section 3.3.

Interspecific comparisons

The caloric value of fish eggs varies little among species (*c.* twofold difference, Table 3.4, Fig. 3.8) as compared with egg weight (several orders of magnitude, Fig. 3.1). Consequently, caloric value may contribute to the variability in the energy content of an egg, although to a lesser degree than egg weight. The high caloric value of *Anguilla anguilla* oocytes (Table 3.4) is probably related to the high energy requirements of the larvae to travel over considerable distances (over 7000 km) on their return from the spawning grounds; it is likely, however, that the oocytes did not complete their vitellogenesis (GSI 32–47%) and thus the proportions of lipids and proteins were shifted towards lipids (see Section 2.1 and Epler *et al.*, 1981b).

A. anguilla is an exception because caloric values of eggs for most marine fish occur lower in the distribution than eggs of species spawning in fresh waters (Fig. 3.8(B)). One should remember that egg size (Fig. 3.1) follows a similar pattern. This was studied in detail by Faustov and Zotin (1967), who summarized data for 50 species of fishes of five ecological groups named after their respective spawning grounds: marine pelagophilous, marine bottom spawning, freshwater pelagophilous, freshwater phytophilous and freshwater lithophilous. After recalculation, the respective mean caloric values are 21.3, 21.9, 22.6, 23.7 and 25.0 J mg^{-1} dry weight, the mean egg dry

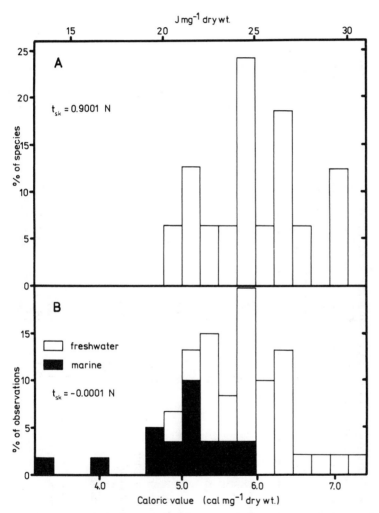

Fig. 3.8 Frequency distribution of caloric value of fish eggs or mature ovaries: interspecific comparisons. (A) Mean values for 16 species listed in Table 3.4 (each species is represented by a single value; erroneous data of Malyarevskaya and Birger, 1965, for two species are excluded); (B) 60 values from 50 teleost species compiled by Wootton (1979). Neither distribution is skewed nor departs from normality (see text).

Table 3.4 Interspecific comparison of caloric values of freshly spawned fish eggs (E)* or mature ovaries (MO)*. Based on 20 sets of data from 18 fish species spawning in fresh waters (f) or in the sea (s). Taxonomy according to Nelson (1976)

Family and species		Caloric value ($J\,mg^{-1}\,dry\,wt$)			Method†	Source
		Mean	95% conf. lim.	Range		
Clupeidae						
1. Clupea harengus pallasi	E, s	25.2	–	–	B	1
2. Sardinops caerulea	MO, s	22.6	21.9–23.2	–	B	2
Anguillidae						
3. Anguilla anguilla	MO, s	30.1	–	–	C‡	3
Esocidae						
4. Esox lucius	MO, f	24.1	–	–	B	4
Salmonidae						
5. Oncorhynchus nerka	MO, f	27.5	–	–	B	5
6. Salmo trutta m. trutta	E, f	27.5	–	25.2–30.4	B	6
7. Oncorhynchus mykiss	E, f	29.5	–	–	C‡	7
8. O. mykiss	E, f	27.8	27.6–28.1	25.9–28.8	B	8
9. Coregonus albula	E, f	27.0	–	23.8–35.3	B	9
Cyprinidae						
10. Rutilus rutilus heckeli	E, f	17.9§	–	–	C	10
11. Abramis brama	E, f	17.7§	–	–	C	10
12. Tinca tinca	E, f	25.5	25.2–25.7	23.4–27.3	B	11
13. Cyprinus carpio	E, f	25.2	24.8–25.5	23.4–26.4	B	12, 13, 14

Oryziatidae					
14. *Oryzias latipes*	E, f	23.6	–	–	B	15
Percichthyidae					
15. *Morone saxatilis*	E, f	30.9	28.5–33.3	26.2–34.5	B‡	16
Centrarchidae					
16. *Micropterus salmoides*	E, f	25.1	23.0–27.1	–	B	17
17. *Lepomis macrochirus*	E, f	21.3	–	–	B	18
18. *L. macrochirus*	E, f	24.4	–	–	B	19
Cichlidae					
19. *Tilapia mossambica*	E, f	26.2	–	–	?	20
Cottidae					
20. *Cottus bairdi*	E, f	22.7	22.3–23.1	21.1–24.7	B	21

*Lasker (1962) showed the similarity of the major organic constituents in mature ovarian tissue and yolk.

†B, caloric value measured directly using a bomb calorimeter; C, computed by the author(s) from chemical composition.

‡Caloric values computed from data reported by the author(s).

§Caloric value under-estimated owing to incomplete lipid extraction.

Sources: 1, Eldridge et al. (1977); 2, Lasker (1962); 3, Epler et al. (1981b); 4, Diana and Mackay (1979); 5, Nishiyama (1970); 6, Kamler (1987); 7, Suyama and Ogino (1958); 8, Kato and Kamler (1983); 9, Kamler et al. (1982); 10, Malyarevskaya and Birger (1965); 11, Stachowiak and Kamler (unpubl.); 12, Kamler (1972a); 13, Kamler (1976); 14, Kamler and Malczewski (1982); 15, Hirshfield (1980); 16, Eldridge et al. (1982); 17, Laurence (1969); 18, Cummins and Wuycheck (1971); 19, Toetz (1966); 20, Mironova (1977); 21, Docker et al. (1986).

weights are 0.06, 0.29, 0.46, 0.63 and 14.59 mg, and the resulting mean energy content is 1.3, 6.7, 10.5, 14.8 and 365.6 J per egg.

Thus, the differences in egg size are paralleled by differences in caloric values. Both of these participate in differences of energy content per egg.

In interspecific comparisons, the distribution of caloric values for fish eggs is unskewed (Fig. 3.8). It does not deviate from normality: for Fig. 3.8(A), $\chi^2_{obs} = 1.104 < \chi^2_{theor} = 6.635$ (d.f. = 1); for Fig. 3.8(B), $\chi^2_{obs} = 7.527 < \chi^2_{theor} = 16.812$ (d.f. = 6). A normal distribution for the frequency of caloric values has been suggested by Paine (1965). Prus (1970) has shown that this distribution is normal for somatic tissues of 64 species of aquatic animals. However, his diagram for seeds of 51 species of terrestrial plants shows skewness to the right, indicating a preponderance of low values. Such a distribution of caloric values would have been favoured under conditions of strong selection because it offers the possibility of producing more progeny for the same reproductive effort. Contrary to what Prus (1970) has shown for the seeds of terrestrial plants, it is evident that caloric values of fish eggs show a normal distribution, just as the bodies of aquatic animals do. This means that some particular combination of proteins, lipids, carbohydrates and minerals would be favoured, and any strong deviations in either direction from this optimum composition would be eliminated (Prus, 1970).

One might intuitively expect that the distribution of organism size would be normal, but that caloric values, being ratios, would have no such distribution. As it turns out, for interspecific comparisons of fish egg size (Fig. 3.1) and their caloric values (Fig. 3.8), just the opposite is true.

Intraspecific comparisons

Differences in caloric value of dry matter for eggs of *Coregonus albula* from different populations are shown in Table 3.5. These differences were not conspicuous. Caloric values of eggs produced by fish from Finnish lakes tend to group at the lower part of the range (Table 3.5, Fig. 3.9(C)). These differences in caloric values depict, although to some small degree, differences in egg weights: vendace eggs from Finland were smaller than those from Poland (Fig. 3.2).

Statistically significant differences in caloric values of eggs were found among collections made in different years (Table 3.5). In three Polish lakes – Pluszne, Maróz and Narie – the caloric values of dry matter of *Coregonus albula* eggs were examined in three subsequent years (1975–77); these values were consistently highest in 1977 for all these populations. This was similar to the situation described for egg weights (Fig. 3.2). A highly significant difference ($P < 0.001$) was found between caloric values of *Salmo trutta* eggs in 1982 and 1983 (Table 3.5); the values for 1983 were higher. Also in this case the changes in caloric value of eggs between years kept pace with changes in egg size (Fig. 3.4).

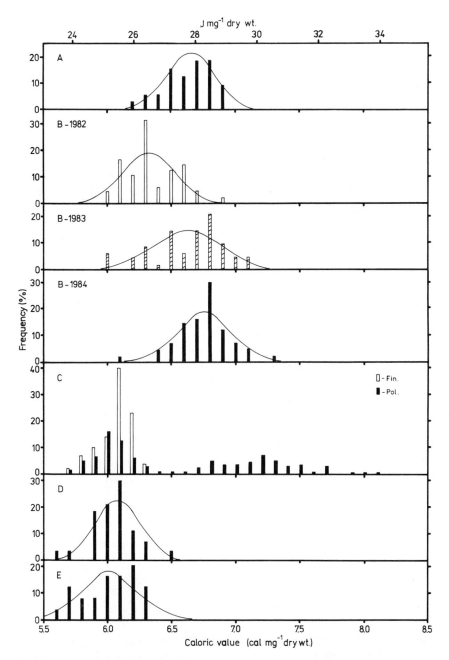

Fig. 3.9 Examples of frequency distributions for egg caloric value in three salmonid species: (A) *Oncorhynchus mykiss*; (B) *Salmo trutta* in 1982–84; (C) *Coregonus albula* in 4 Finnish (open columns) and 5 Polish (filled columns) lakes; and in two cyprinid species: (D) *Tinca tinca*; (E) *Cyprinus carpio*. Further details are given in Table 3.1. Theoretical (normal) distributions are depicted by smooth curves.

Table 3.5 Levels and distributions of caloric value of newly spawned egg dry matter in (A–C) three salmonid and (D–E) two cyprinid species

Species and material*	No. of females	Caloric value ($J\,mg^{-1}$)		Skewness†			Normality‡		
		Mean	95% conf. lim.	sk	t_{sk}	d.f.	χ^2_{obs}		$\chi^2_{(0.01)}$
A. *Oncorhynchus mykiss*									
	33	27.85	27.58–28.12	−0.193	−0.472 NS	2	1.35	<	9.21
B. *Salmo trutta*									
1982	48	26.48	26.23–26.73	0.506	1.474 NS	3	9.81	<	11.35
1983	47	27.74	27.40–28.08	−0.600	−1.730 NS	5	10.24	<	15.09
1984	43	28.29	28.02–28.57	−0.300	−0.830 NS	3	4.42	<	11.35
C. *Coregonus albula*									
Pol. S77	30	30.15	29.48–30.82	0.030	0.071 NS	3	6.72	<	11.35
Pol. H77	30	28.89	28.38–29.41	−0.409	−0.958 NS	2	2.36	<	9.21
Pol. P75	21	25.48	25.27–25.69	0.600	1.197 NS	1	0.76	<	6.54
Pol. P76	26	25.48	25.36–25.59	0.333	0.732 NS	1	5.62	<	6.64
Pol. P77	30	30.09	29.62–30.57	0.434	1.017 NS	2	4.04	<	9.21

	n	mean	range						
Pol. M75	21	25.01	24.57–25.45	2.658	5.305 HS*	1	5.73	<	6.64
Pol. M76	26	24.79	24.20–25.37	0.200	0.439 NS	1	0.87	<	6.64
Pol. M77	26	30.62	30.04–31.20	1.168	2.563 S*	2	1.25	<	9.21
Pol. N75	22	25.37	25.18–25.56	−0.900	−1.833 NS	1	1.54	<	6.64
Pol. N76	23	25.07	24.90–25.25	−0.714	−1.484 NS	1	5.90	<	6.64
Pol. N77	30	29.09	28.47–29.71	−0.263	−0.616 NS	2	4.14	<	9.21
Fin. K76	23	25.71	25.58–25.85	0.750	1.558 NS	1	5.65	<	6.64
Fin. Kg77	30	25.58	25.44–25.73	0.375	0.879 NS	1	13.73	>	6.64
Fin. Ou76	24	25.38	25.19–25.56	−0.500	−1.059 NS	1	2.69	<	6.64
Fin. Pu76	23	24.91	24.66–25.17	0.035	0.073 NS	1	1.66	<	6.64
D. *Tinca tinca*	27	25.44	25.15–25.74	0.087	0.194 NS	2	1.85	<	9.21
E. *Cyprinus carpio*	24	25.17	24.79–25.55	−0.222	−0.470 NS	1	2.11	<	6.64

*A, ponds in central Japan (Kato and Kamler, 1983); B, four near-Baltic rivers in 1982–84 (Kamler, 1987); C, Polish (Pol.) and Finnish (Fin.) lakes in 1975–77 (for abbreviations see Fig. 2.3) (Kamler et al., 1982) (Kamler et al., 1982); D, heated ponds in Poland (Stachowiak and Kamler, unpubl.); E, carp ponds in Poland (Kamler, 1972a, 1976; Kamler and Malczewski, 1982).

†See Fig. 3.1. NS, not significant ($P > 0.05$); S, significant ($P < 0.05$); HS, highly significant ($P < 0.01$).

‡See Table 3.1.

As can be seen from the examples given in Fig. 3.9 and Table 3.5 for five fish species, intraspecific distributions of the frequency of caloric values of eggs are usually unskewed and do not deviate from normality. Although the frequency distribution presented in Fig. 3.9(C) for *Coregonus albula* deviates from normal, this results from the nonhomogeneity of the material: the Figure includes data for many different populations from different years. The same data, recomputed for each population and year separately (Table 3.5), did not show any deviations from normality in 14 cases out of 15.

3.3. EGG COMPOSITION

In earlier Sections, an egg was treated as a single entity. This Section investigates its interior. After a very brief outline of egg structure and properties, which is essential for understanding what follows, the water content and main chemical components in an egg will be discussed.

Egg structure and properties

In a freshly laid egg, there is no water between the outer porous shell (egg capsule, chorion) and the inner protoplasmic egg membrane. The chorion of fresh eggs is soft and possesses an operculum–micropyle through which the sperm cell enters the egg. When the egg is released into water and fertilized, the cortical alveoli present under the chorion burst out and release a colloidal material – mucoproteins – into the perivitelline space, which occurs between the egg membrane and the chorion (Bogucki, 1930). Water is drawn in as a result of swelling of these mucoproteins. The chorion first becomes rigid and smooth, then it becomes hard and the micropyle is closed. The cytoplasm thickens at the pole of the egg holding the nucleus: this is the spot where the embryo develops after fertilization. The interior of the egg is filled with yolk. The eggs of different fish species differ in their structure and physical properties. These problems were examined long ago (reviews: Hayes, 1949, Smith, 1957; Grodziński, 1961).

The liquid phase of yolk is yolk fluid (ichthulin) (Szubińska-Kilarska, 1959; Szubińska, 1961; Grodziński, 1961; Devillers, 1965). In Salmonidae free fat droplets are suspended in ichthulin, whereas in *Perca fluviatilis*, *Gymnocephalus cernuus*, *Misgurnus fossilis* and *Cyprinus carpio* there is little ichthulin but the main component of the yolk is yolk spheres (Szubińska-Kilarska, 1959; Szubińska, 1961). A yolk sphere has a diameter of about 10 μm. It consists of a membrane and a protein–lipid fluid which fills its interior. Membranes of yolk spheres contain proteins and lipid components, which occur in different proportions in various species. In yolk-sphere membranes of *Esox lucius* the lipid components are abundant, membranes of *Gymnocephalus* and *Perca* show intermediate character, and membranes of *Cyprinus* and *Misgurnus* consist largely of proteins (Szubińska, 1961).

Water content of egg matter

The hydration of unswollen eggs just before or just after spawning ranges from 47 to 83% of wet matter (Table 3.6). The average value for the 30 species listed in Table 3.6 is 64.4%. The hydration of fish eggs is lower than that of their soma, which, in general, contains 80 to 85% water (Love, 1957). In Stroganov's (1962) compilation for somatic tissues of sexually mature fish from 17 taxonomic groups (81 species), in 12 groups of 17 the mean water content for the group ranged from 70 to 80%; the findings of Bogucki and Trzesiński (1950) on *Gadus morhua*, Stepanova and Tyutyunik 1973) on *Ctenopharyngodon idella*, Chechenkov (1973) on *Coregonus albula*, Fischer (1976) on *Cyprinus carpio* and Mironova (1977) on *Tilapia mossambica* are within the 70–85% limits. Hydration of eggs of Crustacea (range 15–78%, average 56.7%; based on 95 sets of data for 65 species, Khmeleva and Golubev, 1984) is in general similar to that of fish eggs, although some species have eggs that are less hydrated than fish eggs.

Interspecific comparisons

Interspecific comparisons (Table 3.6, Fig. 3.10) have shown that the hydration of eggs of marine fish species assumes higher values than those of species spawning in fresh waters; see also Wallace and Selman (1981). It is interesting that the eggs of migratory species (*Oncorhynchus* spp., *Salmo salar* and *Salmo trutta* m. *trutta*), who grow mainly in the sea but reproduce in fresh waters, as well as eggs of Acipenseridae, contain less water than those of species spawning in the sea, less than 62%. The hydration of

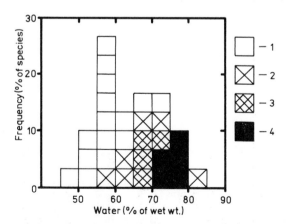

Fig. 3.10 Frequency distribution of percentage water content in fish eggs: interspecific comparisons of mean values for 30 species listed in Table 3.6. (1) to (3), species spawning in fresh waters; (4), species spawning in the sea. (1), Acipenseridae and Salmonidae; (2), Esocidae, Osmeridae, Siluridae, Percidae and Cichlidae; (3), Cyprinidae; (4), Clupeidae, Anguillidae and Gadidae.

Table 3.6 Interspecific comparisons of water content in freshly spawned fish eggs (E) or mature ovaries (MO). Based on 53 sets of data from 30 fish species spawning in fresh waters (f) or in the sea (s)

Family and species		Water (% wet wt)			Mean value for species	Source*
		Mean	95% conf. lim.	Range		
Acipenseridae						
1. Huso huso	E, f	58.9	–	–	58.9	1
2. Acipenser sturio	E, f	53.9†	–	51.4–56.4	53.9	1
3. A. stellatus	E, f	51.4†	–	47.5–55.2	55.8	1
4. A. stellatus	E, f	60.1†	56.5–63.7†	58.5–61.3†	55.8	2
Clupeidae						
5. Clupea harengus	E, s	76.0	–	74.0–78.0	76.0	3
6. Sardinops caerulea	MO, s	70.7	69.7–71.7	–	70.7	4
Anguillidae						
7. Anguilla anguilla	MO, s	77.2	–	76.1–78.2	78.4	5
8. A. anguilla	MO, s	79.5	77.4–81.6	–	78.4	6
Esocidae						
9. Esox lucius	E, f	64.3	–	–	65.7	7
10. E. lucius	E, f	67.0	–	–	65.7	1
Salmonidae						
11. Coregonus lavaretus	E, f	60.8	–	–	60.8	8
12. C. albula	E, f	69.5†	–	67.2–74.8†	69.5	9
13. C. pollan	MO, f	71.6†	67.8–75.4†	–	71.6	10
14. Oncorhynchus tshawytscha	E, f	48.1	–	–	48.1	11
15. O. nerka	E, f	59.0	–	–	59.7	1
16. O. nerka	E, f	60.3†	–	59.9–60.9†	59.7	12
17. O. keta	E, f	53.4†	–	51.3–55.5†	54.6	1

18.	*O. keta*	E, f	55.7†	—	55.1–56.3†	54.6	12
19.	*O. gorbuscha*	E, f	56.6†	—	55.3–57.1†	57.4	12
20.	*O. gorbuscha*	E, f	58.3†	—	56.6–59.4†	57.4	12
21.	*O. kisutch*	E, f	59.9†	—	58.8–61.5†	59.9	12
22.	*O. masu* (1959)	E, f	54.4†	—	52.9–55.3†	54.6	12
23.	*O. masu* (1960)	E, f	54.8†	—	53.3–55.2†	54.6	12
24.	*Salmo salar*	E, f	59.8†	—	58.6–61.2†	59.8	12
25.	*S. trutta m. trutta*	E, f	58.5†	—	52.7–60.8†	59.7	13
26.	*S. trutta m. trutta*	E, f	60.9†	—	59.8–62.0	59.7	14
27.	*O. mykiss* (full ration)	E, f	54.5	53.9–55.1	—	60.3	15
28.	*O. mykiss* (half ration)	E, f	55.7	54.9–56.5	—	60.3	15
29.	*O. mykiss* (fed)	MO, f	58.2	57.8–58.6†	—	60.3	16
30.	*O. mykiss* (fed)	E, f	58.5	—	—	60.3	17
31.	*O. mykiss* (fed)	E, f	59.0	—	—	60.3	18
32.	*O. mykiss* (fed)	E, f	61.8†	61.1–62.5	55.6–65.0	60.3	19
33.	*O. mykiss* (fed)	E, f	62.5	—	—	60.3	20
34.	*O. mykiss* (fed)	E, f	66.2	66.0–66.4	—	60.3	21
35.	*O. mykiss* (fed)	E, f	66.2	—	—	60.3	22
Osmeridae							
36.	*Osmerus eperlanus*	E, f	62.0	—	—	62.0	1
Cyprinidae							
37.	*Leuciscus idus*	E, f	67.0	—	—	67.0	1
38.	*Abramis brama*	E, f	64.8	—	—	66.4	1
39.	*A. brama*	E, f	68.0	—	—	66.4	23
40.	*Rutilus rutilus heckeli*	E, f	67.3	—	—	67.4	23
41.	*R. rutilus caspicus*	E, f	67.6	—	—	67.4	1
42.	*Cyprinus carpio*	E, f	67.0	—	—	70.5	24
43.	*C. carpio*	E, f	69.0	—	68.0–70.0	70.5	25
44.	*C. carpio*	E, f	70.0	—	–	70.5	1
45.	*C. carpio*	E, f	70.6	—	68.7–72.5	70.5	26

Cont'd overleaf

Table 3.6 *cont'd*

Family and species		Water (% wet wt)			Mean value for species	Source*
		Mean	95% conf. lim.	Range		
46. *C. carpio*	E, f	72.9†	68.4-77.4†	70.9-74.5‡	70.5	2
47. *C. carpio*	E, f	73.3	72.3-74.3	70.8-75.6	70.5	27
Siluridae						
48. *Silurus glanis*	E, f	61.0	–	–	61.0	1
Gadidae						
49. *Gadus morhua*	E, s	72.0	–	–	72.0	1
50. *Eleginus navaga*	E, s	77.9†	–	–	77.9	28
Percidae						
51. *Perca fluviatilis*	MO, f	83.0‡	–	–	83.0	29
52. *Stizostedion lucioperca*	E, f	70.6†	–	66.1-75.2	70.6	1
Cichlidae						
53. *Tilapia mossambica*	E, f	58.0	57.9-58.1	–	58.0	30

*Sources: 1, Stroganov (1962); 2. Nikolskij (1974); 3, Blaxter and Hempel (1966); 4. Lasker (1962); 5, Epler *et al.* (1981b); 6, Boëtius and Boëtius (1980); 7, Diana and Mackay (1979); 8, Vasileva and Averyanova (1981); 9, Dąbrowski *et al.* (1987); 10, Dąbrowski (1982a); 11, Heming (1982); 12. Smirnov *et al.* (1968); 13, Kamler (1987); 14. Trzebiatowski and Domagała (1986); 15, Springate *et al.* (1985); 16, Ridelman *et al.* (1984); 17, Ogino and Yasuda (1962); 18. Suyama and Ogino (1958); 19, Kato and Kamler (1983); 20, Springate *et al.* (1984); 21, Satia *et al.* (1974); 22, Smith (1957); 23, Malyarevskaya and Birger (1965); 24. Kamler (1976); 25, Moroz and Luzhin (1973); 26, Semenov *et al.* (1974); 27. Martyshev *et al.* (1967); 28. Lapin and Matsuk (1979); 29. Craig (1977); 30. Mironova (1977).

†Recomputed value.

‡Value read from a graph.

cyprinid eggs falls into a higher range of values than that of acipenserids and salmonids.

In spite of the differences discussed above, the frequency distribution of egg hydration values in 30 species of fish depicted in Fig. 3.10 does not deviate from a normal distribution ($\chi^2_{obs} = 3.966 < \chi^2_{theor} = 11.345$, d.f. = 3).

Intraspecific comparisons

The data gathered in Table 3.6 seem to point to intraspecific differences of egg hydration: for instance, compare rows 3 and 4 for *Acipenser stellatus* and rows 27 and 32 for *Oncorhynchus mykiss*. Calculations made for *O. mykiss* (spawn portions for 33 females, row 32) and for *Tilapia mossambica* (20 females, row 53) have shown that the intraspecific distribution of egg hydration also does not deviate from normality ($\chi^2_{obs} = 6.482 < \chi^2_{theor} = 11.345$, d.f. = 3, and $\chi^2_{obs} = 0.422 < \chi^2_{theor} = 6.635$, d.f. = 1, respectively).

The physical properties of unswollen eggs make them difficult to isolate, and therefore data pertaining to hydration of individual eggs derived from one female pertain to swollen eggs. Intra-individual variability of egg hydration will be discussed shortly but first the amount of water taken in during the process of swelling and the duration of this process will be discussed.

The amount of water drawn into the perivitelline space during swelling will be expressed in terms of percentage of its initial amount. For example, Suyama and Ogino (1958) found 59.4% water in wet weight of unswollen eggs of *Oncorhynchus mykiss* (row 31 in Table 3.6) and 65.02% water in wet weight of eggs after water absorption was completed. Taking into account egg wet weight increase when swollen, the result is 27% of the initial amount of water absorbed by an egg during swelling. For the same species a value of 28.9% was calculated from data given by Ogino and Yasuda (1962): values of 12.4%, 23.3% and 12.3% for eggs obtained from females 2, 3 and 4 years old, respectively, have been calculated from Kato and Kamler (1983). In general, the amount of water taken up during swelling of salmonid eggs forms about 20% of the initial amount (Bogucki, 1928; Leitritz and Lewis, 1976). Eggs of *Cyprinus carpio*, which hold more water when unswollen than do the eggs of Salmonidae (Table 3.6), absorb much more water during swelling: values between 193 and 214% of initial water content were calculated from table 2 in Semenov *et al.* (1974), and a value of 448% was calculated from Kamler (1976). Eggs of *Eleginus navaga* (Gadidae) absorb 803% water (Lapin and Matsuk, 1979). Bathypelagic eggs of *Ctenopharyngodon idella* absorb as much as 160 times the initial amount of water (estimated from Shireman and Smith, 1983). In freshwater pelagophilous fishes, egg buoyancy is achieved by high water content (Balon, 1975a). It is noteworthy that in species whose unswollen eggs are more hydrated (Table 3.6), more water is taken in during swelling. Semenov *et al.* (1974) showed with carp that this dependence is also manifested at the intraspecific level. Kjørsvik and Lønning (1983) and Kjørsvik *et al.* (1984) have shown that the percentage of fertilization was higher in those egg cultures of *Gadus morhua* which possessed harder chorions and absorbed more water into the perivitelline space.

Swelling takes 60 min in Salmonidae (Hayes, 1949), 30 min in *Oncorhynchus mykiss* (Devillers *et al.*, 1953), 60–90 min in *Salmo ischchan* (Ryzhkov, 1966), 30–90 min in *Coregonus* sp. (Zotin, 1954) and 1.5–2.0 h in *Ctenopharyngodon idella* (Shireman and Smith, 1983). In *Gadus morhua* it takes 10–15 min in normal eggs and 30 min in poor ones (Kjørsvik and Lønning, 1983). In *Cyprinus carpio*, swelling is completed within 12 min in fresh water (Kamler, 1976; Renard *et al.*, 1985), but in a saline solution $(150\,\text{mOsm kg}^{-1})$ it takes 18 min (Renard *et al.*, 1985); during swelling the egg fertilizability decreases from 90% at 1 min to 50% at 9 min (Renard *et al.*, 1985).

After termination of swelling the water content in an egg remains unchanged up to the time of hatching (Bogucki, 1928; Hayes, 1949; Moroz and Luzhin, 1973; Semenov *et al.*, 1974; Kamler, 1976; Kjørsvik and Lønning, 1983). In consequence, water exchange between the perivitelline space and the external environment is limited, as shown by Domurat (1956, 1958) and by Winnicki (1960, 1968) for *Salmo trutta*, *Oncorhynchus mykiss*, *Salvelinus fontinalis*, *Esox lucius* and *Rutilus rutilus*.

Intra-individual variability

Intra-individual variability of hydration of swollen eggs was studied by Kato and Kamler (1983) in *Oncorhynchus mykiss*. It was found to be low and to decrease with increasing female age and egg size.

Main constituents of egg matter

Two factors contribute to the amount of a given chemical constituent in an egg: egg size and the amount of the constituent as expressed in terms of its concentration. The size factor was discussed in Section 3.1; here the percentage content of constituents in egg dry matter – protein, lipids, carbohydrates and ash – will be considered.

Total protein

Total protein is a dominant constituent of fish yolk. A large proportion of protein is transformed into embryonic tissue, and part is consumed to provide energy.

Although the percentages of protein in fish egg dry matter (Table 3.7) lie within a broad range, 35–89%, a more representative range is 55–75%, within which fall 19 of 28 average values for various species. The average total percentage of protein in fish egg dry matter is 66.3% (Table 3.7).

Oocytes of *Anquilla anquilla* are distinguished by a low percentage of

Table 3.7 Interspecific comparisons of total protein in freshly spawned fish eggs (E) or mature ovaries (MO). Based on 45 sets of data from 28 fish species spawning in fresh waters (f) or in the sea (b)

Family and species		Protein (% dry matter)			Mean value for species	Source*
		Mean	95% conf. lim.	Range		
Acipenseridae						
1. *Acipenser sturio*	E, f	54.6†	–	50.0–59.3†	54.6	1
2. *A. stellatus*	E, f	58.3†	–	54.9–61.7†	58.3	1
3. *Huso huso*	E, f	63.0†	–	–	63.0	1
Clupeidae						
4. *Sardinops caerulea*	MO, s	71.6	–	–	71.6	2
Anguillidae						
5. *Anguilla anguilla*	MO, s	36.8	35.0–38.7	–	38.9	3
6. *A. anguilla*	MO, s	41.0		40.1–41.9	38.9	4
Esocidae						
7. *Esox lucius*	E, f	81.8†		–	81.8	1
Salmonidae						
8. *Coregonus albula*	E, f	52.3†	–	45.4–59.4	52.3	5
9. *Oncorhynchus nerka*	E, f	66.5	–	66.1–67.3	68.0	6
10. *O. nerka*	E, f	69.5†	–	68.3–70.7†	68.0	1
11. *O. keta*	E, f	66.3	–	58.5–71.3	67.4	6
12. *O. keta*	E, f	68.4†	–	64.3–72.5†	67.4	1
13. *O. gorbuscha*	E, f	69.1	–	64.5–76.9	73.6	6
14. *O. gorbuscha*	E, f	69.8	–	67.5–73.7	73.6	6
15. *O. gorbuscha*	E, f	81.8†	–	74.7–89.0†	73.6	1
16. *O. tshawytscha*	E, f	68.1	–	–	68.1	6
17. *O. kisutch*	E, f	73.5	–	72.4–75.4	73.5	6

Cont'd overleaf

Table 3.7 cont'd

Family and species		Protein (% dry matter)			Mean value for species	Source*
		Mean	95% conf. lim.	Range		
18. O. masu (1959)	E, f	63.7	–	59.5–66.8	64.4	6
19. O. masu (1960)	E, f	65.0	–	63.3–67.2	64.4	6
20. Salmo salar	E, f	71.6	–	66.3–74.4	71.6	6
21. S. trutta m. trutta	E, f	78.8†	–	76.5–81.0†	78.8	7
22. O. mykiss (fed)	MO, f	67.5	67.0–68.0†	–	69.3	8
23. O. mykiss (full ration)	E, f	69.1	61.6–76.6	–	69.3	9
24. O. mykiss (half ration)	E, f	69.8	69.4–70.2	–	69.3	9
25. O. mykiss	E, f	68.6†		–	69.3	10
26. O. mykiss	E, f	71.3†	–	–	69.3	11
Osmeridae						
27. Osmerus eperlanus	E, f	60.5†	–	–	60.5	1
Cyprinidae						
28. Leuciscus idus	E, f	72.7†	–	–	72.7	1
29. Rutilus rutilus heckeli	E, f	60.8	–	–	67.8	12
30. R. rutilus caspicus	E, f	74.7†	–	–	67.8	1
31. Abramis brama	E, f	61.3	–	–	70.0	12
32. A. brama	E, f	78.7†	–	–	70.0	1
33. Ctenopharyngodon idella	E, f	67.8†	–	67.5–68.1	67.5	13
34. Cyprinus carpio	E, f	62.9	–	61.3–64.5	68.9	14
35. C. carpio	E, f	63.5†	61.2–65.8†	58.3–70.1†	68.9	15
36. C. carpio	E, f	66.5†	–	–	68.9	16
37. C. carpio	E, f	70.3†	69.1–71.5†	67.2–73.9†	68.9	17

38. *C. carpio*	E, f	70.4†	68.0–72.8†	69.4–71.2†	68.9	18
39. *C. carpio*	E, f	80.0†	–	–	68.9	1
Siluridae						
40. *Silurus glanis*	E, f	76.9†	–	–	76.9	1
Gadidae						
41. *Gadus morhua*	E, s	82.1†	–	–	82.1	1
42. *Eleginus navaga*	E, s	66.4†	–	–	66.4	19
Exocoetidae						
43. *Hemiramphus sajori*	E, s	61.0†	–	–	61.0	20
Percidae						
44. *Perca fluviatilis*	MO, f	41.0‡	–	–	41.0	21
45. *Stizostedion lucioperca*	E, f	54.7†	–	51.3–58.1†	54.7	1

*Sources: 1, Stroganov (1962); 2, Lasker (1962); 3, Boëtius and Boëtius (1980); 4, Epler *et al.* (1981b); 5, Kamler *et al.* (1982); 6, Smirnov *et al.* (1968); 7, Trzebiatowski and Domagała (1986); 8, Ridelman *et al.* (1984); 9, Springate *et al.* (1985); 10, Smith (1957); 11, Satia *et al.* (1974); 12, Malyarevskaya and Birger (1965); 13, Stepanova and Tyutyunik (1973); 14, Moroz and Luzhin (1973); 15, Kamler (1976); 16, Semenov *et al.* (1974); 17, Martyshev *et al.* (1967); 18, Nikolskij (1974); 19, Lapin and Matsuk (1979); 20, Kimata (1982); 21, Craig (1977).

†Recomputed value.

‡Value read from a graph.

protein (rows 5 and 6 in Table 3.7). This is connected with their prolific supply of high-energy constituents (Section 3.2).

Other consistent, large-scale differences between taxonomic or ecological groups were not revealed by the data listed in Table 3.7. This does not necessarily mean that there are no such differences: the reasons for the apparent lack of differences can be looked for in methods of protein determination.

In an indirect method commonly used, nitrogen determined by the Kjeldahl method is conventionally multiplied by a factor of 6.25 for conversion into protein, because the nitrogen fraction is assumed to be 0.16. A lower conversion factor, 6.025, has been used for fish soma (Fry, 1957; Fischer, 1976). The protein can also be determined directly, e.g. using the Folin reagent (Lowry *et al.*, 1951; Hartree, 1972) or the total amino acid content. The indirect method yields over-estimates for two reasons (Love, 1957; Gnaiger and Bitterlich, 1984). First, the nitrogen fraction in the protein of aquatic organisms is higher, averaging (according to Gnaiger and Bitterlich) 0.173, so a lower conversion factor, 5.8, should be applied. Second, non-protein nitrogen is present, e.g. in eggs of *Coregonus albula* (Kamler and Żuromska, 1979). In spawn portions derived from three females, total nitrogen determined by the Kjeldahl method was 8.79% of dry matter, on average. The amount of protein calculated from the conversion factors 6.25 or 5.8 would be 54.94 or 50.98% dry matter, respectively, i.e. more than the value of 49.94% protein in dry matter determined directly from the Folin reagent. The differences are 5.00 or 1.04% of protein in dry matter, respectively, or after reconversion to nitrogen, 0.80 or 0.18%. Thus, non-protein nitrogen would be 9% or 2% of total nitrogen, respectively. This percentage falls in the lower range of 2–24% reported for aquatic animals (Fischer, 1976; Penczak *et al.*, 1977; Craig *et al.*, 1978; Gnaiger and Bitterlich, 1984). Part of the non-protein nitrogen consists of free amino acid nitrogen. The egg portions examined held, on average, 0.63% free amino acids in dry matter, which corresponds to 0.10 or 0.11% nitrogen in dry matter when using the factors 6.25 or 5.8, respectively. The remaining nitrogen (0.70 or 0.07% dry matter, respectively) was probably present in non-determined, non-protein, N-containing compounds such as oligopeptides, amines, non-protein amino acids, and nucleotides.

No statistical tests can be applied to the data in Table 3.7. Nevertheless, these data seem to suggest that there are intraspecific differences between percentages of protein in egg dry matter (e.g. compare row 5 with 6 for *A. anguilla*, row 9 with 10 for *O. nerka*, row 13 with 15 for *O. gorbuscha*, row 22 with 26 for *O. mykiss* and row 35 with 38 for *C. carpio*). Statistical corroboration of intraspecific differences in percentages of proteins in egg dry matter can be supplied by examples of *Coregonus albula* (Table 3.8). Significant differences were found both between different populations of *C. albula* examined in one year and between egg collections from one popula-

Table 3.8 Intraspecific comparisons of percentages of main chemical constituents in egg dry matter as exemplified by *Coregonus albula*. Spawn portions obtained from *n* females were analysed. Based on data selected from Kamler *et al.* (1982)

Population	Year	n	Protein*		Lipids		Carbohydrates		Ash	
			Mean (%)	CV (%)	Mean (%)	CV (%)	Mean (%)	CV (%)	Mean (%)	CV (%)
			Different populations in 1976							
Oulujärvi	1976	24	53.35	3.42	16.79	7.71	–	–	4.34	11.21
Puruvesi	1976	23†	51.49	2.94	24.52	11.09	–	–	3.87	8.98
Narie	1976	23	51.05	3.52	18.96	8.10	–	–	4.38	9.97
Maroz	1976	26	51.58	5.28	18.90	5.34	–	–	4.32	9.85
Pluszne	1976	26	49.85	4.25	20.96	10.82	–	–	4.79	7.28
F			9.228		57.072		–		15.256	
d.f.			4, 118		4, 117		–		4, 117	
P			< 0.01		< 0.01		–		< 0.01	
			Narie population in different years							
Narie	1975	22	50.76	3.89	22.49	9.12	2.71	8.33	4.73	5.74
Narie	1976	23	51.05	3.52	18.96	8.10	2.89	10.381	4.38	9.97
Narie	1977	30	52.53	5.31	28.42	11.56	2.65	14.95	5.40	29.00
F			4.567		96.666		3.684		6.702	
d.f.			2, 72		2, 72		2, 72		2, 72	
P			< 0.05		< 0.01		< 0.05		< 0.01	

*Measured using the Folin reagent.
†24 for protein.

tion gathered during three years. In spite of this variability, the percentage content of protein is more stable than the percentage content of other chemical components of fish eggs, lipids and ash: this refers both to the interpopulation level (F values were lowest for protein) and to the intrapopulation level (coefficients of variation were lowest for protein). The distribution of percentage protein in dry matter of eggs collected from 389 female *Coregonus albula* (Kamler *et al.* 1982) was unskewed (sk = 0.041, t_{sk} = 0.335) and did not deviate from normality (χ^2_{obs} = 9.605 < χ^2_{theor} = 23.209 for d.f. = 10).

Leucine, alanine, lysine and glutamic acid are the most abundant amino acids in eggs of the three fish species listed in Table 3.9, as well as in eggs of other fishes (Suyama, 1958b; Zdor *et al.*, 1978). Free amino acids in freshwater fish eggs formed only a fraction of 1% of dry matter (Table 3.9 and Kim, 1974a, for *Cyprinus carpio*).

Proteins in fish sexual products and embryos are reviewed in detail by Konovalov (1984).

Table 3.9 Protein amino acids, free amino acids and the four most abundant amino acids occurring in mature eggs of *Abramis brama* (Kim and Zhukinskij, 1978), *Rutilus rutilus heckeli* (Kim, 1981) and *Coregonus albula* (Kamler and Żuromska, 1979). Y, young females; MA, middle-aged females; O, old females

	A. brama		R. rutilus heckeli		C. albula	
	Y	MA	Y	O	Y	O
Total of protein amio acids (% dry matter)	47.70*	46.69*	38.86*	41.62*	36.82	35.11
Alanine	11.53	13.56	15.85	14.05	9.35	9.52
Leucine	12.21	12.03	11.38	13.24	10.14	10.22
Lysine	10.64	11.70	10.75	11.91	10.95	11.90
Glutamic acid	7.50	8.69	9.63	8.76	11.75	12.60
Total of free amino acids (% dry matter)	0.15*	0.14*	0.12*	0.12*	0.63	0.59

*Original data given on a wet weight basis were converted to a dry weight basis using hydration values from Table 3.6.

Total lipids

Total lipids (Table 3.10) are the second component of dry matter of fish eggs (Table 3.8). The major part of yolk fat reserves is used up for energy; the rest is stored in the embryo. Free oil globules in eggs of pelagophilous fishes assure buoyancy (Kryzhanovskij, 1960). In some salmonids, oil globules accumulate near the animal pole; this ensures an uppermost position of the embryo during incubation (Ryzhkov, 1966).

Lipid percentages of egg dry matter (Table 3.10) differ widely from one species to another, showing a vast range from 3 to 54%. Love (1970) reported an exceptionally high lipid content of 67% in muscles of *Salvelinus namaycush siscowet.* However, lipids usually range from 10 to 35% of dry matter in fish eggs: 31 out of 39 species averages in Table 3.10 can be found within these limits. The total overall percentage of lipids in fish eggs, as calculated from the data in Table 3.10, is 19.3%.

The diversity of procedures used in lipid determinations is partly responsible for the diversity of results (Love, 1957). In some earlier papers, very low values of lipid content (less than 10%) were reported for cyprinid eggs (Grodziński, 1961; and Table 3.10 for *Rutilus rutilus* (rows 26 and 27), *Abramis brama* (row 28) and *Cyprinus carprio* (rows 31 and 32 only)). These values were probably under-estimated because solvents could not easily penetrate to lipids locked inside yolk spheres, whose membranes consist mainly of protein components (page 58). More recent investigations (Table 3.10) indicate that the lipid content in dry matter of cyprinid eggs does not drop below 10%. Kim's (1981) values for the lipid content in eggs of *Rutilus rutilus heckeli* and *Abramis brama* (not listed in Table 3.10) amount, after conversion from wet to dry weight using Table 3.6, to 22.5 and 25.8%, respectively. Complete determination of lipids in cyprinid eggs (i.e. determination in which caloric values calculated from the chemical composition and those directly assessed in the bomb calorimeter are alike) requires a lengthy extraction using a chloroform–methanol mixture followed by further purification using a chloroform–petroleum ether mixture (Kamler, 1976). On the other hand, lipids from salmonid eggs are easily extracted (Kamler and Żuromska, 1979; Kamler *et al.*, 1982), as they occur there in the form of free fat droplets (page 58).

The lipid content in oocytes of *Anguilla anguilla* is grouped in the uppper range of values listed in Table 3.10, i.e. more than 40% of dry matter. An explanation for high caloric values in *A. anguilla* is given above (page 50). Second in rank in terms of egg lipid content are the Acipenseridae, whose spawned eggs contain from 26 to 36% lipids, are large (Fig. 3.1 (G)), and are little hydrated (Fig. 3.10). The eggs of most species listed in Table 3.10 that reproduce in the sea (except for *Anguilla* and *Trachurus*) possess very little lipids, most often 11–25% of dry matter. Similarly small quantities of lipids were found in not-fully-mature oocytes of Clupeidae (Kaitaranta and Ackman, 1981). This corroborates the finding (page 50) that marine fish

Table 3.10 Interspecific comparison of total lipids in freshly spawned fish eggs (E) or mature ovaries (MO). Based on 54 sets of data from 39 fish species spawning in fresh waters (f) or in the sea (s)

Family and species		Lipids (% dry matter)			Mean value for species	Source*
		Mean	95% conf. lim.	Range		
Acipenseridae						
1. *Acipenser stellatus*	E, f	30.4†	–	25.7–35.1†	31.9	1
2. *A. stellatus*	E, f	33.4†	30.9–35.9†	32.3–34.2†	31.9	2
3. *A. sturio*	E, f	31.1†	–	24.3–37.8†	31.1	1
4. *Huso huso*	E, f	36.5†	–	–	36.5	1
Clupeidae						
5. *Sardinops caerulea*	MO, s	13.0	12.0–14.0	–	13.0	3
Engraulidae						
6. *Engraulis encrasicgolus ponticus*	E, s	13.1	11.5–14.7	12.0–14.2	13.1	4
Anguillidae						
7. *Anguilla anguilla*	MO, s	47.3	40.2–54.3	–	48.6	5
8. *A. anguilla*	MO, s	49.9	–	49.7–50.1	48.6	6
Esocidae						
9. *Esox lucius*	E, f	4.9†	–	–	4.9	1
Salmonidae						
10. *Coregonus pollan*	MO, f	29.8†	26.6–33.0†	–	29.8	7
11. *C. albula*	E, f	22.8†	–	9.0–30.0	28.6	8
12. *C. albula*	E, f	24.3†	–	14.0–37.8	28.6	9
13. *C. albula*	MO, f	27.8	–	–	28.6	10
14. *C. albula*	E, f	39.5†	–	32.6–51.5	28.6	11
15. *Oncorhynchus gorbuscha*	E, f	9.6†	–	8.2–11.0†	9.6	1
16. *O. keta*	E, f	25.8†	–	24.3–27.2†	25.8	1
17. *O. nerka*	E, f	26.8†	–	24.4–29.2†	26.8	1
18. *Salmo trutta m. trutta*	E, f	14.8†	–	12.2–17.3†	14.8	12
19. *O. mykiss* (half ration)	E, f	7.5	7.0–8.0	–	13.3	13
20. *O. mykiss* (full ration)	E, f	7.3	7.1–7.5	–	13.3	13

21. *O. mykiss* (fed)	MO, f	11.1	10.7–11.5†	—	13.3	14
22. *O. mykiss*	E, f	11.4†	11.1–11.7†	—	13.3	15
23. *O. mykiss*	E, f	29.3†	—	—	13.3	16
Osmeridae						
24. *Osmerus eperlanus*	E, f	31.6†	—	—	31.6	1
Cyprinidae						
25. *Leuciscus idus*	E, f	18.0†	—	—	18.0	1
26. *Rutilus rutilus caspicus*	E, f	5.5†‡	—	3.1–7.9†	6.8?	1
27. *R. rutilus heckeli*	E, f	6.8†	—	—	6.8?	17
28. *Abramis brama*	E, f	6.4‡	—	—	9.6?	17
29. *A. brama*	E, f	12.8†	—	—	9.6?	1
30. *Ctenopharyngodon idella*	E, f	10.9†	—	8.9–12.9	10.9	18
31. *Cyprinus carpio*	E, f	5.6†‡	—	4.8–6.5†‡	16.0?	19
32. *C. carpio*	E, f	6.7†‡	—	—	16.0?	1
33. *C. carpio*	E, f	21.5†	19.7–23.3†	19.3–29.3†	16.0?	20
34. *C. carpio*	E, f	22.5†	21.6–23.4†	19.4–25.0†	16.0?	21
35. *C. carpio*	E, f	23.6†	20.6–26.6†	22.8–25.0†	16.0?	2
Siluridae						
36. *Silurus glanis*	E, f	9.2†	—	—	9.2	1
Gadidae						
37. *Gadus morhua*	E, s	4.6†	—	—	4.6	1
38. *Eleginus navaga*	E, s	20.5†	—	—	20.5	22
Exocoetidae						
39. *Hemiramphus sajori*	E, s	18.8†	—	—	18.8	23
Belonidae						
40. *Belone belone euxini*	E, s	15.2	—	—	15.2	4
Atherinidae						
41. *Atherina mochon pontica*	E, s	12.3	—	—	12.3	4
Syngnathidae						
42. *Syngnathus nigrolineatus*	E, s	12.8†	—	11.4–14.3	12.8	4
43. *S. typhle argentatus*	E, s	15.3†	—	14.5–16.0	15.3	4

Cont'd overleaf

Table 3.10 cont'd

Family and species		Lipids (% dry matter)			Mean value for species	Source*
		Mean	95% conf. lim.	Range		
Gasterosteidae						
44. Gasterosteus aculeatus	E, s	15.1	–	–	15.1	4
45. Pungitius platygaster platygaster	E, s	15.7	–	–	15.7	4
Percidae						
46. Perca fluviatilis	MO, f	16.0†§	–	–	16.0	24
47. Stizostedion lucioperca	E, f	29.9†	–	22.4–37.4†	29.9	1
Carangidae						
48. Trachurus mediterraneus ponticus	E, s	34.2	–	–	34.2	4
Gobiidae						
49. Pomatoschistus microps leopardinus	E, s	16.8	–	–	16.8	4
50. Gobius fluviatilis	E, s	18.9	12.3–25.5	15.4–21.0	18.9	4
51. G. cephalarges	E, s	20.0	15.3–24.7	15.8–22.8	20.0	4
52. G. melanostomus	E, s	26.2	21.4–31.0	24.1–27.9	26.2	4
Bothidae						
53. Scophthalmus maeoticus maeoticus	E, s	20.5†	–	19.8–21.2	20.5	4
Pleuronectidae						
54. Platichthys flesus luscus	E, s	12.1	–	–	12.1	4

*Sources: 1, Stroganov (1962); 2, Nikolskij (1974); 3, Lasker (1962); 4, Vinogradov (1973); 5, Boëtius and Boëtius (1980); 6, Epler et al. (1981b); 7, Dąbrowski (1982a); 8, Potapova (1978); 9, Kamler et al. (1982); 10, Lizenko et al. (1973); 11, Dąbrowski et al. (1987); 12, Trzebiatowski and Domagała (1986); 13, Springate et al. (1985); 14, Ridelman et al. (1984); 15, Satia (1974); 16, Suyama and Ogino (1958); 17, Malyarevskaya and Birger (1965); 18, Stepanova and Tyutyunik (1973); 19, Moroz and Luzhin (1973); 20, Kamler (1976); 21, Martyshev et al. (1967); 22, Lapin and Matsuk (1979); 23, Kimata (1982); 24, Craig (1977).

†Value computed from data reported by the author(s).
‡Value probably under-estimated owing to incomplete extraction.
§Value read from a graph.

eggs have a lower caloric value than freshwater ones. However, the high percentage of lipids in dry matter of eggs of Acipenseridae and *Oncorhynchus* spp. which spend part of their lives in fresh waters and part in sea deviates from the low lipid values observed in typically marine species (Table 3.10). Love (1970) has reported that the body fatty-acid composition of *Acipenser* spp. caught in the North Sea and of *Oncorhynchus kisutch* were typical of freshwater fish. The same can be said about hydration of eggs of these species (Table 3.6).

The lowest egg lipid content found among marine species in Table 3.10 was that of *Gadus morhua* (4.6%). It is plausible because a record protein content was found in these eggs (82.1% dry weight, row 41 in Table 3.7). Similarly in two species spawning in fresh waters, *Esox lucius* and *Oncorhynchus gorbuscha*, low lipid values (4.9% and 9.6%, Table 3.10, rows 9 and 15) were paralleled by exceptionally high protein content (81.8% for each, Table 3.7, rows 7 and 15). High values of lipid content in *Coregonus albula* – up to 51.5% of dry matter – were reported by Dąbrowski *et al.* (1987) (Table 3.10); protein was not determined in this case. According to Balon (1977), fishes with the most advanced (protective) life-history styles, i.e. those that bear fertilized eggs externally (*Labeotropheus* sp.) or internally (*Scyliorhinus caniculus*, *Latimeria chalumnae*) have the highest percentage of lipids in their eggs.

Lipid content is the most labile of the main components of fish eggs (Anukhina, 1968; Smirnov *et al.*, 1968; Kim, 1974b, 1981; Kamler, 1976; Kamler *et al.* 1982). Similarly comparisons between different populations of *Coregonus albula* in one year as well as between collections of eggs from one population in subsequent years (Table 3.8) have shown that there are significant differences in lipid content. Values of F for lipids were considerably higher than those for the remaining components–protein, carbohydrates and ash. Intrapopulation variations (CV, %) of the lipid fraction of egg dry matter were higher than those of protein.

The distribution of total lipid percentage in dry matter of eggs collected from 391 female *Coregonus albula* (Kamler *et al.*, 1982) was unskewed (sk = 0.189, t_{sk} = 1.530). Similarly, Shulman (1972) has demonstrated statistically the normal distribution of lipid percentage in the soma of 100 individuals of *Engraulis encrasicolus maeoticus*.

Yolk lipids of many fish species occur mainly in the form of triglycerides and cholesterol (Terner, 1979) or phospholipids, triglycerides and cholesterol (Kim, 1981). Phospholipids form 72% and 80% of total lipids present in mature eggs of *Abramis brama* and *Rutilus rutilus heckeli*, respectively (Kim, 1981); they are a main lipid component (85%) of mature oocytes of *Sardinops caerulea* (Lasker, 1962) and eggs of *Coregonus albula* (58.2%), whereas triglycerides and cholesterol occur in *C. albula* eggs in smaller amounts (29.0 and 3.5% of total lipids, respectively – Potapova, 1978). In eggs of *Eleginus navaga*, phospholipids, triglycerides and cholesterol form

64, 15 and 13% of total lipids, respectively (Lapin and Matsuk, 1979). Wax esters are the main (80–90%) storage lipid in eggs of *Trichogaster cosby* (Rahn *et al.*, 1977). Studies by Kaitaranta and Ackman (1981) and a literature review (Kim, 1981) point to differences of yolk lipid composition between various fish species.

Total carbohydrates

Total carbohydrates form a small fraction of a fish egg (Table 3.11), constituting 0.6–8.7% of its dry matter (2.6% on average). Erroneous figures of almost 30% of dry matter (rows 7 and 8 in Table 3.11) were disregarded since they are over-estimated (the lipid determination was under-estimated and the carbohydrates were calculated by subtraction of lipid, protein and ash percentages from 100%).

Significant differences were found between the percentage of carbohydrates in dry matter of *Coregonus albula* eggs collected from Lake Narie in various years (Table 3.8), but these differences were smaller than those of protein, ash, and especially of lipids.

Total ash

Total ash percentage in fish egg dry matter is presented in Table 3.12. It ranges from 1% to 17%, but the most typical range is 3–10% (Fig. 3.11), which comprises averages for 31 out of 36 species listed in Table 3.12. The level of ash in fish eggs seems not to differ from that in eggs of aquatic invertebrates (Oligochaeta, Kosiorek, 1979; Crustacea, Khmeleva and Golubev, 1984; Gastropoda, Pandian, 1969). The same can be said of 'soft' somatic tissues in fish, such as muscles, heart, and kidney, where the ash percentage (Love, 1970) is similar to that of eggs. On the other hand, the percentage of ash in the total soma of juvenile and/or mature fishes is higher than in eggs since scales and bones are much richer in ash. Data for total soma of freshwater fish (Fischer, 1970b, *Ctenopharyngodon idella*; Fischer, 1976, *Cyprinus carpio*; Penczak *et al.*, 1976, *Rutilus rutilus*; Craig, 1978, *Perca fluviatilis*; Galicka, 1984, *Oncorhynchus mykiss*) ranges most often from 10 to 20% of ash in dry matter. Although somatic tissues of marine and freshwater fish do not differ in terms of ash content (only the body liquids differ: Love, 1970), the percentage of ash in marine fish eggs (Table 3.12) is twice as high as in fish that spawn in fresh waters (8.2 and 4.0%, respectively; Fig. 3.11). An especially low (most often below 4%) and even ash content was observed in eggs of *Oncorhynchus* spp., although these fish spend only a part of their life in fresh waters and the rest of it in the sea. Thus, the ash content is yet another feature that makes *Oncorhynchus* spp. eggs more of a 'freshwater-type' than eggs of fish living exclusively in fresh waters.

The unusually high ash content in *O. gorbuscha* eggs (row 17 in Table 3.12; Stroganov, 1962) was not corroborated by more recent studies (rows

Table 3.11 Interspecific comparisons of carbohydrates in freshly spawned fish eggs (E) or mature ovaries (MO). Based on 10 sets of data from 9 fish species spawning in fresh waters (f) or in the sea (s)

Family and species		Carbohydrates (% dry matter)			Source*
		Mean	95% conf. lim.	Range	
Clupeidae					
1. *Sardinops caerulea*	MO, s	<1	–	–	1
Anguillidae					
2. *Anguilla anguilla*	MO, s	1.05	–	0.90–1.20	2
Salmonidae					
3. *Coregonus albula*	E, f	2.60†	–	1.83–3.48	3
4. *Oncorhynchus mykiss*	E, f	0.56†	–	0.53–0.59†	4
Cyprinidae					
5. *Cyprinus carpio*	E, f	1.16†	1.01–1.31†	0.84–1.54†	5
6. *C. carpio*	E, f	3.60†	–	3.20–3.90†	6
7. *Rutilus rutilus heckeli*	E, f	28.02‡	–	–	7
8. *Abramis brama*	E, f	27.50‡	–	–	7
Exocoetidae					
9. *Hemiramphus sajori*	E, s	8.70†	–	–	8
Percidae					
10. *Perca fluviatilis*	MO, f	2.40	1.65–3.15	–	9

*Sources: 1, Lasker (1962); 2, Epler *et al.* (1981b); 3, Kamler *et al.* (1982); 4, Smith (1957); 5, Kamler (1976); 6, Moroz and Luzhin (1973); 7, Malyarevskaya and Birger (1965); 8, Kimata (1982); 9, Craig (1977).
†Value computed from data reported by the author(s).
‡Erroneous value (over-estimated).

Table 3.12 Interspecific comparisons of total ash in freshly spawned fish eggs (E) or mature ovaries (MO). Based on 48 sets of data from 36 fish species spawning in fresh waters (f) or in the sea (s)

Family and species		Ash (% dry matter)			Mean value for species	Source*
		Mean	95% conf. lim.	Range		
Acipenseridae						
1. *Huso huso*	E, f	3.6†	–	–	3.6	1
2. *Acipenser stellatus*	E, f	4.3†	–	–	4.3	1
3. *A. sturio*	E, f	5.4†	–	–	5.4	1
Clupeidae						
4. *Sardinops caerulea*	MO, s	7.0	–	–	7.0	2
Engraulidae						
5. *Engraulis encrasicolus ponticus*	E, s	8.4†	7.6–9.2†	8.0–9.1†	8.4	3
Salmonidae						
6. *Coregonus pollan*	MO, f	4.0†	3.6–4.4†	–	4.0	4
7. *C. albula*	E, f	4.5†	–	2.8–9.8†	4.5	5
8. *Oncorhynchus tshawytscha*	E, f	2.5	–	–	2.5	6
9. *O. masu* (1959+1960)	E, f	3.5†	–	3.3–3.7	3.5	6
10. *O. nerka*	E, f	3.6†	–	2.4–4.9†	3.6	1
11. *O. nerka*	E, f	3.7	–	3.7–3.8	3.6	6
12. *O. keta*	E, f	3.5†	–	3.0–3.9†	3.6	1
13. *O. keta*	E, f	3.7	–	3.7–3.8	3.6	6
14. *O. kisutch*	E, f	4.1	–	4.0–4.2	4.1	6
15. *O. gorbuscha* (Kola Peninsula)	E, f	3.4	–	3.3–3.5	3.5	6
16. *O. gorbuscha* (Sakhalin)	E, f	3.6	–	3.4–3.8	3.5	6
17. *O. gorbuscha*	E, f	10.0†‡	–	7.1–13.0†‡	3.5	1
18. *O. mykiss* (half ration)	E, f	3.6	3.4–3.8	–	3.8	7

No.	Species	Type					Ref.
19.	O. mykiss (full ration)	E, f	3.7	3.3–4.1	–	3.8	7
20.	O. mykiss	E, f	3.7	–	–	3.8	8
21.	O. mykiss	E, f	3.8†	–	–	3.8	9
22.	O. mykiss	E, f	3.9†	3.8–4.0†	–	3.8	10
23.	O. mykiss	E, f	3.9	–	–	3.8	11
24.	O. mykiss (fed)	MO, f	4.0	–	–	3.8	12
25.	Salmo trutta m. trutta	E, f	4.3†	–	4.1–4.4†	4.3	13
26.	S. salar	E, f	4.4	–	4.3–4.5	4.4	6
Osmeridae							
27.	Osmerus eperlanus	E, f	5.3†	–	–	5.3	1
Cyprinidae							
28.	Leuciscus idus	E, f	3.0†	–	–	3.0	1
29.	Rutilus rutilus caspicus	E, f	4.3†	–	–	4.4	1
30.	Rutilus rutilus heckeli	E, f	4.4	–	–	4.4	14
31.	Abramis brama	E, f	4.0†	–	–	4.4	1
32.	A. brama	E, f	4.8	–	–	4.4	14
33.	Cyprinus carpio	E, f	4.4†	3.9–4.9†	3.5–6.3†	4.4	15,16
Gadidae							
34.	Eleginus navaga	E, s	2.1†	–	–	2.1	17
Exocoetidae							
35.	Hemiramphus sajori	E, s	7.2†	–	–	7.2†	18
Belonidae							
36.	Belone belone euxini	E, s	6.3†	–	–	6.3	3
Atherinidae							
37.	Atherina mochon pontica	E, s	16.3†	–	–	16.3	3
Syngnathidae							
38.	Syngnathus nigrolineatus	E, s	6.6†	–	–	6.6	3
39.	S. typhle argentatus	E, s	7.6†	–	6.8–8.4†	7.6	3

Cont'd overleaf

Table 3.12 cont'd

Family and species		Ash (% dry matter)			Mean value for species	Source*
		Mean	95% conf. lim.	Range		
Gasterosteidae						
40. *Gasterosteus aculeatus*	E, s	9.1†	–	–	9.1	3
41. *Pungitius platygaster platygaster*	E, s	16.7†	–	–	16.7	3
Carangidae						
42. *Trachurus mediterraneus ponticus*	E, s	9.6	–	–	9.6	3
Gobiidae						
43. *Gobius melanostomus*	E, s	4.4†	2.3–6.5†	3.7–5.3†	4.4	3
44. *G. fluviatilis*	E, s	7.7†	1.2–14.2†	4.7–9.3†	7.7	3
45. *G. cephalarges*	E, s	9.2†	4.0–14.4†	6.3–13.8†	9.2	3
46. *Pomatoschistus microps leopardinus*	E, s	4.9†	–	–	4.9	3
Bothidae						
47. *Scophthalmus maeoticus maeoticus*	E, s	6.2†	–	5.2–7.1†	6.2	3
Pleuronectidae						
48. *Platichthys flesus luscus*	E, s	10.1†	–	–	10.1	3

*Sources: 1, Stroganov (1962); 2, Lasker (1962); 3, Vinogradov (1973); 4, Dąbrowski (1982a); 5, Kamler et al. (1982); 6, Smirnov et al. (1968); 7, Springate et al. (1985); 8, Ogino and Yasuda (1962); 9, Smith (1957); 10, Satia et al. (1974); 11, Suyama and Ogino (1958); 12, Ridelman et al. (1984); 13, Trzebiatowski and Domagała (1986); 14, Malyarevskaya and Birger (1965); 15, Kamler (1972a); 16, Kamler (1976); 17, Lapin and Matsuk (1979); 18, Kimata (1982).

†Value computed from data reported by the author(s).

‡Suspect value, excluded when calculating the mean for the species.

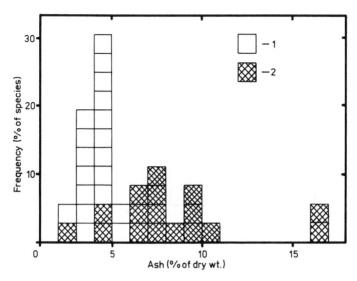

Fig. 3.11 Frequency distribution of percentage ash content in dry matter of fish eggs: interspecific comparisons of mean values for 36 species listed in Table 3.12. (1), species spawning in fresh waters; (2), species spawning in the sea.

15 and 16) and was therefore excluded when calculating the mean for this species. Also, in the percentages of protein (row 15 in Table 3.7) and lipids (row 15 in Table 3.10), Stroganov's (1962) data for *O. gorbuscha* deviated considerably from the remaining data. In general, the percentage of ash in egg dry matter does not vary greatly among different populations of the same species (Table 3.12). Nevertheless, intraspecific differences in ash content of dry matter of *Coregonus albula* eggs, both between different populations within the same year and between egg collections from one population within subsequent years, were highly significant (Table 3.8).

Ogino and Yasuda (1962) reported the following analysis for dry matter of unfertilized *Oncorhynchus mykiss* eggs: P, 1.03%; Ca, 0.182%; Mg, 0.135%; K, 58.7 mg%; Na, 19.2 mg%; Fe, 13.8 mg%; Si, 5.8 mg%; Cu, 0.66 mg%.

3.4 METHODOLOGICAL REMARKS

Proximate analysis

The analysis of the four main constituents – protein, lipids, carbohydrates and ash – is often called **proximate analysis**. The sum of their percentage in dry matter, or their recovery (or **tally**) should not reach 100% (Dowgiałło, 1975) because proximate analysis leaves numerous unidentified minor components, residual water included (Gnaiger and Bitterlich, 1984). In analysis of fish eggs a relatively high tally ($\geqslant 90\%$) is achieved (Fig. 3.12 (A)).

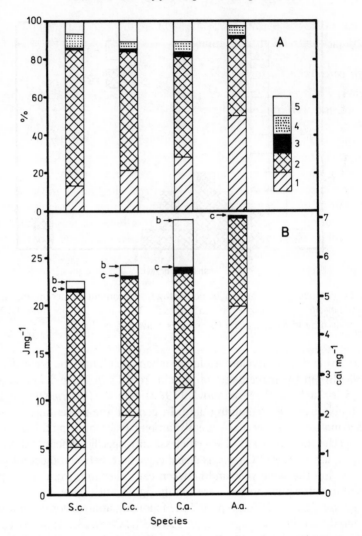

Fig. 3.12 (A) Egg dry matter composition (%) and (B) sources of energy (J mg^{-1} dry weight), as exemplified by mature oocytes of *Sardinops caerulea* (S.c., Lasker, 1962), eggs of *Cyprinus carpio* (C.c., Kamler, 1976), eggs of *Coregonus albula* (C.a., Kamler *et al.*, 1982) and mature oocytes of *Anguilla anguilla* (A.a., Epler *et al.*, 1981b). 1, lipids; 2, protein; 3, carbohydrates; 4, ash; 5, deficit; b and c, caloric values determined using a bomb calorimeter (b) and from chemical composition (c).

In marine invertebrates the tally is at best 85%, and often amounts only to 60–70% (Giese, 1967). The deficit or excess in the tally amounts to 15–20% of ash-free dry matter (Gnaiger and Bitterlich, 1984). Tally fluctuations are comprehensible if we realize the complexity of the biological material. Common reasons for variability in the tally include incomplete extraction, coextraction, and impurities.

Energy conversion factors

The energy conversion factors proposed for the needs of studies within the scope of the International Biological Programme (Winberg and Collaborators, 1971) are $23.0 \, J \, mg^{-1}$ (probable range 22.2–24.1) for protein, $39.8 \, J \, mg^{-1}$ (38.5–41.8)) for lipids and $17.2 \, J \, mg^{-1}$ (15.7–17.6) for carbohydrates. Gnaiger and Bitterlich (1984) reported slightly different factors – 23.9, 39.5, and $17.5 \, J \, mg^{-1}$ for protein, lipids and carbohydrates, respectively (see also Brafield and Llewellyn, 1982; Jobling, 1983); all these conversion factors are within the IBP probable range. Differences between these sets of conversion factors originate from different sets of substances used for their determinations. Highly unsaturated fats are of low caloricity, therefore a lower conversion factor, $36.2 \, J \, mg^{-1}$, is recommended for fish food lipids (Brett and Groves, 1979). Craig *et al.* (1978), as a result of direct determination by bomb calorimetry of the energy content in crude (impurified) lipids extracted from somatic tissues of *Perca fluviatilis*, obtained a very low conversion factor, $35.5 \, J \, mg^{-1}$. The share of low-caloricity impurities in crude lipids extracted from carp eggs was considerable and it increased with increasing lipid extraction time (Kamler, 1976).

Methods for measuring caloric value

Figure 3.12 (B) compares, for four species of fish, caloric values calculated from chemical analysis using the IBP conversion factors and obtained directly by bomb calorimetry. It is evident that the latter values are higher than the calculated values by 3–18%. Similar discrepancies between results of these two techniques are obtained when protein is determined directly and the lipids are purified (Fischer, 1970b; Dowgiałło, 1975; Kamler *et al.*, 1982). When other analytical methods are used the 'proximate energy' can be even higher than that directly determined in the bomb calorimeter (Craig *et al.*, 1978).

Direct calorimetry using an oxygen bomb calorimeter is considered to be the most reliable method for measuring the caloric value of biological material; the two main indirect methods, i.e. estimates from the chemical composition of the material and the dichromate wet oxidation method (Maciolek, 1962) are approximate (Cummins and Wuycheck, 1971; Winberg, 1971; Dowgiałło, 1975; Prus, 1975). More recently Gnaiger and Bitterlich (1984) advocated automatic CHN (carbon–hydrogen–nitrogen) analysis for the purpose of ecological and physiological energetics. This method provides information on proximate chemical composition and simultaneously a basis for calculation of caloric value of analysed material; the measurements are done in a short time.

3.5 CONCLUDING REMARKS

Due to their high caloric value, lipids supply a greater proportion of available energy (Fig. 3.12 (B)) than they contribute to dry matter (Fig. 3.12 (A)). The four species of fish presented in Fig. 3.12 were selected for contrasting chemical composition of their eggs. In a marine species, *Sardinops caerulea*, only about one-fourth of the energy present in an egg originates from lipids (Fig. 3.12 (B)). Energy in eggs of a cyprinid, *Cyprinus carpio*, is one-third derived from lipids, that of a coregonid, *Coregonus albula*, about one-half, but in eggs of *Anguilla anguilla*, whose larvae migrate for long distances, as much as two-thirds of energy is accumulated in the form of lipids. Most remaining energy comes from protein; carbohydrates contribute little to egg energy reserves.

Summing up the considerations in Sections 3.1 to 3.4 on ways of regulation of energy reserves in fish eggs, we will first recall that egg size (mg egg^{-1}) and caloric value of matter (J mg^{-1}) both contribute to the energy content per egg. Table 3.13 summarises these parameters for eggs of some fish groups. It is evident that, as usual, size contributes to the variability of the energy content per individual to a much higher degree than does the caloric value of matter. Caloric value of wet matter depicts both changes in chemical composition of dry matter and changes in hydration of wet matter. Differences in egg energy reserves between fish groups can arise from three causes; increased egg size can be associated with an increase in lipid content, and – simultaneously – by a decrease in hydration. In consequence (Table 3.13) an average egg of *Oncorhynchus* sp. weights about 124 times as much as an average cyprinid egg, but contains about 189 times as much energy. An analogous comparison between eggs of Acipenseridae and an average sea spawner shows that the former have 19 times as much weight and 36 times as much energy as the latter. The 'syndrome' of these three properties was also reported by Balon (1977): in fishes that exhibit advanced protection of their offspring, an increase of egg

Table 3.13 Energy content in fish eggs: contribution of egg size and composition. A summary based on Fig. 3.1(G) and on Tables 3.4, 3.6, 3.7 and 3.10

Group	Wet wt (mg egg^{-1})	Water (% wet wt)	Caloric value (J mg^{-1} wet wt)	Energy content (J egg^{-1})
Sea spawners*	1.3	74	6.0	8
Cyprinidae	1.5	68	8.2	12
Acipenseridae	24.5	56	11.7	287
Salmo	80.0	60	11.2	896
Oncorhynchus	186.0	56	12.2	2269

*Excluding *Anguilla* and *Trachurus*.

size is accompanied by both a decrease of hydration and an increase of lipid content.

FACTORS AFFECTING FISH OFFSPRING

Data presented in Sections 3.1 to 3.5 point to variability in size of fish eggs as well as in the composition of their matter. In Sections 3.6 to 3.10, mechanisms that mediate between properties of fish offspring and various factors will be discussed. Investigations in this field have been developed in connection with man's increasing influence on fish reproductive processes. Factors, especially endogenous ones, affecting variability in early onto-genesis in fish were broadly investigated by Vladimirov (1965a, 1974b) and Romanenko (1981).

The complex pattern of relationships is presented as a simplified scheme in Fig. 3.13. The pathways shown illustrate the direct and indirect effects of various factors on fish offspring. The scheme was partly inspired by the classifications advanced by Brett (1979), Zhukinskij (1981), Żuromska and Markowska (1984) and Trojan (1985). It is neither complete, nor universal; its purpose is only to illustrate the more important factors influencing fish offspring.

In the rest of this Chapter no discussion will be given of quantitative aspects of fecundity, nor of the number of spermatozoa. The main factors controlling fecundity, namely body size (pathways 3–12 in Fig. 3.13), age (4–12), amount of food (6–12), population density (7–12), temperature (9–12), as well as relationships between egg number and their size (16), have been presented in Section 2.2: of these factors, the effect of body size on fecundity is especially conspicuous.

The quality of offspring is influenced by a combined effect both of endogenous properties of spawners, or internal factors, and of external factors (Fig. 3.13). Under natural conditions, the effect of external factors plays a dominant role. They can affect offspring quality indirectly by their impact on spawners during gonad formation and especially during their maturation. These problems will be discussed in detail in this Chapter; in later Chapters, the direct action of external factors will be examined as they affect fish embryos (pathway 28, Chapter 4) and larvae (pathway 29, Chapter 6). The situation with fish culture under controlled conditions is somewhat different. Here the external factors should be maintained within the optimal range. Therefore one should expect that under controlled con-ditions, the internal factors will be manifested more clearly than in the field.

3.6 GENETIC FACTORS

Galkina (1967) and Kato (1975) found that some *Oncorhynchus mykiss* females produced, respectively, highly fertilizable and large eggs, and that

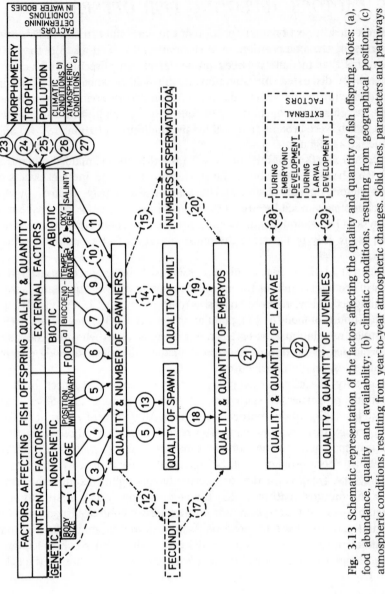

Fig. 3.13 Schematic representation of the factors affecting the quality and quantity of fish offspring. Notes: (a), food abundance, quality and availability: (b) climatic conditions, resulting from geographical position; (c) atmospheric conditions, resulting from year-to-year atmospheric changes. Solid lines, parameters and pathways (numbered) discussed in detail in Sections 3.6 to 3.10; broken lines, those discussed elsewhere and/or briefly mentioned in Sections 3.6 to 3.10. Further explanations in the text.

this feature persisted throughout their reproductive life. Sometimes such features are hereditary. They are used when selecting for commercially important characters such as egg number (pathway 2–12 in Fig. 3.13), egg size (2–13; Section 3.1), and especially body size. Examples for *O. mykiss* are described by, among other authors, Donaldson and Olson (1955) and Kato (1979). A positive relationship between the level of heterozygosity and developmental rate has been found in embryos of *O. mykiss* (Ferguson *et al.*, 1985a, b; Danzmann *et al.*, 1986); faster growth of heterozygotes is associated with depressed oxygen consumption (Danzmann *et al.*, 1987). Genetics are, however, beyond the scope of this work and will receive no further attention here.

3.7 NONGENETIC INTERNAL FACTORS

Two nongenetic internal factors affecting fish offspring are especially important: parental age and body size. Fish are unusual animals, because in many species somatic growth is continued after reaching sexual maturity (Chapter 2), and thus in adult fish also, body size is inseparably coupled with age (pathway 1 in Fig. 3.13). Physiological processes are controlled by age (Craig, 1985) and adjusted to body size. On the other hand, it is known that fish body size can vary at a given age; genetic and/or external factors contribute to this variability. The problem of influence of age and body size in fish on their progeny is of both basic and applied interest, and so has long been studied. Much work was performed in the hope that these easily measurable parameters would allow the quality of offspring to be predicted.

Parental age

Now we will discuss how age affects various properties of fish eggs via quality of spawners (pathway 4–13 in Fig. 3.13).

Effect on egg size

Several studies have shown that female fish spawning for the first time produce the smallest eggs. Egg diameter increases clearly between the first and second spawnings, and more slowly at further spawnings. This pattern has been described for *Rutilus rutilus* and *Abramis brama* (Kuznetsov, 1973), *Coregonus albula* (Potapova, 1978; Kamler *et al.*, 1982), *Oncorhynchus mykiss* (Bartel, 1971; Savostyanova and Nikandrov, 1976; Kato and Kamler, 1983 – see Table 3.2), *Cyprinus carpio* (Tomita *et al.*, 1980) and for *Brachymystax lenok* and *Thymallus arcticus* (Shatunovskij, 1985).

Egg weight is more dependent on age than is egg diameter (Lyagina, 1975). A linear increase in egg weight with female age was found by Zhukinskij (1965) for *Rutilus rutilus* from young spawners (age 3 years) to old ones (8 years), and similarly dry weight of eggs increased in *Cyprinus carpio* aged 3–11 years (Semenov *et al.*, 1974) and 4–9 years (Kamler, 1976). However, more authors have recorded that the age of spawners is associated with a curvilinear increase of egg weight (*Salmo salar*, Galkina, 1970; *Oncorhynchus mykiss*, see Fig. 3.3 and Table 3.2; *Oncorhynchus nerka*, Kato, 1978; *Rutilus rutilus*, Kuznetsov, 1973; *Cyprinus carpio*, Martyshev *et al.* 1967, Semenov *et al.*, 1974 – egg wet weight; *Coregonus albula*, Potapova, 1978 – see Fig. 3.5, Kamler *et al.*, 1982 – See Fig. 3.2; *Brachymystax lenok* and *Thymallus arcticus*, Shatunovskij, 1985; *Cottus bairdi*, Docker *et al.*, 1986; and many others). Only Scott (1962) did not find differences in egg size from *O. mykiss* females aged 3 and 4 years in Lake Paul, whereas in Lake Pennask the differences were evident.

The typical course of dependence of egg size on female age is as follows: young females spawning for the first time produce the smallest eggs, females of average age produce the largest eggs, and old spawners again lay lighter eggs. This relationship is shown for *Coregonus albula* in Fig. 3.2 with an equation which can be used for its approximation (Bartel, 1971; Kamler *et al.*, 1982). This equation also allows the optimal age (τ_{opt}, i.e. the age at which females produce the largest eggs) to be calculated. For *Oncorhynchus mykiss* this age is 5.8 or 6 years (Bartel, 1971); for different populations of *Coregonus albula* it is from 1.97 to 4.49 years, which correspond to age groups from 2^+ to 4^+ (Kamler *et al.*, 1982). In *C. albula*, τ_{opt} depends on maximum egg weight (W_{max}); for Polish lakes $r = 0.77$ (d.f. $= 8$, $P < 0.01$), and for Finnish lakes $r = 0.93$ (d.f. $= 1$,NS). This indicates that in poor trophic conditions *C. albula* produced smaller eggs and exhibited an earlier and shorter (cf. Fig. 3.2) period of production of the largest eggs. The high energy demands for metabolism in the oldest and, at the same time, largest females prevented them from producing high-quality offspring, especially in conditions of low food supply (Kamler *et al.*, 1982).

A very clear drop in egg size of old females *Oreoleuciscus potanini* in Mongolian Lake Dayan–Nur was reported by Shatunovskij (1985); this fish matures at age 5–6 years, and when 10 to 20 years old it produces eggs, whose wet weight decreases evenly from 1.32 to 1.23 mg egg^{-1}. However, in many species no egg size decrease has been observed in old females. This pertains especially to short-lived species, but sometimes to long-lived species such as *Acipenser baeri*, *Esox lucius* and *Eleginus navaga* (Nikolskij, 1974). Probably few fish have a chance to reach old age in exploited populations.

Changes in intra-individual egg size variability occurring with female age were illustrated in Table 3.2 and discussed in Section 3.1. By and large, this variability is smallest at the optimum age and is inversely related to egg size.

Effect on chemical composition of ovaries and eggs

The effect of age on the chemical composition of gonads (pathway 4 in Fig. 3.13) was studied in *Platichthys flesus* by Shatunovskij (1963). The highest percentage of lipids in wet matter of ovaries was found in females of average age. In an earlier-maturing Baltic population, this optimum age came earlier (at age $4^+ - 5^+$) than in longer-lived populations inhabiting the White Sea ($6^+ - 8^+$).

Changes in chemical composition of mature eggs with female age (pathway 4–13) will be exemplified mainly for common carp a fish intensely cultured in Central Europe and Asia. Hydration of eggs derived from young fish is higher than that of eggs produced by females at an average age (Semenov *et al.*, 1974; Nikolskij, 1974), whereas in older females the egg hydration is either unchanged (Semenov *et al.*, 1974) or again increases (Nikolskij, 1974). On the other hand, Kato and Kamler (1983) did not find any differences between hydration of eggs of *Oncorhynchus mykiss* aged 2–4 years. According to Nikolskij (1974), the percentage of total protein and total lipids in wet matter of fish eggs increases with fish age to maximum values and then decreases. These changes can result only from changes in hydration or both from changes of hydration and dry matter composition. There are data suggesting changes in egg dry matter composition with age. In *Cyprinus carpio* the sum of protein amino acids (sum of cysteine, cystine, trypotophan, leucine, valine, glycine, serine, ornithine and aspartic acid), as expressed in terms of concentration in dry matter, was lowest (average 270 mg g^{-1}) in eggs of three-year-old females, highest (306 mg g^{-1}) in females aged 7–8 years and in eggs of old females (11–14 years) it amounted to 280 mg g^{-1}, on average, or returned to the level observed in four-year-old females (Vladimirov, 1973, 1974a). A similar course of changes in the sum of cysteine, cystine, tryptophan and ornithine in dry matter of eggs of *C. carpio* was observed by Kim (1974a), whereas the leucine and isoleucine contents increased continuously with female age, reaching its maximum in old fish. The concentration of phospholipids (mg g^{-1} dry matter) in eggs of *C. carpio* was lowest and most variable in females aged 3–4 years, highest in females 5–8 years old, and dropped again in females 9–14 years old; the course of changes in cholesterol content was reversed, i.e. it decreased to minimum values in eggs produced by females of an average age (Kim, 1974b). By contrast, Potapova (1978) has found an increasing concentration of cholesterol in dry matter of *Coregonus albula* eggs with female age from young ones (2^+) to those of average age (4^+); this increase was accompanied by increased concentration of phospholipids, triglycerides and total lipids. Kamler *et al.* (1982) did not observe changes in percentage of total protein, lipids, carbohydrates and ash in egg dry matter for *C. albula* spawners from 1^+ to 7^+.

Thus changes with age in egg chemical composition, if expressed in terms of concentration, can follow similar courses to changes in egg size, or other patterns, or may not appear at all. In consequence, changes in the amounts per egg of important constituents can often follow trends similar to those in egg size, i.e. they can reach maximum values at the optimal age for reproduction.

Effect on egg metabolism and fertilization

The same can be said of the metabolic rate of ripe, unfertilized eggs (pathway 4–13); oxygen consumption rate, cytochromoxidase activity and rate of glycolysis were highest in eggs stripped from middle-aged *Rutilus rutilus heckeli* and *Abramis brama* (Zhukinskij and Gosh, 1970, 1974; Gosh and Zhukinskij, 1979). Females of optimum age also produce eggs more resistant to postovular overripening (pathway 4–13). During overripening of eggs stripped from young (4–6 years) *Abramis brama*, the concentration of protein amino acids decreased more, and that of free amino acids increased more, than in eggs obtained from females of average age (7–9 years) (Kim and Zhukinskij, 1978). Similarly, young (3–4 years) and old (7–8 years) *Rutilus rutilus heckeli* produced eggs less resistant to overripening than did those aged 3–6 years; amino acid composition of the latter eggs was more stable (Zhukinskij *et al.*, 1981). Percentage egg fertilization was related to both female age (pathway 4–13–18: *Rutilus rutilus heckeli*, Zhukinskij, 1965, and Vladimirov *et al.*, 1965; *Oncorhynchus mykiss*, Gall, 1974; *Cyprinus carpio*, Vladimirov, 1974a) and male age (pathway 4–14–19: *R. rutilus heckeli*, Zhukinskij, 1965, and Vladimirov *et al.*, 1965) in the same way.

Effect on embryos, larvae and juvenilles

Martyshev *et al.* (1967) found a higher embryo growth rate in eggs derived from *Cyprinus carpio* females aged 6–8 years than in those from the younger and older females (see also Nikolskij, 1974) (pathway 4–13–18). Gerasimova (1973) reported an increase of embryo metabolism with increasing age of the parental *Cyprinus carpio* females (pathway 4–13–18). Embryo survival also depends on age of parental females (Martyshev *et al.*, 1967; Zhukinskij and Gosh, 1970; Nikolskij, 1974). Vladimirov *et al.* (1965) fertilized eggs derived from *Rutilus rutilus heckeli* of different ages with sperm from different males. The highest survival at the morula and gastrula stages and at hatching was observed in progeny of average-aged females × average-aged males, and lowest survival in progeny of old and young spawners. The fewest deformed larvae hatched from eggs derived from average-aged females (Vladimirov, 1973, 1974a); this illustrates one of the indirect pathways for an effect of female age on the number of normal larvae (pathway 4–13–18–21). A far-reaching age effect was shown by Vladimirov *et al.* (1965), who found that the highest survival rates among exogenously feeding *R. rutilus heckeli* larvae were those for larvae hatched from eggs that were produced by average-aged females and then fertilized by average-aged males. Similarly the highest survival was observed in one-month-old *Cyprinus carpio* larvae that were offspring of average-aged females (Vladimirov, 1974a). Larger and faster-growing fingerlings of *Oncorhynchus mykiss* originated from three-year-old females than from two-year-olds (Gall, 1974). A still broader range of effects of female age on the quality of offspring was observed by Martyshev *et al.* (1967) in *Cyprinus carpio*. The most rapid increase in subcutaneous fat and the greatest muscle thickness was found in one-year-old juveniles derived from seven-to-nine-year-old female parents; these juveniles used nitrogen more efficiently than those one-year-olds derived from either younger or older females. Nikolskij (1974) believes that biochemical differences between offspring of female spawners of different age are still maintained in two-year-old juveniles (pathway 4–13–18–21–22).

Thus the effect of parental age on the offspring is strong, multidirectional and manifested in a long-lasting way in various aspects of the progeny's life.

Parental size

Peters (1983) has shown from extensive material that in various groups of animals, parental body size determines the offspring's body size. Fish are no exception in this respect; many authors have recorded a positive correlation between egg size and fish size (pathway 3–13 in Fig. 3.13).

This was reported by Laurence (1969) for *Micropterus salmoides*, Pope *et al.* (1961) and Thorpe *et al.* (1984) for *Salmo salar*, Alderdice and Forrester (1974) for *Hippoglossoides elassodon*, Lyagina (1975) for *Rutilus rutilus*, Rogers and Westin (1981) for *Morone saxatilis*, DeMartini and Fountain (1981) for *Seriphus politus*, Fujita and Yogata (1984) for *Seriola aureovittata*, Mann and Mills (1985) for *Leuciscus leuciscus*, Beacham and Murray (1985) for *Oncorhynchus keta*, McFadden *et al.* (1965) and Szczerbowski (1966) for *Salmo trutta*, in which Trzebiatowski and Domagała (1986) have observed an increase in egg size with increased body length and weight, and a decrease in egg hydration and increase in lipid concentration (% wet weight). Similarly an increase in egg diameter and weight with increasing body length was observed by Chełkowski *et al.* (1985, 1986) in *S. trutta*; these authors conclude that the most practical method of selection of female spawners for reproduction should be based upon the body length. Garcia and Brana (1988) measured *S. trutta* oocyte diameters at the final stages of vittelogenesis and found them significantly correlated with female body length. Wootton (1984) quotes data collected by Scott and Crossman for 162 species of freshwater fishes; egg diameter was significantly correlated ($P < 0.05$) with fish length at maturity. The increase in wet weight of eggs collected from two- and three-year-old females of *Oncorhynchus mykiss* together with the increase in body weight has been described by a linear equation (Shimma *et al.*, 1978); the relationship was highly significant.

Although Galkina (1970) found a significant relationship between egg weight and female body length in *S. salar* and *Oncorhynchus mykiss*, she did not always find that egg weight depended upon female weight. Similarly Suzuki (1983) found egg diameter in two-year-old *Misgurnus fossilis* females to be positively correlated with body weight, but no such relationship in three-year-old females. Islam *et al.* (1973) applied linear equations to describe the relationship between egg diameter and female weight in *O. mykiss*. In some series of their investigations these relationships were significant, but not in others. These relationships were usually stronger within the series with young females, whereas egg diameter was almost independent of body weight in older females.

Egg size was found not to increase significantly with female size in *Gadus morhua* (Oosthuizen and Daan, 1974), *Cichlasoma nigrofasciatum* (Townshend and Wootton, 1984), *Etheostoma spectabile* (Marsh, 1984), and *Coregonus albula* (Dąbrowski *et al.*, 1987); in *C. albula* no changes were observed in the percentage of lipids in egg dry matter as female size increased. No significant relationship between egg size and female size was found in *Oncorhynchus mykiss* belonging to the same year-class in a single population (Scott, 1962; Bartel, 1971; Kato, 1975, 1979).

Summing up, the data presented here suggest that female body size per se does not always affect egg size, and that the observed effect of body size can often result from the age (pathway 1–3–13).

Oocyte position within the ovary

In contrast to the two previous internal factors, oocyte position within the ovary directly influences the offspring's properties (pathway 5 in Fig. 3.13).

Studies by MacGregor (1957) have indicated that the right and left ovarian lobes of *Sardinops caerulea* are at the same stage of maturity. Toetz (1966) showed this to be true for the ovarian lobes of *Lepomis macrochirus*: dry weight, energy and nitrogen content per egg as well as fertilizability and hatchability of eggs from the right and the left lobes were alike. Size-frequency distributions of oocytes in the right and left ovary of *Seriphus politus* did not differ significantly (DeMartini and Fountain, 1981). Similarly, no significant differences in hydration, dry weight, caloric value of dry matter, or in energy content per egg were found for the right and left ovarian lobes of *Salmo trutta* (Kamler, 1987). Hence, the right and left ovarian lobes can be regarded as equivalent.

Dry weight, energy content, nitrogen content and fertilizability of eggs were alike in *Lepomis macrochirus* regardless of the antero-posterior position of eggs within the ovary (Toetz, 1966). Nevertheless, the hatchability of eggs taken from the most posterior part of an ovary (i.e. situated near to the urogenital vent) was significantly higher than that of the more anterior ova. Toetz (1966) compared also the quality of peripheral eggs and those situated in a medial position in relation to the longer axis of an ovary. The medial eggs were somewhat heavier and contained a little more energy per egg, but the differences were not significant. On the other hand, the percentage of small (immature) eggs was clearly lower in medial eggs. Fertilizability rate and hatchability of medial eggs were significantly higher than in eggs collected from the periphery of an ovary.

The investigations described above have therefore revealed no marked differences in egg quality in various parts of an ovarian lobe. One should remember, however, that these investigations were run on a 'macro scale': large numbers of eggs (up to 200, Toetz, 1966), were sampled from a given part of an ovary. The considerations presented earlier (Section 3.1) point to the existence of clear differentiation in egg size within an ovary.

3.8 BIOTIC EXTERNAL FACTORS

External factors, both biotic and abiotic, can affect fish offspring directly during embryonic and larval development – pathways 28 and 29 in Fig. 3.13; these pathways are discussed later (Chapters 4 and 6). Here, indirect pathways for the action of external conditions on fish offspring via the parents will be discussed.

Food

In this section we shall consider how food conditions prior to spawning, such as food abundance, food quality and food availability, can affect fish offspring.

Abundance

It is believed that fish with a more abundant food supply generally produce larger eggs than the same species receiving less food (Brown, 1957; Nikolskij, 1974) (pathway 6–13). The eggs of *Coregonus albula* from Finland's ultraoligotrophic Lake Puruvesi were smaller than those from two oligo-dystrophic lakes, Oulujärvi and Kangosjärvi; females from a Polish mesotrophic lake (Narie) produced eggs which were systematically smaller than those from the eutrophic Lake Maróz (Fig. 3.2) (pathway 24–6–13). However, the increase in trophic status of Polish lakes affected in a catastrophic way the survival of *C. albula* embryos, and especially of *C. lavaretus* on spawning grounds (pathway 28) (Żuromska, 1982; Wilkońska and Żuromska, 1982).

Referring again to the influence of food supply on egg properties, Townshend and Wootton (1984) showed experimentally with *Cichlasoma nigrofasciatum*, and Springate *et al.* (1985) with *Oncorhynchus mykiss*, that egg size was positively correlated with food ration. By contrast, Bartel (1971) found no such relation in the latter species. Similarly Wootton (1985) reported that in *Gasterosteus aculeatus*, egg size changed only a little over a wide range of food rations. Also Scott (1962) and Ridelman *et al.* (1984) recorded that experimentally induced starvation during gonad maturation did not affect *O. mykiss* egg size. The same can be said of the chemical composition of egg matter. Springate *et al.* (1985) found no changes in hydration of swollen eggs, nor changes in protein, lipid and ash percentages in egg dry matter of

O. mykiss fed with restricted rations (Tables 3.6, 3.7, 3.10 and 3.12, respectively). Similarly Wootton (1985) did not find any changes in chemical composition of eggs of *G. aculeatus* receiving different rations; in *O. mykiss*, Ridelman *et al.* (1984) did not observe that any changes in egg proximate composition resulted from starvation of the females before spawning. No effect was ascertained of food limitation during ovary maturation on offspring viability, as measured by percentage fertilization of eggs, percentage embryo survival at eyeing (pathway 6–13–18) (Springate *et al.*, 1985), and percentage hatching (pathway 6–13–18–21) (Ridelman *et al.*, 1984).

Thus although when food is abundant for female spawners they can produce larger eggs, a lack of effect of food abundance on egg size, chemical composition and offspring viability has also been observed. Any effect of food limitation on egg quality is counterbalanced by the fact that a fish can maintain egg quality at the expense of their numbers (pathway 16) (Scott, 1962; Wootton, 1985), and lipids present in gonads can be used for metabolic purposes under the conditions of extreme food shortage only (Nikolskij, 1974).

Another aspect of this problem is the abundance of food that could be adequate for the larvae. Marsh (1984) supports the genetic hypothesis that under poor food conditions, selection favours females producing large eggs (K strategists), whereas females producing a greater number of small eggs (r strategists) would be confined to conditions of abundant larval food. In the latter case a disadvantage resulting from the lower fitness of small larvae (size is a measure of fitness, Begon, 1984) would be balanced by their greater numbers. However, recently Daoulas and Economou (1986) did not find in *Sardina pilchardus* any negative relationship between egg size and plankton biomass at the time of fish sampling, as the genetic explanation predicts; they are of the opinion that intraspecific variation of egg size is a phenotypic phenomenon.

Quality

The effect of food quality on egg properties (pathway 6–13) will be illustrated with examples for *Oncorhynchus mykiss* (size of eggs) and *Cyprinus carpio* (composition of eggs). Bartel (1971) observed that *O. mykiss* females fed with food pellets produced smaller eggs than females fed with wet food; Shimma *et al.* (1978) observed a positive effect on egg size of a diet based on methanol-grown yeast, as compared with a diet based on fish meal. In *C. carpio* fed different feeds, alterations in egg chemical composition expressed in terms of concentration were reported for amino acids by Vladimirov (1973) and Kim (1974a), and for cholesterol and phospholipids by Kim (1974b).

Availability

The effect of food availability to spawners during gonad formation on the quality of offspring will be illustrated by our studies (Kamler *et al.*, 1982) on autumn-spawning populations of *Coregonus albula* of three Finnish and five

Polish lakes. Of the morphometric features of a lake which affect food availability (pathway 23), the development of shoreline will be considered:

$$D_L = L/(2\sqrt{\pi Sa}) \tag{3.1}$$

where L is length of the shoreline and Sa is surface area. In lakes where the shoreline is strongly developed, i.e. having numerous bays, peninsulas and islands (and thus a large D_L value), wind-driven mixing is obstructed and nutrients are swiftly eliminated from the upper layers (Patalas, 1960a,b). This leads to decreased food availability for a planktivorous fish such as *C. albula*. Of the three Finnish lakes, the most developed shoreline is in Lake Puruvesi ($D_L = 8.1$, v. 2.6–7.7), and among the five Polish lakes, Narie ($D_L = 4.3$, v. 1.2–2.8). In fact, the populations in Lake Puruvesi and Lake Narie, when compared with the remaining populations, were characterized by a low rate of fish growth, and thus by low somatic production (P_g), low egg size (amount of dry matter, main chemical constituents and energy per egg), low reproductive production (P_r) and low percentages of fertilization and embryo survival to morula stage (data for 1975–77 in Kamler *et al.*, 1982). Moreover, the above indices were usually higher in the Polish lakes than in the Finnish lakes, whose shorelines were more developed. Some of these data are depicted in Fig. 3.2 and Tables 2.3, 3.1, 3.5, and 3.14. Later studies performed in 1981–85 by Wilkońska and Żuromska (1988) have confirmed that the growth of *C. albula* in Lake Narie is exceptionally low as compared with that in other Polish lakes; egg and sperm quality is also low in this population. The poor growth of *C. albula* in Lake Narie as compared with the other Polish lakes has long been well known (Radziej, 1965). Radziej's (1973) experiment has shown that it is a phenotypic phenomenon, rather than a genotypic one: *C. albula* fry transferred from Lake Narie to Lake Wierzbiczany exhibited satisfactory growth. Thus one can expect that morphometric properties of lakes, acting via food conditions for spawners and spawner quality, can modify the quality of spawn and embryos (pathway 23–6–13–18).

The flexibility of salmonid reproductive strategies has been reviewed by Thorpe (1990). Initiation of development can be attributed to abiotic factors (temperature and light), but its completion depends on a biotic factor: the trophic conditions.

Biocoenotic factors

Studies by Suzuki (1974) have indicated that population density exceeding a biomass of 300 g m^{-2} decreased the percentage of fertilization and of hatching in *Misgurnus angullicaudatus* (pathway 7–13–18–21). Fish producing a small number of large eggs (i.e. exhibiting a K strategy) are favoured under conditions of weak competition, whereas under more intense competition, selection for a large number of small eggs (r strategy) occurs, thus interpopulation, intraspecific variations in egg size may result from natural selection (Svärdson, 1949; Scott, 1962).

The presence of fish eggs in stomachs of Coregonidae was reported by Jacobsen (1982) and in *Fundulus heteroclitus* by Penczak (1985); Wootton (1985) observed egg cannibalism in *Gasterosteus aculeatus*. Vendace, *Coregonus albula*, is an autumn-spawning species, but in some deep Scandinavian lakes which were situated outside the ice margin during the time of the Baltic Ice Lake in the Younger Dryas Period, winter- or spring-spawning forms exist (Airaksinen, 1968; Lind and Turunen, 1968). They spend summer in cold deep waters; in late autumn, during the period of intense growth of their oocytes they move to shallower waters, which are the spawning grounds of the autumn-spawning vendace, and eat the latter's eggs (Airaksinen, 1968). These large amounts of very valuable food, consumed without large energy costs and at the right moment, permit eggs spawned in winter in Lake Kajoonjärvi to be 3–7 times larger (dry weight) than the eggs of the autumn-spawning form (Section 3.1). Winter-spawning vendace hatch later and are larger than autumn-spawning vendace (K. Salojärvi, pers. comm., 1979). In consequence the type of relationship between weight (W) and body length (L) in the winter-spawning *Coregonus albula* from Lake Kajoonjärvi is different from that in the autumn-spawning *C. albula* in other Finnish lakes: the intercept a in the equation $W = aL^b$ is much higher (0.302 instead of 0.004–0.009), whereas the slope b is much lower (1.772 v. 2.980–3.242), with the rate of growth in the winter-spawning form being higher than in the autumn-spawning form (Kamler *et al.*, 1982).

In a recent review of egg size in 71 cyprinid species, Coburn (1986) concluded that larger eggs were observed in territorial and/or nesting species than in species lacking these behaviours. One possible explanation is that these behaviours may select for fish with larger body size, which is often positively correlated with egg size (Section 3.7). On the other hand, parental care decreases the net reproductive effort and shifts the reproductive strategy towards a K strategy (Section 2.2).

3.9 ABIOTIC EXTERNAL FACTORS

Temperature

Climatic conditions

Larger eggs produced by fish in waters situated in colder climates, i.e. in higher latitudes (pathway 26–9–13), have been described at the intra-specific level for *Engraulis anchoita* (Ciechomski, 1973), *Etheostoma spectabile* (Marsh, 1984), *Noemacheilus barbatulus* (Mills and Eloranta, 1985) and *Sardina pilchardus* (Daoulas and Economou, 1986), as well as for a crus-tacean, *Palaemon paucidens* (Nishino, 1980). The same can be said of egg size considered at the interspecific level (data of Rass in Hoar, 1957; Zalewski and Naiman, 1985; Zalewski, 1986). From a review of data for 71 North American cyprinids, Coburn (1986) found that larger eggs were produced by species inhabiting cooler waters at higher altitudes. An exponential increase of egg diameter with decreasing temperature during the peak of spawning was reported by Ware (1975) for 23 North-West Atlantic species. However, Smirnov *et al.* (1968) did not observe any increase in egg size of *Salmo* spp. and *Oncorhynchus* spp. in higher latitudes. In northern populations of *Coregonus albula* (Kamler *et al.*, 1982) and of *Fundulus heteroclitus* (Penczak, 1985), egg size was smaller than in more southerly populations. Thus the size increase of fish eggs is often a response to low temperatures, but other responses have also been observed.

It is commonly accepted that long development of embryos in cold water may be a factor promoting selection for large eggs in low temperatures. However, other aspects of this relationship have also been considered. Ciechomski (1973) found slower growth and lower fecundity of *Engraulis anchoita* in colder spawning areas; according to her opinion, the larger eggs found in the colder areas can be a consequence of lowered fecundity (pathway 16). Ware (1975) summarized data on the interdependence between temperature during peak of spawning, egg size, and duration of incubation. Moreover, Ware showed that mortality rate is inversely proportional to egg or larval size; low temperatures select for large eggs. Kamler and Kato (1983) showed with *Oncorhynchus mykiss* that efficiency of yolk utilization for embryonic growth was lower in colder water. Marsh (1984) believes that warm temperatures select for small eggs because large eggs have a smaller surface-to-volume ratio, which is unfavourable in warm waters holding less oxygen (pathway 8 in Fig. 3.13). Although embryos of oviparous fish have large stores of yolk, they can also use exogenous energy supplies taken up from the water (Terner, 1979). Zalewski (1986) concluded that this can be one reason why fish inhabiting boreal waters, which are poor in dissolved organic matter, produce large eggs. According to Mills and Eloranta (1985), a greater egg size in a northern population of *Noemacheilus barbatulus* than in a southern one is probably an adaptation to reduce larval mortality under conditions of food scarcity, and to achieve a certain weight before the long first winter. Thus potential reasons for producing large eggs in cold waters can be diverse, and it is difficult to judge which of them are at work at a given moment and in what combinations. However, their effect can sometimes be blurred by a masking factor. In the case of *Coregonus albula*, which produce smaller eggs in Finnish lakes than in warmer Polish ones (Kamler *et al.*, 1982), such a masking factor can be food for spawners; in cold, oligotrophic Finnish lakes they grew more slowly, were less fecund, and at the same time produced smaller eggs.

The problem of egg size was explained in a most complex way by Zalewski and Naiman (1985) and Zalewski (1986), as based on the concept of a continuum of abiotic and biotic factors. In their opinion, there are ecosystems where the pressure of abiotic factors is higher than that of biotic ones, and there are other ecosystems where biotic factors are decisive for the structure of fish communities. Between these two types of ecosystem lies a continuum (Fig. 3.14(A)). Zalewski (1986) has observed that the graph of egg size on water temperature during spawning does not follow a simple falling line, as thought hitherto, but is U-shaped (Fig. 3.14(B)). Eggs that are large in relation to the body size of spawners are observed not only in the zone of low temperatures but also at high temperatures; however, in both cases they were found in ecosystems with a prevalence of abiotic regulation (cf. Figs 3.14(A) and (B)), i.e. in rivers of boreal regions and in trout rivers (left side of Fig. 3.14(B)) and in desert rivers (right side). In the latter habitats, food is not a limiting factor, but periodic deficiencies of oxygen and

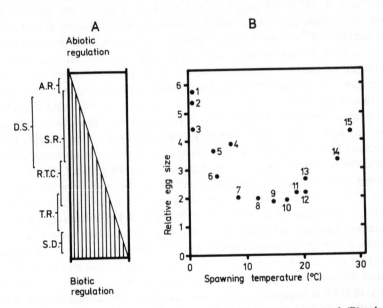

Fig. 3.14 (A) Abiotic and biotic regulation in riverine ecosystems and (B) relative size of eggs spawned at different temperatures. A.R., arctic river; D.S., desert stream; S.R., salmonid river; R.T.C., rivers in temperate climate; T.R., tropical river; S.D., desert source. The relative egg size is the ratio of egg diameter to L_∞, where L_∞ is a parameter from the von Bertalanffy equation describing growth of fish, calculated using the Ford–Walford method. Spawning temperature is the lower limit of the optimal temperature. Species: 1, *Salmo salar;* 2, *Salmo trutta;* 3, *Salvelinus fontinalis;* 4, *Thymallus thymallus;* 5, *Cottus poecilopus;* 6, *Esox lucius;* 7, *Phoxinus phoxinus;* 8, *Rutilus rutilus;* 9, *Noemacheilus barbatulus;* 10, *Gobio gobio;* 11, *Misgurnus fossilis;* 12, *Tinca tinca;* 13, *Silurus glanis;* 14, *Cyprinodon nevadensis* (desert streams); 15, *Poecilopsis monacha* (small desert streams in which water, amount of food and temperature change abruptly). Fish exhibiting special strategies of early ontogenesis, e.g. Cichlidae, are excluded. (Reproduced with permission from Zalewski, 1986.)

oscillations in mineral salt concentration are observed. Under conditions of low oxygen availability, the share of energetically less efficient anaerobic processes increases; in other words, the management of energy stored in the yolk is more wasteful. There are also added energy expenditures for osmoregulation, but they are probably not very important. The adaptive significance of large eggs in abiotically controlled ecosytems from the two extremes of the temperature scale was suggested by Zalewski (1986). Supporting evidence is supplied by Nishino (1981) for a crustacean, *Paratya compressa*, which displays an increase in egg size towards both the southern and the northern borders of its range along the Japanese archipelago. Therefore extreme abiotic conditions would favour K selection, and competition would favour r selection, as discussed in Section 3.8.

Atmospheric conditions

The other aspect of an influence of temperature on aquatic organisms is connected with atmospheric conditions which produce year-to-year differences in temperature patterns within the same water body. Examples of changes in egg size within a population among collections made in different years have been given in Section 3.1, so we will restrict our considerations to two populations of autumn-spawning *Coregonus albula* (Table 3.14). Diversified atmospheric conditions were observed in the years 1975–77 in the Masurian Lake District of Poland (Kamler *et al.*, 1982). The period of gonad formation by *C. albula* was very warm in 1975. High temperatures in the epilimnion in summer forced the cold-water fish to stay for longer periods in the deeper, cooler layers of the lake, which have less abundant food. The following growing season was generally cooler. On the other hand, in 1977 the temperature curve was flattened. An early, warm spring with abundant food permitted a faster recovery of fish from the winter depletion. Summer was cool, but in autumn, at the last stage of vitellogenesis, temperatures higher than the long-term average occurred again and the abundant zooplankton was maintained for a longer period than usual. In the year with a flattened air temperature curve, 1977, therefore, the eggs of *C. albula* were larger and less variable than in the two previous years (Table 3.14) (see also Wilkońska and Żuromska, 1988) the caloric value of their dry matter was higher, a smaller proportion of eggs had traces of resorption, and there were higher percentages of fertilization and survival to the morula stage (pathway 27–6–13–18).

Salinity

Information on the effect of salinity on fish egg properties (pathway 11–13) is presented in Sections 3.1–3.3 and will be only briefly summarized here. There is a tendency to produce smaller eggs in sea water (Fig. 3.1), although the egg size ranges for marine fish and for those spawning in fresh waters

Table 3.14 Coregonus albula egg properties and embryo viability. The years 1975–77 differed in atmospheric conditions (see text)*

Lake and year	W_{max}† (mg egg⁻¹)	Egg diam. variability, CV (%)	Mean caloric value of egg dry matter (J mg⁻¹)	$C.e._{max}$† (J egg⁻¹)	Mean % of eggs without resorption	Mean % fertilization	Mean % survival to morula
Narie, 1975	0.771	–	25.35	19.51	–	–	–
Narie, 1976	0.746	6.19	25.06	18.75	66.9	68.4	65.1
Narie, 1977	0.779	4.05	29.07	23.05	95.8	74.3	74.3
Pluszne, 1975	0.798	–	25.47	20.36	–	–	–
Pluszne, 1976	0.843	4.17	25.46	21.56	80.8	78.5	78.5
Pluszne, 1977	0.880	3.74	30.08	26.04	97.3	85.1	85.1

*Source, Kamler et al. (1982).

†Maximum weight (W) or caloric equivalent ($C.e.$) of an egg at the optimal age: compare Fig. 3.2.

overlap clearly. The same holds for the caloric value of egg dry matter (Fig. 3.8) owing to the lower lipid percentage in marine eggs (Table 3.10), whereas hydration (Table 3.6, Fig. 3.10) and mineral salt content (Table 3.12, Fig. 3.11) are higher in eggs of marine fish (see also Table 3.13).

Some internal and external factors influencing fish offspring have been presented in Sections 3.6–3.9. A summarizing concept was given by Daoulas and Economou (1986). A positive relationship between egg size and female parent size, and inverse relationships between egg size and temperature, as well as between egg size and food availability for spawners, were tentatively explained by an inverse relationship between egg size and adult growth rate. The weak point of the concept, as its authors admit, is lack of direct evidence relating egg size to female growth rate. This is not surprising, as egg size is only one component of reproductive growth (for negative relations between reproductive and somatic growth see Section 2.2).

3.10 EFFECT OF EGG QUALITY ON FISH OFFSPRING

In previous Sections, 3.6–3.9, attention was paid to factors that modify fish egg size and chemical composition. Now egg quality will be considered from another point of view: if and how it affects the fish offspring that subsequently develop (pathways 18, 18–21, or 18–21–22 in Fig. 3.13). In their recent review of criteria for the determination of quality of marine fish eggs Kjørsvik *et al.* (1990) define egg quality as their potential to produce viable fry.

Offspring properties related to egg size

The evaluation of the effect of any factor on early developmental stages of fish is most often based on two main indices: growth (or size attained) and survival.

Size

Greater size of larvae derived from large eggs than those derived from smaller eggs is a well-known phenomenon (Kryzhanovskij, 1940). It is manifested at both the intra- and interspecific levels, with the exception of viviparity occurring in some teleosts. For brevity, only a few primary works providing intraspecific comparisons will be mentioned here. A positive relationship between larval size and egg size was recorded for *Salmo salar* (Hayes and Armstrong, 1942; Thorpe *et al.*, 1984), *Oncorhynchus mykiss* (Gall, 1974; Kamler and Kato, 1983 – see Fig. 3.15; Escaffre and Bergot, 1984, 1985, intra-female studies; Springate *et al.*, 1985; Springate and

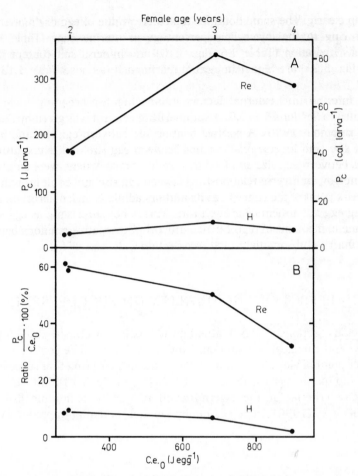

Fig. 3.15 Relationship of *Oncorhynchus mykiss* larval size (A) and efficiency of yolk utilization (B) to initial egg size ($C.e._0$). P_c, total amount of energy in the larval tissues at hatch (H) and at the end of yolk sac resorption (Re). The age of females producing eggs of a given size is shown on the upper horizontal axis. Based on data from Kamler and Kato (1983).

Bromage, 1985), *O. keta* and *O. kisutch* (Beacham *et al.*, 1985), *O. tshawytscha* (Rombough, 1985), *Coregonus albula* (Dąbrowski *et al.*, 1987; Wilkońska and Żuromska, 1988), *Leuciscus leuciscus* (Mann and Mills, 1985), *Clupea harengus* (Blaxter and Hempel, 1963, 1966), *Morone saxatilis* (Eldridge *et al.*, 1981a, 1982), *Gadus morhua* (Knutsen and Tilseth, 1985) and *Etheostoma spectabile* (Marsh, 1986). Also a direct positive relationship between nauplius size and egg size was found in the calanoid copepod *Diaptomus clavipes* (Cooney and Gehrs, 1980). A study on 71 species of North American Cyprinidae by Coburn (1986) supplies an explanation for this phenomenon. He found the number of vertebrae to increase significantly with egg size; his opinion is that somite formation is dependent upon space, which can be related to egg size.

The adaptive significance of larval size is obvious. Predation, which is recognized as one of the major causes of mortality of fish larvae in the wild (Mills, 1982), operates in a size-dependent way. In contrast to juvenile and adult fish, the smaller members of a *Salmo trutta* larval brood suffered greater mortality than larger ones (Hansen, 1985). Knutsen and Tilseth (1985) have shown with *Gadus morhua* larvae that the size of mouth gape just before the onset of external feeding was strongly ($P<0.0001$) correlated with the mean dry weight of eggs from which the larvae had hatched; the percentage of larvae having prey in their gut during the first days of feeding was related to mouth gape. The adaptive significance of the larval size is also connected with the broader feeding spectra.

The initial size differences between the larvae hatched from large eggs and those hatched from small ones were observed to persist in *Salmo salar* for almost 5 weeks (Hayes and Armstrong, 1942) and in *Oncorhynchus mykiss* for 2, 8 and 16 weeks (Springate and Bromage, 1985; Escaffre and Bergot, 1985; Springate *et al.*, 1985, respectively). However, the initial size advantage conferred upon larvae hatched from large eggs may be obscured during subsequent development, or even lost. In *Salmo salar* it did not persist after the first 5 weeks of growth (Hayes and Armstrong, 1942) or the first year (Thorpe *et al.*, 1984); in *Oncorhynchus mykiss* it was lost over a 16 week period (Springate *et al.*, 1985) or a 4 week period (Springate and Bromage, 1985). The relation between larval size and egg size in *O. keta* was modified by temperature (Beacham and Murray, 1985). Hayes and Armstrong (1942), Eldridge *et al.* (1982) and Dąbrowski *et al.* (1987) documented compensatory growth of small larvae hatched from small eggs; hence these larvae were not at a disadvantage to larger ones.

An explanation of the possible bioenergetical regulatory mechanisms can be based on the data shown in Fig. 3.15. The amount of energy accumulated in tissues of *O. mykiss* larvae increased with increasing egg energy content. Although Ware (1975) thinks that the weight of a newly hatched larva is proportional to egg volume, in the case presented in Fig. 3.15 these values are not proportional. Eggs derived from a four-year-old female contained three times as much energy as eggs produced by two-year-olds, whereas the amount of energy in larvae hatched from these respective eggs differed only by about 1.5 times (Fig. 3.15(A)). This results from a decrease in the efficiency of yolk energy utilization for growth with increasing amounts of available yolk (Fig. 3.15(B)): at the end of yolk resorption, larvae developing from the largest eggs have transformed into their bodies only one-third of the energy initially present in the egg, those from medium-sized eggs about one-half, and those from the smallest eggs almost two-thirds. This means that the small larvae hatching from small eggs manage the meagre energy resources that they have at their disposal more economically (further examples, Section 4.5).

Survival

Turning now to the viability of progeny, higher survival of larger individuals resulting from larger eggs has been demonstrated for *Salmo salar* (Peslyak, 1967), *S. trutta* (Bagenal, 1969) and *Oncorhynchus mykiss* (Gall, 1974; Pchelovodova, 1976; Kato and Kamler, 1983; Escaffre and Bergot, 1984). Ware (1975) proposed a survivorship equation for marine pelagic fish which shows that the egg and/or larval mortality rate is inversely proportional to egg size. When there is no external food, larger larvae from larger eggs survive longer than those derived from small eggs; this has been shown for *Clupea harengus* (Blaxter and Hempel, 1963). *Oncorhynchus mykiss* (Escaffre and Bergot, 1985), *Coregonus albula* (Wilkońska and Żuromska, 1988), *Leuciscus leuciscus* (Mann and Mills, 1985) and *Etheostoma spectabile* (Marsh, 1986). In contrast, no effect of egg size on viability of progeny was found by many authors: Ciechomski (1966) – *Engraulis anchoita*; Satia *et al.*, (1974), Kato and Kamler (1983), Springate *et al.*, (1985) and Springate and Bromage (1985) – *Oncorhynchus mykiss*; Kamler (1987) – *Salmo trutta*; Thorpe *et al.* (1984) – *S. salar*; Docker *et al.* (1986) – *Cottus bairdi*. Beacham and Murray (1985) observed higher survival of *O. keta* embryos from small females than from large ones.

Żuromska and Markowska (1984) in *Tinca tinca* and Wilkońska and Żuromska (1988) in *Coregonus albula* studied the paternal effect. Throughout embryonic development in *T. tinca*, the quality of spermatozoa, as defined from morphological features, was decisive for survival. The effect of egg quality, as expressed by the energy stored in an egg, on survival became noticeable in the postovular period of embryonic development, but was less pronounced than the effect of milt quality. The effect of egg size became dominant later, during larval development. Similar results were obtained for *C. albula*, with the difference, however, that in this fish the effect of egg size became manifest later than in *T. tinca*, as late as larval development. The survival of larvae deprived of exogenous food depended only on the amount of energy contained in an egg and was not related to the quality of sperm. Thus, the paternal effect on fish viability in early ontogenesis is revealed earlier than the maternal effect. The effect of gametes becomes less important in further ontogenesis (Zhukinskij and Nedyalkov, 1980).

From considerations presented in this Chapter on properties of offspring as related to egg size, the counterbalancing advantages of reproductive r and K strategies arise. In an r strategy, from the same total biomass of eggs a larger number of smaller individuals develops; they utilize food resources more effectively and exhibit a potential for rapid growth. In a K strategy, fewer but large larvae capture bigger prey; they are at an advantage if food is scarce and more easily avoid predators and achieve territorial dominance; they have a better chance of survival under unfavourable conditions in the wild.

Offspring viability related to chemical properties of eggs

It is logical to expect that not only the quantity but also the quality of matter transferred by a mother to her progeny should have important consequences for its vitality. This expectation is corroborated by numerous data from the literature.

In *Oncorhynchus mykiss* a significant negative correlation was found between embryo survival and the moisture content (%) in egg matter (Satia *et al.*, 1974; Kato and Kamler, 1983). In consequence, positive correlations were observed between offspring vitality and the percentage of protein in wet matter: in *O. mykiss* this dependence was shown to persist until yolk sac resorption (Satia *et al.*, 1974), and in *Cyprinus carpio* to the 16th day of larval life (Semenov *et al.*, 1974). Positive correlations between fertilizability of eggs, percentage of hatched individuals that are normal and survival of larvae during one month of culture, on the one hand, and the concentration of total protein amino acids in egg dry matter, on the other hand, have been reported for carp by Vladimirov (1974a). Fertilizability and hatchability of carp eggs, and survival of larvae over 16 days of exogenous feeding, were positively correlated with total protein concentration and with concentration of sulphydryl groups (–SH) in eggs at gastrulation (Konovalov, 1979; review: Konovalov, 1984). In *Acipenser stellatus*, however, fertilizability of eggs was not dependent upon percentage content of protein in dry matter (Arutyunova and Lizenko, 1985). A positive relationship between viability of offspring and total lipid concentration in eggs was shown by Vladimirov (1965b) for *Rutilus rutilus heckeli* and *Abramis brama*, Arutyunova and Lizenko (1985) for *Acipenser stellatus*, and Dąbrowski *et al.* (1987) for *Coregonus albula*. One should therefore expect a positive relationship between viability of offspring and caloric value of egg matter. In *Cottus bairdi*, survival from fertilization to yolk resorption was reported by Docker *et al.* (1986) to be significantly correlated with caloric value, but the relationship was almost horizontal. In *Oncorhynchus mykiss*, survival at hatch was independent of caloric value of egg matter (Kato and Kamler, 1983). Higher amounts of carbohydrates were found in eggs of *Cyprinus carpio* containing developing embryos than in dead eggs (Kamler, 1976); the viability of starved larvae of *Rutilus rutilus heckeli* and *Abramis brama* was dependent upon the eggs' carbohydrate contents (Vladimirov, 1965b).

Offspring properties related to egg ripeness

Precocious maturity of eggs, either evoked by heated water in *Tinca tinca* (Morawska, 1986) or provoked in *Cyprinus carpio* by hormone treatment in November and February (i.e. in a period clearly deviating from the natural spawning season, May/June: Kamler and Malczewski, 1982), yielded eggs that were poorly supplied with energy resources and offspring of low vitality.

One important factor controlling egg quality in farmed fish is the time during which ovulated eggs are retained in the abdominal cavity. Ripe, unfertilized eggs are not able to exist autonomously as they are not supplied by blood and are exposed to a reduced oxygen supply, and accordingly, after a certain time has elapsed, their degradation and degeneration take place. Examining the cytolysis of eggs, their fertilizability, hatchability and percentage of deformed embryos, Suzuki (1975) concluded that overripening of eggs of *Misgurnus anguillicaudatus* occurred after 6–8, 5–6 and 3–4 h after ovulation at 20, 25, and 30 °C, respectively. An inverse, exponential relationship between the time of egg fertilizability and temperature (within the range 9.3–16 °C with $Q_{10} = 2.9$) was found in *Oncorhynchus masu* by Kawajiri (1927d). By artificially stripping eggs from *O. mykiss*, Springate *et al.* (1984) determined the date of ovulation, and confirmed this from hormone profiles. Fertilization of underripe eggs was low; 100% fertilization, and simultaneously the highest survival of eyed stage, to hatch, and to swimming-up, was observed 4–6 d after ovulation, whereas during overripening (6–20 d after ovulation), the percentage fertilization decreased and was followed by a reduced success at each subsequent development stage. Nedyalkov (1981) showed with *Ctenopharyngodon idella* that in overripe eggs the mortality at critical periods (in embryos at gastrulation and in larvae from hatching to swimming-up) was greater than in ripe eggs. Kjørsvik and Lønning (1983) found *in vitro* that unchanged fertilizability of eggs of *Gadus morhua* was maintained longer when they were kept dry than when in sea water.

Zhukinskij *et al.* (1981) described changes in overripe eggs of *Abramis brama* retained in the abdominal cavity for 7–8 h. Oxygen consumption rate decreased by one-half in overripe eggs in relation to ripe eggs, anaerobic glycolysis rate decreased by one-half, oxidative phosphorylation (P/O coefficient) decreased by 85–88%, and ATPase activity decreased by about 30% with simultaneous mortality rates twice as high as usual for embryos at the morula, gastrula and tail-bud stages. A decrease of protein amino acids and an increase of free amino acids, cholesterol and free fatty acids indicate structural decomposition. Gosh (1985) has summarized the literature concerning metabolic changes in fish eggs during overripening. She supports the opinion that overripening is associated with an accumulation of metabolic inhibitors of unknown nature.

Chapter four

Endogenous feeding period

Yolk is the main energy source for most fishes during the endogenous feeding period,* which begins at fertilization (or any other triggering stimulus that begins cell division) and ends with the onset of exogenous feeding by the hatched larvae (Fe to S in Fig. 4.1). There is an obvious difference between the early development of fish and that of some other oviparous animals, such as invertebrates, reptiles and birds: the former hatch before their yolk is fully resorbed (H in Fig. 4.1).

4.1 DEVELOPMENT

General remarks

Large, transparent fish eggs, developing externally, are a convenient subject for embryological studies. Therefore since about 1870 substantial information has accumulated concerning fish embryonic development, in particular that of salmonids. For brevity, only a few examples will be discussed here. The morphogenetic development and various other features of the embryonic development of salmonid fishes were described by Knight (1963), Vernier (1969), Ballard (1973), Ryzhkov (1966, 1976), Leitritz and Lewis (1976), Dąbrowski (1981) and Velsen (1987), of *Acipenser* by Grodziński (1961), of *Cyprinus carpio* by Balon (1958a), Matlak (1972), and Peňáz *et al.* (1983), of *Ctenopharyngodon idella* by Shireman and Smith (1983), of *Abramis brama* by Dziekońska (1956), of *Xiphister atropurpureus* by Wourms and Evans (1974), of *Esox lucius* by Lindroth (1946), of Clupeidae by

---

*The exception is the viviparous fishes, whose embryos receive part or all of their nutrition directly from the mother (review: Balon, 1975a).

FEEDING

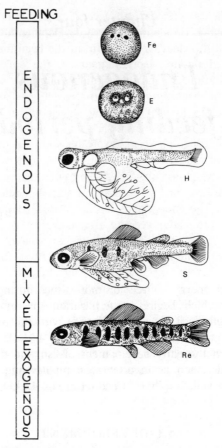

Fig. 4.1 Schematic representation of the early development of a salmonid fish. Fe, fertilization; E, eyeing (eyes of embryo are visible through egg cases); H, hatching; S, free-swimming (beginning of food-seeking movements and initiation of external feeding); Re, completion of yolk sac resorption. (Modified with permission from Raciborski, 1987.)

Kryzhanovskij (1956), of *Scophthalmus maximus* and *S.rhombus* by Jones (1972), of *Eleginus navaga* by Lapin and Matsuk (1979), of *Labeotropheus* sp. by Balon (1977), of *Vimba vimba* by Pliszka (1953), of *Acanthopagrus cuvieri* by Hussain *et al.* (1981), and of *Anoplopoma fimbria* by Alderdice *et al.* (1988).

Five developmental **periods** – embryonic, larval, juvenile, adult and senescent – are recognized in fish. The ecomorphological approach to fish development was founded on Vasnetsov's (1953) concept of developmental **etaps** (or stanzas) and on Kryzhanovskij's (1949) theory of ecological groups of fishes. The ecomorphological approach was further developed by many authors; only a few will be listed here. Balon (1958a, b, 1975b, 1986) divided developmental periods into several **phases, steps** and **stages**, arranged hierarchically. He identified three developmental phases within the embryonic period: the cleavage phase (from the beginning of development until the commencement of organogenesis), the embryonic phase (organogenesis within the egg) and the eleutheroembryonic phase (free embryo, from hatch to the beginning of external feeding). Lange *et al.* (1974) recognized nine etaps in fish embryonic development; development from hatching to the beginning of external feeding is included in the embryonic period according to their classification. In his study on salmonids, Ryzhkov (1976, 1979) synthesized the ecomorphological and the ecophysiological approaches; the latter, known as 'the theory of critical periods', is presented in Section 5.2. He described seven etaps in embryonic development, which ends at hatching. The endogenous feeding period comprises another four larval etaps prior to active feeding (Ryzhkov, 1976, 1979; Lapin and Matsuk, 1979). Some other classifications are based on morphometric and/or gravimetric criteria. Bams (1970) described the development of *Oncorhynchus gorbuscha* in hatcheries by the developmental index, $K_D = 10 \, W^{0.33} \, L^{-1}$, where W is weight (mg) and L is length (mm). Marr (1966) and Thorpe *et al.* (1984) defined developmental stages of *Salmo salar* by the ratio of embryo weight to the combined weight of the embryo and any remaining yolk. The theory of saltatory ontogeny (Balon, 1986) assumes early development to be a series of consecutive stabilized states (steps) during which various structures differentiate and grow in such a way that they are complete at the end of a step. The transition from one step to another is via a rapid threshold.

A word about terminology must be added here. The term **alevin** is usually applied to hatched, endogenously feeding salmonids; they are called **fry** after they begin to feed externally. Balon (1975b) proposes the term alevin for much older, actively feeding salmonids, mouthbrooding cichlids and other fish having no larval period; he discourages the term 'fry'.

Bioenergetical studies are not usually accompanied by a deep analysis of development. For staging the literature data I used the classification shown in Fig. 4.1. This very general classification is applied to salmonids, both in hatchery practice and in Japanese scientific fisheries. As in the great majority of bioenergetical and ecological studies, a fish developing inside an egg will be called an **embryo**, whereas for a hatched fish the term **larva** will be used (cf. Osse, 1989). Hatching is not an ideal boundary between the two developmental periods. In different species it occurs at different levels of development. Within a species, precocious hatching can result from high temperature (Gray, 1928b; Kowalska, 1959; Kokurewicz, 1969a; Peňáz, 1974; Heming, 1982; Peňáz *et al.*, 1983), low temperature (Kokurewicz, 1969a,b; Peňáz *et al.*, 1983; Eckmann, 1987), low oxygen concentration (Buznikov, 1955; Dziekońska, 1956; Alderdice and Forrester, 1974; Hamor and Garside, 1979) and from electric shock (Łuczyński and Dettlaff, 1985).

However, growth and metabolism both change virtually at hatch (Sections 4.3, 4.4).

Factors affecting the developmental rate during the endogenous feeding period

Temperature

The influence of temperature $(t, °C)$ on development can be expressed by changes in the time (τ) from fertilization to the mass appearance of any development stage. Developmental rate (V) is the inverse of τ. Fractions of the incubation periods occurring daily at a given temperature are tabulated (Alderdice and Velsen, 1978; Łuczyński and Kirklewska, 1984) to predict hatching time. The duration of a single mitotic cycle in early cleavage (τ_m) was proposed by Detlaf and Detlaf (1960) as a unit of duration of development in transparent eggs.

The rate of development is closely related to temperature: it is slow at low temperatures and increases with increasing temperature. Temperature is the major factor controlling the developmental rate of *Salmo salar* embryos (Hamor and Garside, 1976). Velsen (1987) compiled 783 data points, mostly from North America, on the time to 50% hatch at constant and ambient temperatures for six *Oncorhynchus* species, including *O. mykiss*; the time to 23 developmental stages is also shown.

The relationship between developmental time $(\tau$, days) and constant water temperature $(t, °C)$ within the restricted optimum range is given for *Oncorhynchus tshawytscha*, *O. mykiss* and *S. salar* (Fig. 4.2) by an exponential equation:

$$\tau = ae^{-bt} \qquad (4.1)$$

where a and b are constants. This equation forms a straight line on a semi-log scale. In most cases the regressions were a good fit to the data (Fig. 4.2). The values of b from the regression equation were not significantly different $(P > 0.05)$, neither between hatching, maximum tissue weight and yolk resorption time in *O. tshawytscha* (Fig. 4.2(A)), nor between eyeing, hatching and resorption time in *O. mykiss* (Fig. 4.2(B)). Thus, in the different sections of endogenous and mixed feeding, temperatures within the optimum range accelerated development uniformly. The same can be said of all embryonic stages of *Acipenser stellatus* and *A. gueldenstaedti* (Detlaf and Detlaf, 1960), of *Cyprinus carpio* (Peňáz et al., 1983), and of *Coregonus albula* (Łuczyński and Kirklewska, 1984), as well as of the second part of *Salmo salar* embryonic development (from eyeing to hatching, Fig. 4.2(C)), of hatching and emergence of *Oncorhynchus keta* (Beacham and Murray, 1985) and of eyeing, hatching, swimming-up and yolk resorption of *Salmo trutta* (Raciborski, 1987); four *Rana* (Amphibia) species were no exception to the rule (Detlaf and Detlaf, 1960). The dimensionless expression of developmen-

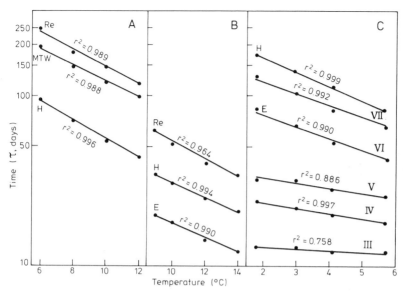

Fig. 4.2 Effect of constant temperatures (*t*, °C) on the time (τ, days after fertilization) required to attain different developmental steps (E, eyeing; H, hatching; MTW, maximum tissue weight; Re, complete yolk resorption; III–VII, etaps after Ryzhkov, 1976). Equation 4.1 was used, and near-optimum incubation temperatures were applied. (A) *Oncorhynchus tshawytscha*, 6–12 °C, initial egg wet weight 490 mg, b±95% confidence intervals: H, 0.128±0.015; MTW, 0.115±0.022; Re, 0.121±0.023 (computed from table 1 in Heming, 1982); (B) *O. mykiss*, 9–14 °C, 30 mg wet wt egg^{-1}, b±95% CI: E, 0.104±0.033; H, 0.102±0.019; Re, 0.118±0.067 (compiled from tables I and III in Kamler and Kato, 1983); (C) time of the beginning of embryonic etaps in *Salmo salar m. sebago*, 1.8–5.7 °C, 85 mg wet wt egg^{-1}, b±95% CI: etaps I and II, no influence of temperature; III, 0.024±0.021; IV, 0.075±0.000; V, 0.068±0.037; VI, 0.171±0.026; VII, 0.170±0.026; H, 0.193±0.006 (computed from table 41 in Ryzhkov, 1976). Semi-log scale; note change of horizontal scale in (C). Slopes within sections (A) or (B) do not differ significantly (see text).

tal advancement, the τ_n / τ_m ratio (Detlaf and Detlaf, 1960), is based on the assumption that the relationship between developmental rate and temperature remains the same in the different embryonic stages. The same assumption was made by Gnaiger (1980). Ferguson *et al.* (1985a) confirmed a good agreement between hatching time, ontogeny of enzyme expression and morphological development. It seems, then, that hatching time can be used as a measure of the effect of temperature on developmental rate. **Hatching time** is defined as the time by which 50% of eggs have hatched.

Data on the relationship between developmental time to hatch and constant temperature over the almost complete range of temperatures that can be tolerated are shown for *Oncorhynchus mykiss* in Fig. 4.3. These data,

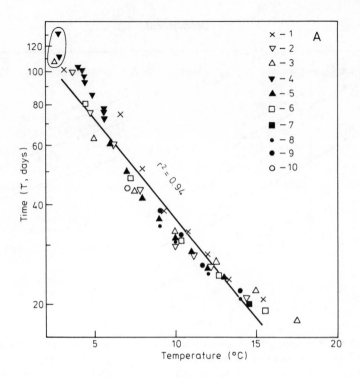

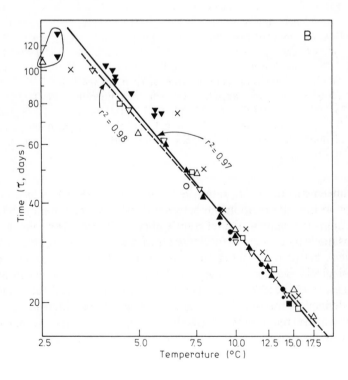

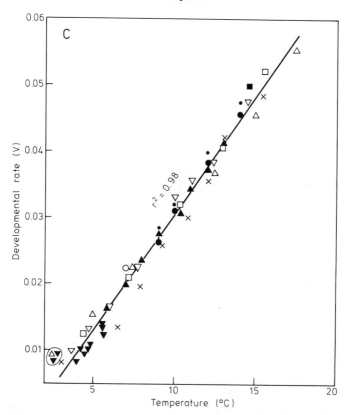

Fig. 4.3 Comparison of three models describing the relationship between embroynic development of *Oncorhynchus mykiss* and constant temperature over a broad temperature range (almost the complete range for viable hatching), 3–17.5 °C; $n = 53$; results obtained at 2.5–2.8 °C (ringed) are excluded from computations (cf. Winberg, 1987) because of high mortality (Goryczko, 1972). (A) and (B) Hatching time (τ, days after fertilization): (A) Equation 4.1. $a = 145 \overset{\times}{\div} 1.10$, $b = 0.139 \pm 0.01$; (B) power-law Equation 4.2, solid curve computed from 1–10, $a = 498 \overset{\times}{\div} 1.13$, $b = 1.18 \pm 0.05$, broken curve, regression line given by Humpesch (1985) (50% of eggs hatched, pooled data). (C) Developmental rate ($V = \tau^{-1}$), linear Equation 4.3, $a = -0.00446 \pm 0.00151$, $b = 0.00353 \pm 0.00015$. Variability of a and b is expressed as 95% confidence intervals. Combined data from: 1, Kawajiri (1927a); 2, Embody (1934); 3, Garside (1966); 4, Goryczko (1972); 5, Tashiro *et al.* (1974); 6, Leitritz and Lewis (1976); 7, Calamari *et al.* (1981); 8 and 9, Kamler and Kato (1983) (small (10 mg dry weight) and large (24 mg) eggs, respectively); 10, Escaffre and Bergot (1984). Semi-log scale in (A), log–log scale in (B), arithmetic scale in (C).

collected during last 60 years in Japan, USA and Europe, do not show any clear intraspecific variations. No differences between populations of the same species were reported for many fishes (Humpesh, 1985; Herzig and Winkler, 1985), *Cyrinus carpio* being one of the rare exceptions. Although the relationship between hatching time and temperature was well described by an exponential function (Equation 4.1) over the restricted range of favourable temperatures (Fig. 4.2(B)), the exponential model was found to be less adequate for a complete range of temperatures over which development occurs (Fig. 4.3(A); for nonlinearity of exponential plots see also Baxter, 1969). A better fit was obtained using both, a power-law equation:

$$\tau = at^{-b} \qquad (4.2)$$

which forms a straight line on a log–log scale (Fig. 4.3(B)), and a model in which the relationship between the developmental rate (V) and temperature (t) is given by a linear equation:

$$V = a + bt \qquad (4.3)$$

which forms a straight line on an arithmetic scale (Fig. 4.3(C)). A deviation from the line occurs at low temperatures ($< 3\,°C$, ringed points in Fig. 4.3). Such deviations at the lowest and the highest end of viable temperatures have been discussed by Hayes (1949), Detlaf and Detlaf (1960), Garside (1966), Kokurewicz (1970) and many others. According to Winberg (1971, 1987) and Peňáz *et al.* (1983), the curve of Equation 4.3 is S-shaped (sigmoidal) when obtained over a wide temperature range. Its lower left concave and upper right convex sections are those of asynchronous development, high percentage of deformities and low viability. The oxygen supply that can penetrate into an egg capsule is probably too low to cover the embryo's augmented demand for oxygen at high temperature (Section 4.4), and so is probably the reason for the slowed development in the upper right section. For the middle range of temperatures (tolerated temperatures, Winberg, 1987; optimum temperatures, Peňáz *et al.*, 1983), the linear equation can be used and the rule of thermal summation, credited to Réaumur, can be applied. The threshold temperature (t_0), at which development is theoretically arrested, can be obtained from Equation 4.3: $t_0 = -a/b$. This is the point where the extrapolation of the curve intersects the abscissa and is only a rough estimation of the true threshold temperature. The effective temperature, t_{eff}, can be obtained from $t_{eff} = t - t_0$. The number of day-degrees ($D°$) above the threshold temperature, the effective day-degrees, $D°_{eff} = \tau(t - t_0)$ can also be obtained from Equation 4.3: $D°_{eff} = 1/b$, therefore the duration of development is related hyperbolically to temperature:

$$\tau = D°_{eff}/(t - t_0) \qquad (4.4)$$

In the model attributed to Bělehrádek, a power b is applied,

$$\tau = a/(t - c)^b \qquad (4.5)$$

Here $c \neq 0$ and $b \neq 0$ is understood. When $c = 0$ then Equation 4.5 is identical to the power-law model (Equation 4.2); when $b = 1$, c is the threshold temperature and $a = D°_{eff}$, then Equation 4.5 is identical to the hyperbolic model (Equation 4.4). Thus, the general equation (4.5) summarizes Equations 4.2 and 4.4 (Humpesch and Elliot, 1980; Humpesch, 1985). The sigmoidal Vt curve can be described by Leiner's (1932) equation:

$$\tau = a + b^{(c-t)} \qquad (4.6)$$

which has not been in use for decades. Different polynomials such as:

$$\tau = a + bt + ct^2 \qquad (4.7)$$

$$\tau^{-1} = V = ab^t c^{t^2} \qquad (4.8)$$

$$\log \tau = a + b \log t + c(\log t)^2 \qquad (4.9)$$

have been used to obtain the best fit of the data for aquatic crustaceans (Bottrell, 1975; Kankaala and Wulff, 1981) and fishes (Table 4.1).

The popular models are illustrated for fish in Table 4.1 along with the temperature ranges over which they were obtained. More than one model could be successfully fitted to the hatching time of a given species. Even the same set of data from one species can be satisfactorily described by different models (Fig. 4.3(B) and (C)) (see also Alderdice and Velsen, 1078). The parameters of the equations are interrelated (Table 4.1, Fig. 4.4; see also Kamler *et al.*, 1982; Winberg, 1987). The use of polynomials has been discouraged (Humpesch and Elliott, 1980; Winberg, 1987), but they allow corregonid hatching times to be predicted with considerable accuracy (Łuczyński and Kirklewska, 1984; Eckmann, 1987). Equations 4.2 to 4.5 have been recommended as suitable for many fish (Alderdice and Velsen, 1978; Jungwirth and Winkler, 1984; Humpesch, 1985; Herzig and Winkler, 1985, 1986) and aquatic insect species (Humpesch and Elliott, 1980; Brittain *et al.*, 1984). Recently Winberg (1987) has focused attention on the linear relationship between developmental rate and temperature (Equation 4.3) as well as on the old rule of temperature summation. On the basis of a review of vast quantitative data he finds that Equation 4.3 adequately describes the relationship between developmental rate of many aquatic poikilotherms (fish embryos included) and temperature within the range of temperature tolerance. Studies by Gnaiger (1980) on embryonic development of a crustacean, *Cyclops abyssorum tatricus*, provided supporting evidence for the rule of temperature summation.

Further examples of the application of models 4.3 and 4.4 to hatching times in fishes are given in Fig. 4.4, in which the calculated threshold temperatures t_0 are compared with the ranges of viable temperatures for incubation of some common and/or commercially important fishes. The values of t_0 (which is calculated from the relationship between developmental rate and temperature) and the lower lethal temperature (which is estimated directly in embryo survival tests) are in most cases close to one another, in which circumstances 'biological zero' (i.e. the true threshold

Table 4.1 Popular models* relating developmental time (τ) of poikilotherms to constant temperature (t): examples of their application to fish hatching times (50% hatch, days after fertilization)

Eqn no.	Taxa	Temp. range (°C)	Parameter/Variable						Source†
			a	b	c	D°_{eff}	t_0		
4.1	Salmo salar m. sebago	1.8–5.7	255	0.196	–	–	–		1
	Oncorhynchus mykiss	3.1–15.4	186	0.157	–	–	–		2
	Salmo ischchan typicus	2.0–11.0	167	0.143	–	–	–		1
	S. ischchan gegarkuni	2.0–11.0	158	0.134	–	–	–		1
	Oncorhynchus masu	6.1–16.1	155	0.105	–	–	–		3
	O. mykiss	3.0–17.5	145	0.139	–	–	–		4
	O. mykiss	4.0–10.0	126	0.113	–	–	–		1
	Salmo ischchan aestivalis	6.0–14.0	125	0.113	–	–	–		1
4.2	Oncorhynchus mykiss	3.0–17.5	498	1.18	–	–	0		4
	Thymallus thymallus	3.5–16.2	459	1.37	–	–	0		5
	O. mykiss	3.4–18.9	456	1.14	–	–	0		5
	Hucho hucho	5.1–16.0	432	1.28	–	–	0		5
	Ammodytes personatus	6.5–15.5	372	1.09	–	–	0		6
	Salmo trutta	1.4–15.2	281	0.84	–	–	0		5
	Salvelinus fontinalis	1.4–12.8	241	0.73	–	–	0		5
	S. alpinus	1.4–8.0	206	0.63	–	–	0		5
4.3 and 4.4	Oncorhynchus mykiss	3.0–17.5	−0.00446	0.00353	–	283	1.3		4
	Esox lucius	9.0–21.0	−0.07463	0.02073	–	48	3.6		7
	Cyprinus carpio	15.1–26.7	−0.37433	0.03376	–	30	11.1		8

Further examples in Fig. 4.4

4.5	*Leuciscus idus*	7.0–22.0	222 775	3.055	—	—	−11.98	9,10
	Abramis brama	9.8–22.0	15 440	2.746	—	—	−1.68	9,10
	Cyprinus carpio	12.7–30.4	3 491	2.198	—	—	−3.27	9,10
	C. carpio	12.5–30.0	2 790	2.284	—	—	2.79	9,10
	Chalcalburnus chalcoides mento	9.8–22.8	2 507	2.203	—	—	1.20	9,10
	Carassius carassius	12.1–31.1	1 499	2.040	—	—	3.50	9
	Rutilus rutilus	7.8–23.9	1 447	1.898	—	—	0.42	9,10
	Chondrostoma nasus	10.0–17.3	1 258	1.994	—	—	2.62	9
	Vimba vimba	12.8–23.7	100	1.355	—	—	6.90	9,10
	Tinca tinca	14.5–30.2	6.8	0.392	—	—	14.06	9,10
4.6	*Gasterosteus aculeatus*	8.0–27.0	4	1.26	—	—	22	11
	Esox lucius	8.0–18.5	4	1.26	—	—	19	12
	Onchorhynchus mykiss	3.1–15.4	15	1.26	—	—	23	12
	Coregonus wartmanni and *C. fera*	1.6–6.9	15	1.26	—	—	20.7	13
4.7	*Leuciscus leuciscus*	6.7–16.5	96.6	−9.30	—	—	0.254	9
	Scardinius erythrophtalmus	11.0–19.5	63.0	−5.44	—	—	0.127	9
4.8	*Coregonus albula*	0.1–9.9	0.00497	1.162	—	—	1.001	14

*(4.1) Exponential model, $\tau = ae^{-bt}$; (4.2) power-law model, $\tau = at^{-b}$; (4.3) linear model, $\tau^{-1} = V = a + bt$; (4.4) hyperbolic model, $\tau = D_{eff}^{\circ}/(t - t_0)$; (4.5) Bělehrádek's model, $\tau = a/(t - c)^b$; (4.6) Leiner's model, $\tau = a + b^{(c - t)}$; (4.7) polynomial, $\tau = a + bt + ct^2$; (4.8) polynomial, $\tau^{-1} = V = ab^t c^{t^2}$. See text for further explanation.

†Sources. 1, Ryzkhov (1976); 2, Kawajiri (1927a); 3, Kawajiri (1927b); 4, data from many literature sources (Fig. 4.3) computed herein; 5, Humpesch (1985); 6, Yamashita and Aoyama (1985); 7, Winberg (1987); 8, data from Peňáz *et al.* (1983) computed herein; 9, Herzig and Winkler (1985); 10, Herzig and Winkler (1986); 11, Leiner (1932); 12, Lindroth (1946); 13, Braum (1964); 14, Luczyński and Kirklewska (1984).

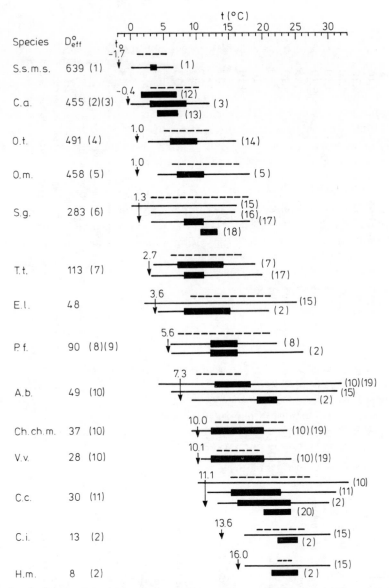

Fig. 4.4 Effective day-degrees (D_{eff}^{o}) and threshold temperatures (t_0, arrows) as compared with viable constant temperatures (lower to upper lethal temperature, solid horizontal lines) and optimum constant temperatures (thick horizontal bars) for embryonic development of 14 fish species (the determination coefficients r^2 for Equation 4.3 are given in parentheses): S.s.m.s., *Salmo salar morpha sebago* (0.98); C.a., *Coregonus albula* (0.97); O.t., *Oncorhynchus tshawytscha* (1.00); O.m., *O. masu* (0.98); S.g., *O. mykiss* (0.98); T.t., *Thymallus thymallus* (0.92); E.l., *Esox lucius* (–); P.f., *Perca fluviatilis* (0.97); A.b., *Abramis brama* (0.98); Ch.ch.m. *Chalcalburnus chalcoides mento* (0.98); V.v., *Vimba vimba* (0.95); C.c., *Cyprinus carpio* (0.98); C.i., *Ctenopharyngodon idella* (0.97); H.m., *Hypophthalmichthys molitrix* (0.99). Broken line, range of constant temperatures for which calculations were made (the range within

temperature at which development is arrested) is better approximated than by t_0 or survival test alone.

Fish species differ greatly in the optimal temperature for embryo development. Chernyaev (1981) reported that a constant temperature of 0.5 °C was optimal for Baikal omul (Coregonus autumnalis migratorius) embryogenesis and that at a constant temperature of 3.0 °C, a high percentage of embryo deformities appeared. Cyprinodon spp. reproduce in thermal streams reaching 43.8 °C (Smith and Chernoff, 1981). Apart from the obvious interspecific differences, with Salmo and Coregonus at the cold end and cyprinids at the warm end of the temperature scale, Fig. 4.4 also reveals intraspecific variability in the viable and optimum temperature ranges. Geographical location, season, parental factors, as well as experimental conditions and different criteria used to define these ranges may contribute to this variability; examples are given by Humpesch (1985) and are summarized by Zhukinskij (1986).

The acceleration of the developmental rate by temperature is shown by the van't Hoff temperature coefficient Q_{10} for a 10 °C difference in temperature:

$$V_2/V_1 = \tau_1/\tau_2 = Q_{10}^{(t_2-t_1)/10} \tag{4.10}$$

or by the coefficient Q_1 for 1 °C of temperature difference: $Q_1 = Q_{10}^{0.1}$. In the case of exponential relationship 4.1, the temperature coefficients remain constant over the temperature range, $\ln Q_{10} = 10\,b$, $\ln Q_1 = b$. Most frequently, however, the values of temperature coefficients decline with increasing temperature (examples in Blaxter, 1969 and Ignateva, 1979). When the developmental rate V is linearly related to temperature (Equation 4.3), temperature coefficients depend only on the effective temperature (t_{eff}) (Winberg, 1987):

$$Q_1 = 1 + 1/t_{eff} \tag{4.11}$$

This is shown for Oncorhynchus mykiss and Cyprinus carpio in Fig. 4.5(A) and (B), respectively. In the cold-water species, O. mykiss, the Q_{10} values are

---

which the developmental rate was linearly related to temperature). The values of D_{eff}° and t_0 for E. lucius are given by Winberg (1987); for the remaining species they were calculated in the present work. Sources of data for calculating D_{eff}° and t_0 as well as those for the viable and optimum temperature ranges are shown in parentheses: (1) Ryzhkov (1976); (2) Kokurewicz (1971); (3) Łuczyński and Kirklewska (1984); (4) Heming (1982); (5) Kawajiri (1927b); (6) many sources, shown in Fig. 4.3; (7) Kokurewicz et al. (1980); (8) Kokurewicz (1969b); (9) Swift (1965); (10) Herzig and Winkler (1986); (11) Peňáz et al. (1983); (12) Lebedeva (1981); (13) Łuczyński (1987); (14) Yarzhombek (1986); (15) Zhukinskij (1986); (16) Kawajiri (1927a); (17) Humpesch (1985); (18) Kamler and Kato (1983); (19) Herzig and Winkler (1985); (20) Sarig (1966).

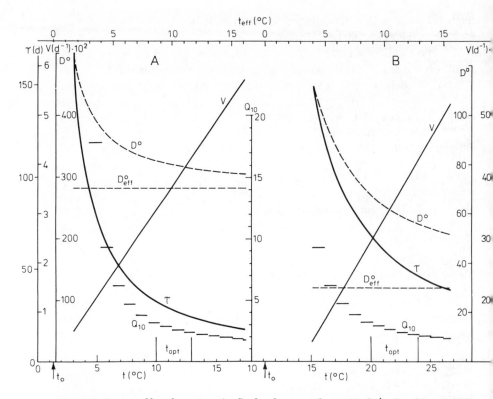

Fig. 4.5 Course of hatching time (τ, d), developmental rate (V, d^{-1}), day-degrees ($D°$), effective day-degrees ($D°_{eff}$) and the temperature coefficient (Q_{10}) over the temperature (t, °C, bottom scale) and the effective temperature (t_{eff}, °C, top scale) ranges within which the V to t relation is linear. Threshold temperature (t_0) and the ranges of optimum constant temperature (t_{opt}) are shown. (A) *Oncorhynchus mykiss*, references in Fig. 4.3 and 4.4; (B) *Cyprinus carpio* (Fig. 4.4). Of the vertical scales, only that for Q_{10} is common to both species.

much lower than in the warm-water *C. carpio* at the same temperatures, 15–17 °C; however, within the respective optimum temperature ranges (*O. mykiss*, 10–13 °C, Kamler and Kato, 1983; *C. carpio*, 20–24 °C, Sarig, 1966) the Q_{10} values are similar, i.e. between 2 and 3 in each species (Winberg, 1983).

The product of the time to attain a given developmental stage and the temperature, the day-degrees ($D°$), changes less with temperature than does the developmental time (Fig. 4.5), and was originally thought to be independent of temperature. Examples of the decrease of day-degrees with increasing temperature can be found in Tatarko (1965), Kokurewicz (1970, 1971), Vovk (1974), Tashiro *et al.* (1974), Leitritz and Lewis (1976) and Peňáz *et al.* (1983), to quote only a few; an increase of $D°$ at a narrow range of low temperatures, followed by a decrease with increasing temperatures, was reported by Kawajiri (1927 a,b), Kokurewicz (1969b) and Kokurewicz *et al.* (1980). The day-degrees ($D°$) of cold-water species are, within the optimum temperature range, less dependent on temperature than those of warm-water species. The effective day-degrees ($D°_{eff}$) are independent of temperature (Fig. 4.5).

The **length of hatching period** is the period over which the eggs hatch, i.e. the number of days between the first hatch and the end of hatching. In most studies it is represented by the period over which 90% or 80% of the eggs hatched (i.e. number of days between the points at which 5% and 95%, or 10% and 90%, respectively have hatched). Synchronous hatching is encouraged by high temperature, e.g. in *Chalcalburnus chalcoides mento* 80% of the eggs hatched over 15.7, 4.3, 2.8 and 1.4 days at 10.3, 12.6, 16.6 and 22.8 °C, respectively (Herzig and Winkler, 1986). The inverse relationship between the length of hatching period and temperature was described in *Ammodytes personatus* by an exponential equation (Yamashita and Aoyama, 1985), and in five species of salmonids and *Thymallus thymallus* by a power-law equation (Humpesch, 1985). Deviations from the models occur at extreme temperatures.

The influence of temperature on embryonic developmental rate has frequently been studied at constant temperatures, which are only rarely found in the field. Studies by Kawajiri (1927c) on *Leuciscus hakuensis* and *Carassius auratus* have shown that embryonic developmental rate was positively related to the amplitude of twice-a-day temperature fluctuations around a constant mean. Temperature variations within the optimal range generally accelerate development of aquatic invertebrates (Galkovskaya and Sushchenya, 1978). On the contrary, fluctuations of temperature prolong embryonic development in *Leuciscus idus* (Florez, 1972) and *Thymallus thymallus* (Kokurewicz *et al.*, 1980). Long-term temperature changes that imitate the natural course of temperature (i.e. a decrease in temperature from fertilization to hatch in autumn-spawning species, and an increase in spring-spawning ones) accelerate development (Chernyaev, 1981; Lebedeva, 1981) and increase survival (Kokurewicz *et al.*, 1980; Lebedeva, 1981).

Temperature regulation during incubation of eggs is used in aquaculture. In this way the first growth season of *Salmo salar* and *Oncorhynchus mykiss* was extended by 1.5–2 months (Ryzhkov, 1976). An earlier first intake of dry diet, lower mortality, and more advanced developmental state at hatch were observed by Rösch (1989) in *Coregonus lavaretus* larvae whose hatch was delayed by low temperature. A low-temperature-induced delay of coregonid hatching was a technique used to obtain larvae just when higher temperatures and sufficient zooplankton were available in the lakes (Flüchter, 1980; Łuczyński *et al.*, 1986; Łuczyński and Kolman, 1987). The resulting faster development and growth are believed to shorten the vulnerable larval period.

Turning to the mathematical relationships between duration of embryonic development and constant temperature (Table 4.1), it must be remembered that these are empirical models. However, the log–log relationship between biochemical reactions and temperature may provide a theoretical basis for the power-law model (Humpesch and Elliott, 1980). A bioenergetical approach to the model was presented by Winberg (1987). The temperature coefficient Q_{10} for developmental rate is inversely related to effective temperature; the same can be said of Q_{10} for growth rate. In contrast, metabolic Q_{10} in fully acclimated animals is constant over the range of tolerated temperatures. This explains why the lowest values for the efficiency $K_2 = P/(P+R)$ are usually found at low temperatures, and maximum efficiencies at the middle, optimum temperature range (for explanation of efficiencies see Section 4.5). According to Winberg (1987), his hypothesis needs to be verified, in particular the assumed-equal final weight of animals developing in different temperatures (Section 4.3). Thus, no universal theory explaining the relationship between embryonic development of poikilotherms and temperature has yet been accepted, although the problem has been much explored.

Remaining abiotic factors

Oxygen Reduced oxygen content retards embryonic development (Skadovskij and Morosova, 1936, *Acipenser stellatus*; Hayes *et al.*, 1951, *Salmo salar*; Alderdice *et al.*, 1958, *Oncorhynchus keta*; Carlson and Siefert, 1974, *Salvelinus namaycush* and *Micropterus salmoides*; review: Zhukinskij, 1986). A synergic effect of low oxygen content and high temperature was reported by Garside (1966) for *Salvelinus fontinalis* and *Oncorhynchus mykiss* and by Hamor and Garside (1976) for *S. salar*. Oxygen deficit initiates hatching by stimulation of chorionase secretion (Röthbard, 1981, various carps; Shireman and Smith, 1983, *Ctenopharyngodon idella*; DiMichele *et al.*, 1986, *Fundulus heteroclitus*). Retardation of development in supraoptimal oxygen concentrations has also been recorded (Zhukinskij, 1986).

pH In *Perca fluviatilis*, time to hatch at pH 5.0 was 11% longer, and at pH 4.0 29% longer, than at pH 6.4 (Rask, 1983). In *Salmo salar* also, low pH (4.0–5.5) delayed hatching of embryos exposed after the eyed stage (Peterson *et al.*, 1980). In *Brachydanio rerio* the longest time to hatch was at both the lowest (4.2) and the highest (8.5) pH, whereas the shortest time was at pH 6,2 (Dave, 1985). Inhibition of chorionase activity and reduction of embryo mobility was ascertained by these authors as the cause of delayed hatching at extreme pH levels.

Salinity Salinity within the range 1–10‰ did not modify the hatching time of *Abramis brama* and *Stizostedion lucioperca* from the Dnieper River estuary. In the former species, neither mortality nor embryo deformities occurred at these salinities, while in the latter, 13–28% and 50% of embryos were abnormal at salinities 2‰ and >6‰, respectively (Belyi, 1967). In *Fundulus heteroclitus*, salinity had no effect on the time to hatching (DiMichele and Taylor, 1980).

Light Acceleration of embryonic development at high light intensity was found in *Lepomis macrochirus* by Toetz (1966) in *Oncorhynchus mykiss* by MacCrimmon and Kwain (1969) and in *O. nerka* by Leitritz and Lewis (1976); in *O. nevka* the resulting larvae were smaller and less viable. Also, in *Salmo ischchan*, morphogenetic rate was accelerated by light, especially at early gastrulation and blastopore closing; hatching occurred 15–20 d earlier in light-exposed eggs than in eggs incubated in darkness (Ryzhkov, 1976). Light accelerates the development of older, pigmented embryos of the Baikal coregonids; thus hatching is synchronized with ice thawing and increase of plankton density (Chernayaev, 1981). In contrast to the embryonic developmental rate, however, the posthatch development of *Salmo salar* was retarded by light (Ryzhkov, 1976).

Egg size

Ware (1975) found a highly significant positive relationship between incubation time and size of eggs derived from different pelagic fish species. Egg weight affects hatching time of rotifers (Duncan, 1983) and crustaceans (Khmeleva, 1988) at the interspecific level. A tendency for embryos in larger eggs to have a longer development when compared at the same temperature is shown in Fig. 4.6B. The same tendency for the entire endogenous feeding period is illustrated in Table 4.2. Turning to Fig. 4.6, an exception is an autumn-spawning , cold-water stenotherm, *Coregonus albula*, whose small eggs develop slowly. In a cyprinid, *Leuciscus leuciscus*, both habitat and the response of embryonic development rate to temperature are not typical of Cyprinidae. Thus, conclusions regarding the effect of egg size on development rate at the interspecific level are clouded by other influences relating to spawning temperature and life history.

Intraspecific comparisons revealed no significant effects of fish egg size, neither on the time to eyeing, nor on hatching time (Blaxter and Hempel,

Table 4.2 Combined effect of temperature (t, °C) and egg size (W, mg) on the time (τ, days) required to reach the eyed stage (E), to hatch (H) and to reach maximum alevin wet weight (MAWW), as described by the multiple regression equation $\tau = ae^{bt}e^{cW}$

Species	Time to	Parameter			Source*
		a	b	c	
Oncorhynchus mykiss	E†	50.8	−0.104	0.00145	1
Oncorhynchus mykiss	H†	87.6	−0.103	0.00095	1
Oncorhynchus tshawytscha	MAWW‡	357.8	−0.152	0.00051	2

*Sources: 1, Kamler and Kato (1983); 2, Rombough (1985).
†Where W is dry weight.
‡Where W is wet weight.

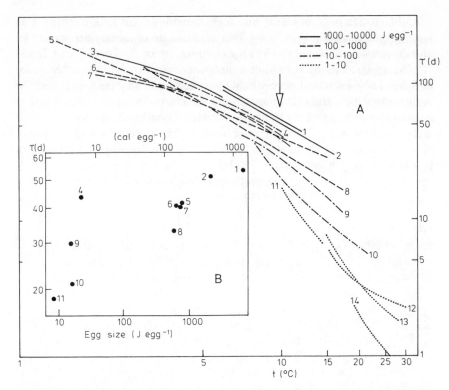

Fig. 4.6 Hatching time (τ, days after fertilization) as related to temperature (t, °C) and egg size (J egg^{-1}); interspecific comparison. (A) Relationships between hatching time and temperature; (B) cross-section through 10 °C (arrow in (A)). Species depicted: 1, *Oncorhynchus tshawytscha*; 2, *O. masu*; 3, *Salmo salar m. sebago*; 4, *Coregonus albula*; 5, *S. trutta*; 6, *S. ischchan typicus*; 7, *S. ischchan gegarkuni*; 8, *O. mykiss*; 9, *Leuciscus leuciscus*; 10, *Perca fluviatilis*; 11, *Abramis brama*; 12, *Tinca tinca*; 13, *Cyprinus carpio*; 14, *Ctenopharyngodon idella*. Pooled data from Fig. 3.1, Tables 3.4 and 3.6 (egg size), Table 4.1 and Fig. 4.4 (hatching time v. temperature). Log–log scales.

1963, 1966; Kamler and Kato, 1983; Escaffre and Bergot, 1984; Beacham and Murray, 1985; Beacham *et al.*, 1985; Knutsen and Tilseth, 1985; Marsh, 1986), likewise the duration of embryonic development of a crustacean, *Boeckella symmetrica*, was almost independent of egg size (Woodward and White, 1981). On the contrary, the subsequent developmental events – maximum larval weight for the yolk supply, end of yolk resorption and 50% mortality by starvation after yolk resorption – were delayed in larvae derived from larger eggs (Smith, 1958; Blaxter and Hempel, 1963, 1966; Kamler and Kato, 1983; Escaffre and Bergot, 1984; Rombough, 1985; Knutsen and Tilseth, 1985). The rare exceptions are the studies by Beacham *et al.* (1985) and Beacham and Murrary (1985) on *Oncorhynchus keta* and *O. kisutch*, in which eggs differing very little in weight were used. In general, large eggs contribute to survival through both a greater size of larvae (Section 3.10) and their longer survival prior to first feeding.

In summary, then, the rate of embryonic development is different in various fish species and is controlled environmentally. Temperature is the major abiotic variable; egg size exerts a less pronounced effect. This is exemplified in Table 4.2 – compare the temperature coefficient b to the weight coefficients c.

4.2 YOLK ABSORPTION

General remarks

Yolk is the major source of energy and materials for developing embryos of oviparous species, although some exogenous supplies can be taken from water (Terner, 1979). Embryos of viviparous species use nutritional substances provided by ovarian fluid (Section 4.5).

Mobilization of yolk reserves in teleostean embryos occurs through the vitelline syncytium (periblast), a thin (c. 0.09 mm in *Sebastes schlegeli*, Shimizu and Yamada, 1980) tissue which envelops the whole yolk mass after the closure of the blastopore. Yolk absorbed (C_Y) by a developing embryo or endogenously feeding larva provides materials to be deposited in the newly formed or growing tissues (P) and supplies energy for metabolism (R). No faeces are produced by fish larvae prior to initiation of exogenous feeding (the exception is acipenserid larvae, which expel a melanin button just before external feeding, Gershanovich *et al.*, 1987); only metabolites (U) are excreted. Therefore the energy consumed from yolk is partitioned as follows:

$$C_Y = P + R + U = D \qquad (4.12)$$

where D is digestible energy (Brett and Groves, 1979; Krasnoper, 1985). Yolk absorbed is estimated from:

$$C_Y = Y_0 - Y.r. \qquad (4.13)$$

where Y_0 is the initial energy content in the yolk sac and $Y.r.$ denotes energy in the remaining yolk.

Decrease in yolk weight can be measured directly in formalin-hardened yolk sacs dissected from larval bodies at appropriate time intervals. This method has been applied to yolk-sac salmonids (Hayes *et al.*, 1951; Gunnes, 1979; Kamler and Kato, 1983; Escaffre and Bergot, 1984; Raciborski, 1987), to eggs of *Morone saxatilis* (Rogers and Westin, 1981) and to embryos and yolk-sac larvae of *Oreochromis niloticus* (De Silva *et al.*, 1986). The caloric value of the yolk dry matter can be also directly determined in these sacs. The most popular calorimetric methods are listed in Section 3.4. In fish producing small eggs, yolk volume, yolk weight and caloric value of its matter were measured at the beginning of development and the energy decrease was calculated from the decrease in volume at each sampling period (Lasker, 1962; Blaxter and Hempel, 1963; Toetz, 1966; Laurence, 1969; Howell, 1980; Quantz, 1985). Two assumptions have to be made in this method: that (a) the caloric value of yolk dry matter, and (b) that of yolk specific gravity, remain constant over the period of endogenous feeding.

The validity of the former assumption was demonstrated by Lasker (1962) on yolk of *Sardinops caerulea* and by Raciborski (1987) on *Salmo trutta*. Similarly, caloric value of yolk decreased only by 5% between fertilization and hatch of *Oncorhynchus mykiss* (Smith, 1957; Kamler and Kato, 1983). In contrast, changes of yolk chemical composition were reported by Hayes (1949) and Toetz (1966). Energy-rich oil globules, found in the high-protein yolk of some species, remained unabsorbed at the final absorption of the remaining yolk (*Morone saxatilis*, Eldridge *et al.*, 1981a; *Ammodytes personatus*, Yamashita and Aoyama, 1985; *Coregonus fera*, Loewe and Eckmann, 1988) and even at the point of death by starvation (*Morone saxatilis*, Rogers and Westin, 1981). Yolk hydration increased significantly from 49% to 53–59% in *Oncorhynchus mykiss* between hatching and swimming-up (Escaffre and Bergot, 1984). Thus, yolk caloric value and specific gravity may remain unchanged or may change as development proceeds.

In fishes the yolk substances are absorbed through the vitelline syncytium, which envelopes the central mass of yolk. Yolk is absorbed by phagocytic activity of the inner part of the syncytial layer (the vitellolysis zone), is degraded into lower-molecular-weight substances and is transported into the blood; the role of the incomplete alimentary tract seems to be minimal (Walzer and Schönenberger, 1979; Shimizu and Yamada, 1980; review: Buddington and Doroshov, 1986).

Factors affecting the yolk absorption rate

The decrease in the amount of yolk energy is illustrated in Table 4.3 for nine fish species which differ by four orders of magnitude in their initial yolk energy. No external food was given to larvae in these experiments. Two types of absorption rate were calculated.

The amount of energy consumed daily from yolk by an embryo or larva (the absolute rate of yolk absorption, $J \, indiv^{-1} \, d^{-1}$) is positively related to the initial yolk energy (Y_0) (see also Blaxter and Hempel, 1963; Rombough, 1985). In contrast, the relative rate of yolk absorption (% d^{-1}) is lowest in the largest eggs ($1.4\% \, d^{-1}$ in *Oncorhynchus keta*) and highest in the smallest eggs ($50.2\% \, d^{-1}$ in *Lepomis macrochirus*). Although this conclusion is clouded by the differences in temperatures at which the tests were performed (Table 4.3), supporting evidence is provided in Table 4.4 for an intraspecific comparison of the effect of egg size at one temperature.

Table 4.3 Yolk absorption by embryos and endogenously feeding (unfed) larvae: interspecific comparisons

Species	Temp. (°C)	Energy remaining (J indiv⁻¹) at stage:*								Source†
		Fe		H		S		Re		
		τ (d)	Y_0	τ (d)	Y.r.	τ (d)	Y.r.	τ (d)	Y.r.	
Oncorhynchus tshawytscha	10	0	4784	55.0	3658	107.0[a]	715	147.0	0	1
O. keta[b]	4.8	0	3981	101.8	2268	–	–	164.0	400[f]	2
O. kisutch[b]	4.6	0	3238	103.0	1547	–	–	143.6	288[f]	2
O. mykiss	12	0	674	26.0	553	44.5	138	54.0	0	3
S. trutta	3–11[c]	0	625	119.0	515	152.0	281	168.0	0	4
Micropterus salmoides	12–18	0	9.56	4.0	5.92	10.0	1.21	13.0	0	5
Clupea harengus pallasi	12.5–13.5	0	5.44	7.0	1.98	–	–	12.0	0	6
Lepomis macrochirus	23.5	0	3.90	1.8	2.94	6.4	0.36	8.5	0	7
Scophthalmus maximus	15	0	0.853	$\tau_H + 0.17$[d]	0.198	$\tau_H + 2$[e]	0.017	$\tau_H + 5.2$	0	8

Cont'd overleaf

Table 4.3—*contd.*

Mean yolk absorption rates‡

Species	Absolute (J indiv.$^{-1}$ d^{-1})				Relative (% d^{-1})			
	Fe–H	*H–S*	*S–Re*	*Fe–Re*	*Fe–H*	*H–S*	*S–Re*	*Fe–Re*
O. tshawytscha	20.5	56.6	17.9	32.5	0.5	3.1	10.7	4.2
O. keta	16.8	–	–	21.8	0.6	–	–	1.4
O. kisutch	16.4	–	–	20.5	0.5	–	–	1.7
O. mykiss	4.65	22.4	14.5	12.5	0.8	7.5	24.9	7.0
S. trutta	0.924	7.09	17.6	3.72	0.2	1.8	18.3	2.1
M. salmoides	0.910	0.785	0.403	0.735	12.0	26.5	49.5	25.5
C. harengus pallasi	0.494	–	–	0.453	14.3	–	–	18.3
L. macrochirus	0.533	0.561	0.171	0.459	15.7	45.6	102.3	50.2
S. maximus	–	0.099	0.005	–	–	134.2	65.8	–

*Stages: Fe, fertilization; H, hatching; S, free-swimming; Re, completion of yolk resorption.

†Sources: 1, Heming (1982); 2, Beacham *et al.* (1985), both species: female no. 2, small eggs; 3, Kamler and Kato (1983), three-year-old female; 4, Raciborski (1987), unfed larvae in 1984; 5, Laurence (1969); 6, Eldridge *et al.* (1977) control larvae; 7, Toetz (1966); 8, Quantz (1985).

Absolute rate $= (Y.r._1 - Y.r._2)/(\tau_2 - \tau_1)$.

Relative rate $= 100(\ln Y.r._1 - \ln Y.r._2)/(\tau_2 - \tau_1)$; in the case of final yolk resorption, calculations were made for day τ_{Re-1}.

^{a} Time of maximum wet weight of endogenously feeding larvae.

^{b} Converted from wet weight to dry weight using hydration values from Table 3.6 and from dry weight to energy using caloric value 27.5 J mg^{-1}.

^{c} Ambient temperature, 3–6 °C before hatching, then gradual increase to 11 °C.

^{d} Time to hatching (τ_H) not stated.

^{e} Time of maximum ash-free dry weight of endogenously feeding larvae.

^{f} Completion of exogenous yolk resorption ('button up').

Table 4.4 Effect of egg size and temperature on mean relative rate of yolk energy absorption, from fertilization to hatching (Fe–H) and from hatching to completion of yolk resorption (H–Re) in endogenously feeding larvae: intraspecific comparisons

Species	Initial yolk size ($J\,indiv^{-1}$)	Temp. (°C)	Absorption rate (% d^{-1})	
			Fe–H	H–Re
Oncorhynchus mykiss*	276	12	0.8	18.7
	280	12	0.8	18.7
	674	12	0.8	13.0
	871	12	0.4	8.0
O. mykiss*	674	9	0.3	–
	674	10	0.6	7.2
	674	12	0.8	13.0
	674	14	1.2	15.5
O. tshawytscha†	4 784	6	0.2	4.7
	4 784	8	0.3	5.9
	4 784	10	0.5	6.4
	4 784	12	0.4	7.6

*Absorption rates computed from Kamler and Kato (1983).
†Absorption rates computed from Heming (1982).

The mean relative yolk absorption rate (% d^{-1}) increases with time as development proceeds (Tables 4.3 and 4.4). The amount of energy consumed daily by an individual ($J\,indiv^{-1}\,d^{-1}$) is low in embryos (Fe–H in Table 4.3) and is highest in larvae (H–S), which are larger and have better-developed blood vessels in the yolk sac; the subsequent decrease of the absolute rate after swimming-up (S–Re in Table 4.3) results from the limited amount of remaining yolk (see also Hamor and Garside, 1977; Escaffre and Bergot, 1984).

Larvae of *Salmo trutta* that were fed consumed yolk less rapidly than starved ones during the final period of yolk resorption (Raciborski, 1987).

Yolk absorption rate is positively related to temperature within its optimum range, as shown in Table 4.4 for *Oncorhynchus tshawytscha* and *O. mykiss*, and found in *Scophthalmus maximus* by Jones (1972), *Salmo salar* by Hamor and Garside (1977), *Gadus morhua* and *Melanogrammus aeglefinus* by Laurence (1978), and *Limanda ferruginea* by Howell (1980).

The absolute rate of yolk utilization by *Salmo salar* declines with decreasing oxygen level at 5 and 10 °C (Hamor and Garside, 1977).

In *Scophthalmus maximus*, high salinity of seawater (28%) reduced the yolk absorption rate in comparison with brackish water (17%), in which the larvae were below their point of neutral buoyancy and had to expend extra energy to keep from sinking (Quantz, 1985).

Hansen and Møller (1985) in *Salmo salar* and Hansen (1985) in *Salmo trutta* observed a slower absorption of yolk in larvae incubated on a flat surface than in larvae reared on an artificial, corrugated substratum (Astro-turf).

Light delayed yolk resorption in larvae of *Salmo salar* (Ryzhkov, 1976).

4.3 BODY GROWTH

General remarks

Expression of growth

Growth is an increase of body size, resulting from production of new tissues. Body size and its growth may be represented in terms of length, wet weight, dry weight or content of energy (or carbon, nitrogen etc.) per individual, in order of increasing validity for comparisons (Blaxter, 1969; see also Section 3.1).

Various numerical descriptions of growth are used. They are shown for weight in Fig. 4.7. Body weight (Fig. 4.7(A)), after conversion to energy and after subtraction of the minute amount of energy contained in the germinal disc, represents the energy deposited in the body from the beginning of development to the end of a period of life (cumulative production, P_c; Section 2.2). The absolute growth rate, dW/dt, can be approximated for the period $\tau_2 - \tau_1$ by:

$$dW/dt = (W_2 - W_1)/(\tau_2 - \tau_1) \tag{4.14}$$

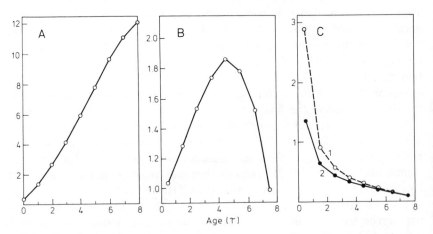

Fig. 4.7 Different expressions of growth. (A) Weight; (B) absolute growth rate, $(W_2 - W_1/(\tau_2 - \tau_1)$; (C) relative growth rates: 1, Minot's curve, $(W_2 - W_1)/W_1(\tau_2 - \tau_1)$; 2, specific growth rate, $G = (\ln W_2 - \ln W_1)/(\tau_2 - \tau_1)$. Units: weight unit in (A), weight unit per time unit in (B), (time unit)$^{-1}$ in (C).

(Fig. 4.7(B)). The absolute growth rate, if expressed in terms of energy, represents the instantaneous production (P_i). Relative growth rates (Fig. 4.7(C)) are related to the body size, $G = dW/W \, dt$. At the end of the 19th century Minot proposed an approximation:

$$G = (W_2 - W_1) / W_1(\tau_2 - \tau_1) \qquad (4.15)$$

which gives results that are very high in the youngest stages (curve 1 in Fig. 4.7(C)).

In the exponential type of growth, described by the equation:

$$W_\tau = W_0 \, e^{G\tau} \qquad (4.16)$$

W_0 is the size at time 0, and the specific growth rate, G, is constant. For a period $\tau_2 - \tau_1$ it is usually estimated from:

$$G = (\ln W_2 - \ln W_1) / (\tau_2 - \tau_1) \qquad (4.17)$$

(curve 2 in Fig. 4.7(C)). For a unit time interval $(\tau_2 - \tau_1 = 1)$, G is sometimes approximated by:

$$G = 2(W_2 - W_1) / (W_2 + W_1) \qquad (4.18)$$

The specific growth rate G is commonly used for all types of growth of aquatic animals, including fish (e.g. Brett *et al.*, 1969; Winberg, 1971; Zaika, 1972; Ricker, 1979). Winberg's (1956) formula:

$$G = 10^{1/\tau(\log W_2 - \log W_1)} - 1 \qquad (4.19)$$

gives relative growth rates that are higher than those computed from Equation 4.17, especially when growth is fast. When $\tau_2 - \tau_1 = 1$, the values from Equation 4.19 approach those from Equation 4.15. Later Winberg (1971) abandoned Equation 4.19 but it has remained in use by Russian authors for fish embryos and larvae (e.g. Baranova, 1974; Ryzhkov, 1976; Ostroumova *et al.*, 1980).

Turning to the values of G from Equations 4.16 and 4.17, they can be computed for all measures of body size. Then we obtain specific growth rate for length (G_L), wet weight (G_{W_w}), dry weight (G_{W_d}) or for energy $(G_{C.e.})$. The specific growth rates for length and for weight are interrelated:

$$G_L = G_W / b \qquad (4.20)$$

where b is the slope from the equation relating weight (W) to length (L):

$$W = aL^b \qquad (4.21)$$

Specific growth rates for wet weight are the most commonly used in larval aquaculture but the rates for dry weight are more reliable. It has to be borne in mind that the values of G_{W_w} and G_{W_d} are the same only when hydration of the body remains unchanged over the period under study. An increase in the water percentage during that period results in G_{W_w} values higher than those of G_{W_d}. Analogously, the relations between the values of G_{W_d} and $G_{C.e.}$ reflect the changes of caloric value of dry matter over the period studied (see also Section 6.3).

Growth during the endogenous feeding period

A multifold increase of body size over a short time is a characteristic feature of embryonic growth. Two approaches have been adopted in studies of growth during the early life of fishes.

Samples collected over short time intervals during endogenous and mixed feeding of *Salmo salar morpha sebago* by Ryzhkov (1976) show periodical changes of G_{W_d} closely related to the discontinuity of development (Fig. 4.8)). Acceleration of growth was observed at the beginning of any developmental step; afterwards the growth rate slowed down. The lowest growth rates were associated with a transition from one developmental step to another, and with the intensification of differentiation processes. A larger-scale periodicity of growth rate was superimposed on the former one, with maximum growth rates at the beginning of development periods: embryonic (after fertilization) and larval (after hatch) (Fig. 4.8). The same pattern was obtained by Ryzhkov (1976) for *Salmo ischchan typicus*, *S. ischchan aestivalis*, *S. ischchan gegarkuni* and *Oncorhynchus mykiss*. This approach is based on Schmalhausen's theory (reviews: Trifonova, 1949; Ryzhkov, 1976). Doubts have been raised about the accuracy of tissue growth determinations at early stages of fish embryonic development (Devillers, 1965; Gosh, 1985) because the weight changes are small and the tissues are closely connected with yolk (Abramova and Vasileva, 1973). Sudden accelerations of growth at the beginning of developmental steps were confirmed by Kamler and Mandecki (1978) for postembryonic development of an aquatic gastropod, *Physa acuta*. Growth curves shown in Fig. 4.9 (A), (B) and (D1) also reveal accelerated growth after hatch.

An alternative approach to the problem is to neglect the short-term accelerations and to describe growth with a smooth model. Many growth curves for postlarval fishes have been proposed (review: Ricker, 1979). Growth in energy of endogenously fed *Clupea harengus pallasi* (embryos and prefeeding larvae) was found to increase linearly with age ($r = 0.88$–0.94, Eldridge *et al.*, 1977), similarly to growth in dry weight of endogenously fed *Morone saxatilis* ($r = 0.90$, Eldridge *et al.*, 1981b) and growth in length and ash-free dry weight of prefeeding larvae of *Limanda ferruginea* (Howell, 1980). The exponential curve (Equation 4.16) is a popular model describing growth of fish embryos for short time intervals (Hayes and Armstrong, 1943; Hayes, 1949; Winberg, 1971; Ryzhkov, 1976; Ricker, 1979; Lindsey and Arnason, 1981). The body weight of *Perca flavescens* increased exponentially after hatch (Henderson and Ward, 1978). However, growth rate usually decreases with increasing body size. A log–log relationship:

$$W_\tau = W_0 \tau^b \tag{4.22}$$

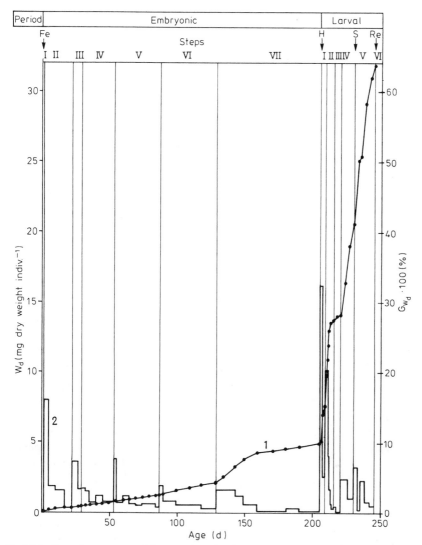

Fig. 4.8 Growth in terms of dry weight of *Salmo salar morpha sebago* for periods of endogenous and mixed feeding. Larvae were fed after swimming-up. Data compiled from Ryzhkov's (1976) tables 1 and 3; each point is the result of 16–20 measurements of dissected tissues. 1, Dry weight of tissues, W_d; 2, specific growth rate, G_{W_d}. Original data were recomputed using Equation 4.17. Developmental steps of the embryonic period (summarized classification of Ryzhkov, 1976): I, formation of perivitelline space and germinal disc; II, cleavage; III, gastrulation; IV, organogenesis; V, differentiation of the tail bud and formation of subintestinal and yolk blood vessels; VI, formation and functioning of hepatic and yolk blood circulation system; VII, differentiation of unpaired fins and hatching. Developmental steps of the larval period (following Ryzhkov, 1976): I, gill breathing; II, formation of protonephros; III, complete differentiation of fins from the fin fold and beginning of formation of the first intestinal loop; IV, formation of pyloric caeca; V, mixed feeding; VI, yolk sac resorbed, exogenous feeding. For explanation of Fe, H, S, and Re see Fig. 4.1.

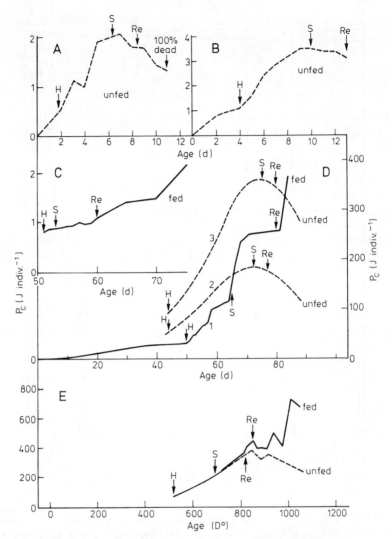

Fig. 4.9 Comparison of energy gain by embryonic or larval tissues of five fish species. Fertilization, age 0; H, S, and Re, see Fig. 4.1. (A) *Lepomis macrochirus*, initial yolk size 3.90 J indiv^{-1}, 23.5 °C, larvae unfed, recomputed from Toetz (1966); (B) *Micropterus salmoides*, 9.56 J indiv^{-1}, 18–21 °C, larvae unfed, recomputed from Laurence (1969); (C) *Ammodytes personatus*, 6.5 °C, larvae fed, data recomputed from Yamashita and Aoyama (1985); (D) *Oncorhynchus mykiss*, curve 1, 592 J indiv^{-1}, 7.5–8.9 °C, larvae fed, recomputed from Ryzhkov (1976), curves 2 and 3, small eggs (<3.7 mm in diameter) and large eggs (>4.2 mm), respectively, 7 °C, larvae unfed, recomputed from the equation given by Escaffre and Bergot (1984) for female no. 2; (E) *Salmo trutta*, 625 J indiv^{-1}, ambient temperature increasing from 3 to 11 °C (hence age expressed in D°), fed and unfed groups of larvae, mortality 4% and 75% at 1050 D°, respectively, recomputed from Raciborski (1987).

where W_0 is the initial weight when $\tau = 1$ and b is a constant, was applied to embryonic growth by Schmalhausen (cited in Ozernyuk, 1985) and to *Morone saxatilis* embryos (Eldridge *et al.*, 1982). MacDowell's modification of Equation 4.22:

$$W_\tau = W_0(\tau - \tau_a)^b \qquad (4.23)$$

where τ_a is the time from fertilization to the establishment of the embryonic axis, was also used (Hayes, 1949; Ricker, 1979). Equations 4.16, 4.22 and 4.23 describe only the left part of the *S*-shaped curve occurring during endogenous feeding. This curve (examples, Fig. 4.9) can be described by a single line using a model proposed by Escaffre and Bergot (1984) for dry weight of unfed *Oncorhynchus mykiss* larvae from hatch to a point in time between yolk absorption and 50% mortality:

$$W_\tau = 4W_{max} Z(1-Z) \qquad (4.24)$$

with:

$$Z = 1 / [1 + \exp \lambda (\tau_{max} - \tau)] \qquad (4.25)$$

where: W_{max} is maximum dry weight, τ_{max} is time needed for weight to reach W_{max} and λ is a growth coefficient. This approximation is presented in terms of energy in Fig. 4.9(D), curves 2 and 3, for larvae hatched from small and large eggs, respectively.

Although growth patterns in the early life of fish show considerable heterogeneity, the following sequence of stages, which differ in absolute growth rate (Table 4.5), can be described in very general terms:

1. embryos grow slowly (fertilization to hatching, Fe to H)
2. growth accelerated after hatching (yolk-sac larvae)
3. larvae reach the maximum size that their yolk reserves will support, then their growth rate decreases (coincident with beginning exogenous feeding, S)
4. between the start of exogenous feeding and final yolk resorption (S to Re):
   (a) larvae that have not fed show an energy deficit and tissue absorption
   (b) fed larvae grow
5. after final yolk absorption:
   (a) unfed larvae show negative growth due to tissue absorption, then die from prolonged starvation
   (b) fed larvae show rapid growth

The slow specific growth rate of embryos as compared with yolk-sac larvae was confirmed by Aguirre's (1986) measurements of protein synthesis rate in *Coregonus schinzi palea* using ^{14}C arginine. Stages 2–4 were described by Farris (1959) for growth in length of four marine fish species.

In unfed larvae, an energy deficit (4(a) above) occurs when metabolic requirements cannot be covered by energy consumed and have to be made

Table 4.5 Interspecific comparisons of mean growth rates, computed in terms of energy for fish embryos and endogenously feeding larvae receiving no external food

Species	Temp (°C)	Initial yolk size (J indiv⁻¹)	Absolute growth rate $(J\ indiv^{-1})$* at stage:				Source[†]
			Fe–H[‡]	H–S	S–Re	Fe–Re[‡]	
Oncorhynchus tshawytscha[a]	10	4 784	8.2	28.9	−26.4	11.0	1
O. keta[ab]	4.8	3 981	2.9	–	–	10.1	2
O. kisutch[ab]	4.6	3 238	3.3	–	–	11.9	2
Salmo ischchan aestivalis	7.8–15[c]	1 112	0.7	7.9	–	–	3
Salmo trutta	3–11[d]	625	0.6	3.9	8.5	2.0	4
O. mykiss	7.5–11[e]	592	0.6	5.8	–	–	3
Micropterus salmoides	18–21	9.56	0.26	0.40	−0.13	0.24	5
Clupea harengus pallasi	12.5–13.5	5.44	0.37	–	–	0.18	6
Lepomis macrochirus	23.5	3.90	0.28	0.33	−0.11	0.21	7
Scophthalmus maximus	15	0.853	–	0.059	−0.028	–	8

Specific growth rate ($\% d^{-1}$)§ at stage:

			Fe–H‡	H–S	S–Re	Fe–Re‡	
Oncorhynchus tshawytscha[a]	10	4 784	5.3	2.5	–1.4	2.9	1
O. keta[ab]	4.8	3 981	5.6	–	–	4.5	2
O. kisutch[ab]	4.6	3 238	5.6	–	–	5.2	2
Salmo ischchan aestivalis	7.8–15[c]	1 112	7.0	10.0	–	–	3
Salmo trutta	3–11[d]	625	3.6	3.1	3.2	3.5	4
Salmo trutta	3–4	?	4.1–5.9[f]	–	–	–	9
O. mykiss	7.5–11[e]	592	6.9	9.0	–	–	3
Micropterus salmoides	18–21	9.56	1.4	19.7	–3.9	8.6	5
Morone saxatilis	18?	8.52	182.2[g]	8.0[g]	–	–	10
Clupea harengus pallasi	12.5–13.5	5.44	13.8	–	–	6.4	6
Lepomis macrochirus	23.5	3.90	>100?	30.1	–6.1	6.7	7
Scophthalmus maximus	15	0.853	–	19.0	–8.6	–	8

*Calculated from Equation 4.14. Stages: Fe, fertilization; H, hatch; S, free-swimming (first feeding); Re, yolk resorption (see Fig. 4.1).

†Sources: 1, Heming (1982); 2, Beacham et al. (1985), both species: female no. 2, small eggs; 3, Ryzkhov (1976); 4, Raciborski (1987); 5, Laurence (1969); 6, Eldridge et al. (1977) control larvae; 7, Toetz (1966); 8, Quantz (1985); 9, Zalicheva (1981), 10, Eldridge et al. (1982).

‡Computed according to Eldridge et al. (1982).

§Calculated from Equation 4.17.

[a]Converted from dry wt to energy using caloric value 23.8 $J mg^{-1}$ (Kamler and Kato, 1983).

[b]Converted from wet wt to dry wt using hydration 84% and 79%, respectively, at hatch and resorption (Ryzhkov, 1973).

[c]7.5–8.9 °C from fertilization to hatch, 5.5–15 °C from hatch to first feeding.

[d]Ambient temperature (3–6 °C during egg incubation and then gradual increase to 11 °C).

[e]7.5–8.9 °C from Fe to H, 11 °C from H to S.

[f]Values given by the author. Computation method unknown; probably Equation 4.17 or 4.19 was used.

[g]The value for Fe–H is given by the authors, that for H–S is recomputed.

up by the resorption of tissue. This has been found in the absence of exogenous food in larvae able to take such food while still utilizing endogenous yolk (Gray, 1926; Smith, 1957; Lasker, 1962; Blaxter and Hempel, 1966; Toetz, 1966–Fig. 4.9(A) herein; Laurence, 1969–Fig. 4.9(B); Ehrlich and Muszynski, 1981; Hunter and Kimbrell, 1981; Heming, 1982; Escaffre and Bergot, 1984 – curves 2 and 3 in Fig. 4.9(D); Rombough, 1985; Quantz, 1985). The energy deficit can begin before the larvae become capable of exogenous feeding (*Sardinops caerulea*, Lasker, 1962) or coinciden-tally (*Tautoga onitis*, Laurence, 1973). In contrast, larvae of some species continue to grow until yolk energy reserves are exhausted (*Oncorhynchus nerka*, Hurley and Brannon, 1969; *Clupea harengus*, Eldridge *et al.*, 1977; *Salmo trutta*, Raciborski, 1987 – Fig. 4.9(E)). In *Clupea harengus*, Blaxter and Hempel (1963) report tissue resorption before yolk exhaustion only for larvae hatching from larger eggs; the same was suggested for *Oncorhynchus mykiss* by Kamler and Kato (1983). Probably the depressed amounts of consumed yolk (see the absolute yolk absorption rates in Table 4.3) were too low to meet the energy expenditures of larger larvae. Thus, an energy deficit before the yolk sac disappears is not universal among fish. A possible explanation is that yolk composition may remain unchanged, or may change over the period of yolk sac resorption (Section 4.2). In unresorbed yolk, some essential component may be lacking.

Tissue resorption is the only way to meet the metabolic demands of larvae if starvation continues after yolk resorption (5(a) above). Larvae of *Oncor-hynchus mykiss* that were unfed until 50% mortality had occurred lost 30–40% of dry weight from the tissues that had been formed from their yolk supply (Escaffre and Bergot, 1984). In *Salmo trutta*, such losses at 75% mortality were 32% of energy and 42% of protein (Raciborski, 1987), and in *Lepomis macrochirus* at 100% mortality they came to 36% of energy (Toetz, 1966). Hence, about one-third of the body's energy or matter can be consumed before mass mortality in larvae that do not obtain external food.

The variability in dry weight of embryonic tissues dissected from five salmonid species at the beginning of development was low (coefficients of variations ranged from 4 to 10%, Ryzhkov, 1976) as compared with that of intact, unfertilized fish eggs (CV 8.5–17%, Table 3.1). Not surprisingly, yolk is more variable than early embryos. After gastrulation, the variability in weight of embryos, and then of larvae, increases (Ryzhkov, 1976), especially when food is limited.

Factors affecting growth during the endogenous feeding period

Temperature

A well-known tendency towards increased growth rate with increasing temperature within the optimum range is illustrated in Fig. 4.10 for the embryonic and larval stages of the endogenous feeding period. However,

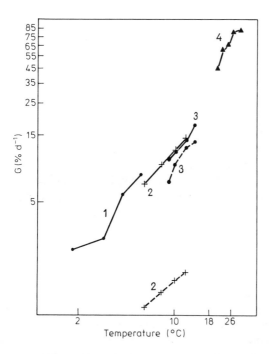

Fig. 4.10 Effect of temperature on specific growth rate (G) of fish embryos (from fertilization to hatching, solid curves) and endogenously feeding larvae (from hatching to the end of yolk resorption, unfed larvae, broken curves). 1, *Salmo salar morpha sebago* dry weight, initial yolk 1692 J indiv^{-1}, computed by Ryzhkov (1976); 2, *Oncorhynchus tshawytscha* dry weight, initial yolk 4784 J indiv^{-1}, recomputed from Heming (1982); 3, *O. mykiss*, energy, initial yolk 280 J indiv^{-1}, recomputed from Kamler and Kato (1983); 4, *Cyprinus carpio* standard length, initial egg 8.3 J indiv^{-1}, recomputed from Peñáz *et al.* (1983). Log–log scale.

aquaculturists are more concerned with the final effect of temperature on growth: what size will larvae attain on their yolk reserves at different temperatures?

No apparent influence of incubation temperature upon larval size at hatching was reported for *Salmo salar* (Ryzhkov, 1976) and for *S. trutta* (Zalicheva, 1981), at first feeding for *Cottus bairdi* (Docker *et al.*, 1986) and at complete yolk resorption for *Scophthalmus maximus* (Quantz, 1985). Larvae of *Eopsetta jordani* (Alderdice and Forrester, 1971) and *Tinca tinca* (Kokurewicz, 1981) converted more of their yolk to body tissue as incubation temperature increased. On the contrary, larvae of *Tautoga onitis* (Laurence, 1973), *Ctenopharyngodon idella* (Vovk, 1974), *Salmo salar* (Hamor and Garside, 1976, 1977) and *Oncorhynchus tshawytscha* (Rombough, 1985) showed a negative correlation between weight and incubation temperature (see also Blaxter, 1969). In *Salmo trutta* (Kowalska, 1959),

Perca fluviatilis and *Stizostedion lucioperca* (Kokurewicz, 1969b), and *Anoplopoma fimbria* (Alderdice *et al.*, 1988), the maximum size resulted from incubation at intermediate temperatures, implying that an optimum temperature existed. Further examples are shown in Fig. 4.11 and Table 4.6. In summary, then, these data do not support Winberg's (1987) hypothesis of equal final size of poikilotherms incubated at different temperatures (see p. 122).

These conflicting results can be tentatively explained in three different ways. First, unusually small larvae are known to be those which hatched precociously. Precocious hatching has been observed at high temperatures, as the effect of increased embryo mobility. It has also been observed at both ends of the viable temperature range, resulting from a desynchronization between the secretion of hatching enzymes and the growth of larvae (Section 4.1).

Second, Lindsey and Arnason (1981) proposed a model (the atroposic model) of vertebral formation in embryos of various fish species. According to these authors, the number of vertebrae is determined by two processes: A (differentiation) and L (growth), both proceeding exponentially at the same time but independently of each other. Attainment of the critical level by process A causes a sudden termination of process L; the final number of vertebrae is fixed at this moment. The number of vertebrae is temperature-sensitive throughout the entire period of their formation up to the point at which their number is fixed. The atroposic model explained the contradictory results of experimental studies on the effect of constant and variable temperatures on the number of vertebrae in fish. However, it has proved to be less applicable to variations in fin-ray number.

The third explanation is a bioenergetical one. It should be borne in mind that yolk represents a predetermined amount of energy; yolk energy is partitioned between tissue production and respiration, the main and competing components of the endogenous energy budget (Equation 4.12). Increased metabolism will inevitably result in a decrease of growth. All the rates – developmental rate, yolk absorption rate, growth rate and metabolic rate – are positively related to temperature, but the extent of their acceleration by temperature is not necessarily the same. A combination of a high developmental Q_{10} with a low metabolic Q_{10} will cause the amount of cumulative energy expended in metabolic processes (R_c) over a developmental period to decrease with increasing temperature. The remaining energy will be invested in growth of tissue (Table 4.6, *Oncorhynchus mykiss*). The same mechanism explains why Dąbrowski (1981) found that the point of no return (see page 177) was reached earlier in *Coregonous pollan* larvae tested at the same temperature as others but obtained from eggs that had experienced a prolonged incubation at lower temperature. An opposite combination (lower developmental Q_{10} than metabolic Q_{10}, increase of

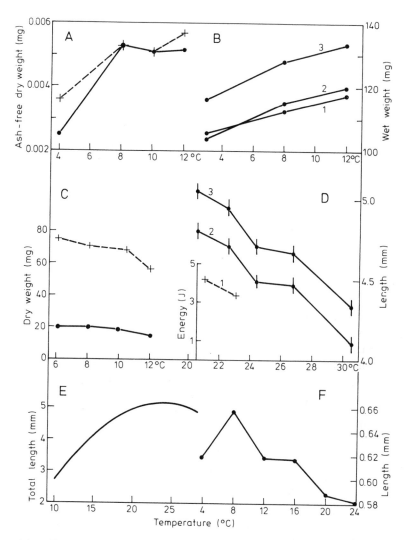

Fig. 4.11 Effect of incubation temperature on size (length or tissue weight) of larvae at hatching (continuous curves) or at complete yolk resorption (unfed larvae, broken curves). (A) *Limanda ferruginea* (Howell, 1980) (ordinate at left); (B) *Oncorhynchus keta*: 1, 2, and 3, larvae from small, medium and large eggs, respectively (Beacham and Murray, 1985) (ordinate at right); (C) *O. tshawytscha* (Heming, 1982) (ordinate at left); (D) *Cyrinus carpio*: 1, energy (Kamler, 1976) (ordinate at left), 2, standard length and 3, total length (Peñáz *et al.*, 1983) (ordinate at right); (E) *Morone saxatilis* 24 h after hatching (computed from the equation given in Morgan *et al.*, 1981) (ordinate at left); (F) *Capnia atra* (Plecoptera) (Brittain *et al.*, 1984) (ordinate at right).

Table 4.6 Change of developmental rate and metabolic rate with temperature, and resulting differences in size of newly-hatched individuals

Species*	Initial egg size ($J\,indiv^{-1}$)	Incubation temp. (°C)	Hatching time (d)	Q_{10} developmental	Mean metabolic rate ($J\,indiv^{-1}\,d^{-1}$)	Q_{10} metabolic	Cumulative metabolism ($J\,indiv^{-1}$)†	Body tissues at hatch ($J\,indiv^{-1}$)‡
Oncorhynchus mykiss								
mykiss	681	9	38		0.330		12.56	27.5
O. mykiss	681	14	22	2.98	0.342	1.07	7.53	56.7
O. mykiss	285	9	35		0.334		11.69	25.7
O. mykiss	285	14	21	2.78	0.339	1.03	7.12	32.0
Physa acuta	0.33	22	8		0.0113		0.090	0.239
P. acuta	0.33	26	7	1.40	0.0153	2.13	0.107	0.223

*Sources: *O. mykiss*, Kamler and Kato (1983); *P. acuta* (Gastropoda) Kamler and Mandecki (1978).

†Note decrease with incubation temperature in *O. mykiss*, but increase in *P. acuta*.

‡Note increase with incubation temperature in *O. mykiss*, but decrease in *P. acuta*.

cumulative metabolism and decrease of the size at hatch with increasing temperature) is illustrated in Table 4.6 for a gastropod, *Physa acuta*.

Fish egg incubation is aimed at both large and viable larvae. However, in the three species listed in Table 4.7, the temperature optimum for survival was lower than the temperature at which the largest larvae were obtained. This is not surprising, as both survival and the final size attained on the yolk reserves depend upon various processes differing in the degree of temperature dependence.

Oxygen

Many authors have observed smaller larvae hatched from eggs incubated at reduced oxygen levels (Winnicki, 1967, 1968, *Salmo trutta* and *Oncorhynchus mykiss*; Carlson and Siefert, 1974, *Salvelinus namaycush*; Hamor and Garside, 1976, 1977, 1979, *Salmo salar*; review: Zhukinskij, 1986). This may be a result of precocious hatching (Section 4.1). Also, with a decrease in oxygen supply, an increasing portion of metabolic costs is thought to be covered by the less efficient anaerobic processes.

Egg size

Absolute growth rates of embryos and endogenously feeding larvae are greater in species producing larger eggs (Table 4.5); the same applies to 'negative growth' (tissue resorption). An inverse relationship between specific growth rate and egg size can be expected, but is clouded by the variability of results. Intraspecific size comparisons for larvae resulting from eggs of different size are presented in Section 3.10.

Table 4.7 Comparison of incubation temperatures at which maximum survival or maximum size was attained

Species	Temp. studied (°C)	Temp. of maximum survival at hatching (°C)	Temp. of maximum size (°C)	Source*
Morone saxatilis	10–28	18	23.8†‡	1
M. americana	6–26	14.1	17.6†	2
Oncorhynchus mykiss	9–14	10–12	⩾14§	3
O. mykiss	9–14	–	⩾14**	3

*Sources: 1, Morgan *et al.* (1981); 2, Morgan and Rasin (1982); 3, Kato (1980) and Kamler and Kato (1983).
†Maximum size was measured as length at stage H (see Fig. 4.1).
‡Obtained from the equation $L = -0.013t^2 + 0.62t - 2.22$, where L is length and t is temperature.
§Maximum size was measured as dry weight at stage H (see Fig. 4.1).
**Maximum size was measured as dry weight at stage Re (see Fig. 4.1).

4.4 METABOLISM

General remarks

Expression of metabolism

Metabolic processes provide energy for physicochemical work. The main subcomponents of the respiratory metabolism (R) are defined by Brett (1962):

$$R = R_r + R_a + R_f \qquad (4.26)$$

R_r is the resting (standard) metabolism, which is recorded in unfed animals that are at rest; below this level the physiological functions are impaired. R_a is the activity metabolism ('the scope for activity', Fry, 1947), i.e. additional energy used for swimming or other activities. R_f is the feeding metabolism, i.e. additional energy expended on foraging, ingestion and SDA (specific dynamic action, costs associated with food conversion, from digestion to synthesis of excretory products); at least part of R_f may reflect the energetic costs of tissue synthesis in fed animals (Jobling, 1985).

Metabolism is measured in terms of energy output. A complete investigation, indirect calorimetry, requires determination of non-faecal nitrogen excretion, oxygen consumption and carbon dioxide production; the amounts of oxidized protein, carbohydrates and fats are computed therefrom (for further details see Brafield, 1985). Small differences ($< 2\%$) were found, however, between energy output computed from complete indirect calorimetry and from oxygen consumption alone, measured in juvenile *Ctenopharyngodon idella* and in a beetle, *Tribolium castaneum*, and converted to energy with a composite oxycalorific coefficient (Kamler, 1970). In fish embryos, buffering carbon dioxide in perivitelline fluid Smith, 1957) and the retention of nitrogen metabolites in it (Blaxter, 1969; Kaushik *et al.*, 1982), make complete indirect calorimetry a questionable procedure.

The main components of eggs at fertilization are fats and protein (Section 3.3). In *Cyprinus carpio*, total energy lost during the endogenous feeding period amounted to 6 J indiv^{-1}: of this, 2 J indiv^{-1} was the energy contained in the matter rejected during hatching and the remaining 4 J indiv^{-1} was metabolized (Kamler, 1976). Most of this energy originated from fats and protein (Fig. 4.12), but this does not indicate in what proportion these substances are actually used as the substrate of metabolism. There is some ambiguity about the role of different compounds in energy production during embryonic development (Needham, 1931; Hayes, 1949; Smith, 1957; Devillers, 1965; Milman and Yurowitzky, 1973; Terner, 1979; Boulekbache, 1981; Kaushik *et al.*, 1982; Gosh, 1985); carbohydrates are controversial as the substrate of metabolism, as to the possibility of gluconeogenesis. However, the general tendency is that carbohydrates act as a main substrate of respiration during very early developmental stages, whereas after gastrulation the utilization of lipids (and proteins?) increases.

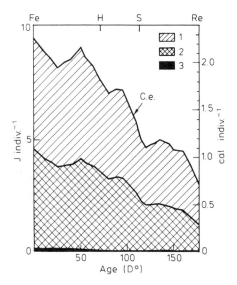

Fig. 4.12 Decrease of energy stored in (1) protein, (2) fats and (3) carbohydrates plus lactic acid of carp, *Cyprinus carpio*, embryos (Fe to H) or unfed yolk-sac larvae (H to Re) during development from fertilization (Fe) to final yolk resorption (Re) at 23 ± 2 °C. The upper line represents the caloric equivalent of one individual (*C.e.*). (Modified from Kamler, 1976).

Hence, for fish early life a general, composite oxycalorific coefficient has to be used. This problem has been considered and reviewed by Elliott and Davison (1975), Brett and Groves (1979), Brafield and Llewellyn (1982) and Brafield (1985) for postembryonic fish. To summarize the discussion, the oxycalorific coefficient given by Kleiber (1961) for mammals, which excrete urea, is too high for fish in which the main component of excreta is ammonia. For starving or carnivorous postembryonic fish catabolizing chiefly fats and protein, the most appropriate composite value is 13.6 J per mg of oxygen consumed, whereas for herbivorous fish having a high proportion of carbohydrates in the diet, a composite value of 14.1 J per mg O_2 can be used (Elliott and Davidson, 1975).

In summary, then, the oxycalorific coefficient for the endogenous feeding period of fish can be accepted as a value between 13.6 and 14.1 J per mg of oxygen. The resulting possible error will be low in comparison with the inaccuracies of oxygen consumption measurements and the inherent variability of respiration.

Metabolism during egg maturation and the endogenous feeding period

Oxygen consumption increases during oocyte growth (Nakano, 1953; Ozernyuk, 1970; reviews: Devillers, 1965; Ozernyuk, 1985; Zhukinskij, 1986); an example is given in Fig. 4.13 in terms of the absolute rate (oxygen consumed per oocyte and per hour). That increase reflects the metabolic

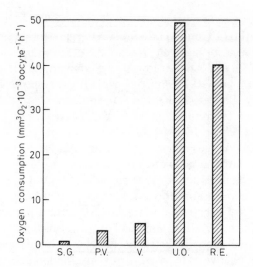

Fig. 4.13 Oxygen consumption by *Misgurnus fossilis* oocytes during slow growth (S.G.), rapid growth (previtellogenesis (P.V.) and vitellogenesis (V.)) and maturation (unripe oocytes (U.O.) and ripe eggs (R.E.)). (From table 12 in Ozernyuk 1985.)

costs of formation of structural and reserve compounds under conditions of good oxygen and nutrient supply via the ovarian blood vessels. During egg maturation, metabolic rate decreases (Figs 4.13 and 4.14). The energy-saving low metabolic expenditure of mature, ovulated eggs is of adaptive significance in the case of poor oxygen supply (see below) and delayed fertilization.

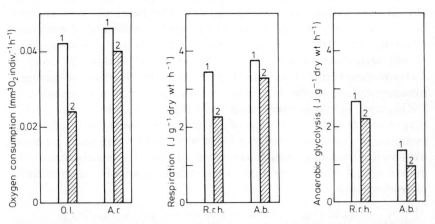

Fig. 4.14 Decrease of metabolic rates during oocyte maturation of *Oryzias latipes* (O.l.). (Nakano, 1953), *Acipenser ruthenus* (A.r.) (Ozernyuk, 1985), *Rutilus rutilus heckeli* (R.r.h) and *Abramis brama* (A.b.) (Gosh and Zhukinskij, 1979). 1, unripe oocytes (stage IV); 2, ripe eggs.

No sudden changes of oxygen consumption on fertilization were found in *Oryzias latipes* (0.0243 and $0.0245\,mm^3\,O_2$ $egg^{-1}\,h^{-1}$ in unfertilized and freshly fertilized eggs, respectively, Nakano, 1953), in *Cyprinus carpio* (0.051 and $0.056\,mm^3\,O_2$ $egg^{-1}\,h^{-1}$, respectively, Kamler, 1976) or in *Acipenser ruthenus* (0.130 and $0.129\,mm^3\,O_2\,egg^{-1}\,h^{-1}$, respectively, Ozernyuk, 1985). Similarly, Philips (1940) did not find any change of oxygen consumption by eggs on fertilization in *Fundulus heteroclitus*, for which Boyd (1928) had reported an increase of respiration just after fertilization. The absence of abrupt changes of respiratory rates and ATP levels on fertilization is typical of fishes and amphibians, in contrast to sea urchins, in which an increase of respiration is found; that depends upon the egg maturity stage at which fertilization occurs (review: Ozernyuk, 1985). Thus, the oxygen consumption of fish eggs remains unchanged until a particular time after fertilization (Nejfakh, 1960). Examples of oxygen consumption by unfertilized or freshly fertilized eggs are listed in Table 4.8. Interspecific comparison shows that the relative rate of oxygen consumption (per g wet weight) fluctuates haphazardly, whereas the absolute rate (per egg) decreases with decreasing egg size.

Teleostean embryos begin to increase their oxygen consumption rate (per individual and per unit time) at the onset of cleavage. Oxygen consumption increases slowly during early development (cleavage, morula) and then accelerates. For example a four-fold increase from fertilization to hatching was found in *Hypophthalmichthys molitrix* (Yarzhombek, 1986), eightfold in *Misgurnus fossilis* (Nejfakh, 1960), 29-fold in *Cyprinus carpio* (Kamler, 1976), 20-, 27- and 37-fold in *Oncorhynchus mykiss* (Yarzhombek, 1986, Kamler and Kato, 1983, and Ryzhkov, 1976, respectively), 14–50-fold in three species of *Acipenser* (Korzhuev *et al.*, 1960) and an 85-fold increase in *Gadus morhua* (Davenport and Lønning, 1980); Fig. 4.15.

Two complementary types of interpretation of changes in absolute oxygen consumption rate during early life of fishes are similar to those used for growth studies (Section 4.3).

One of them is to analyse short-term fluctuations of the absolute oxygen consumption rate and their relation to the key events in the embryo's life. For example, in *Cyprinus carpio* (Fig. 4.15(A)), a temporary acceleration of oxygen consumption at the age of $36\text{--}44\,D°$ coincides with haemoglobin formation in the blood. The subsequent acceleration before hatching arises from an increase in the amount and functional activity of tissues as well as from embryo mobility. Active embryos of *Salmo salar*, *Sardinops caerulea* and *Clupea harengus* consumed three, one to three, and two times as much oxygen, respectively, as quiescent embryos (Hayes *et al.*, 1951; Lasker and Theilacker, 1962; Holliday *et al.*, 1964). Further increase of oxygen consumption occurs at hatching. A 2.5-fold increase of oxygen consumption rate (per mg dry weight and hour) was found at hatching in *Clupea harengus* (Holliday *et al.*, 1964), four- to sevenfold increases (in terms of $mm^3\,O_2$

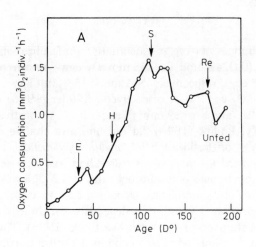

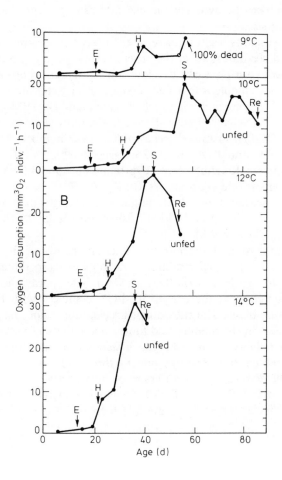

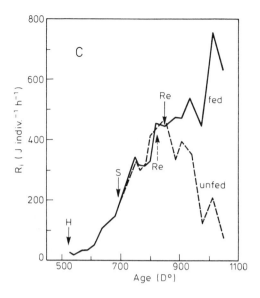

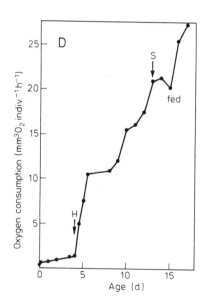

Fig. 4.15 Course of absolute instantaneous metabolic rate (per individual and hour) during embryonic and early larval development of four fish species. Fertilization, age 0; E, H, S and Re as in Fig. 4.1. (A) *Cyprinus carpio* at 23 °C, initial egg size 8.3 J indiv^{-1}, larvae unfed (modified from Kamler *et al.*, 1974); (B) *Oncorhynchus mykiss* at four temperatures, offspring of a three-year-old female, initial yolk 674 J indiv^{-1}, larvae unfed (reproduced with permission from Kamler and Kato, 1983); (C) *Salmo trutta*, incubation performed and respiration measured at ambient temperatures increasing from 3 to 11 °C, initial yolk 625 J indiv^{-1}, fed and unfed groups (recomputed from Raciborski, 1987); (D) *Acipenser gueldenstaedti* at 20 °C, fed (Korzhuev's data, from table 46 in Ozernyuk, 1985).

Table 4.8 Oxygen consumption of unfertilized or freshly fertilized fish eggs

Species	Temp. (°C)	Egg size (mg wet wt)	Oxygen consumption rate Relative* ($mm^3 g^{-1} h^{-1}$)	Oxygen consumption rate Absolute ($mm^3 egg^{-1} h^{-1}$)	Source[†]
Trematomus borchgrevinki	−1.0	840	c. 15.5	c. 13	1
T. borchgrevinki	6.0[‡]	840	c. 15.5	c. 13	1
Salmo salar	–	229.00	12.2	2.7900	2
Oncorhynchus mykiss	–	72.40	14.0	1.0100	2
O. mykiss	10	–	–	0.1750	3
O. mykiss	7.5–8.9	c. 60	c. 2.0	0.1205	4
Huso huso	–	25.80	11.7	0.3020	2
Esox lucius[§]	12	10.80	5.6	0.0600	5
Misgurnus fossilis	21	–	–	0.0420	3
M. fossilis	–	2.38	14.7	0.0350	2
Cyprinus carpio	23	1.00	51.0	0.0510	6
Brachydanio rerio	–	0.48	15.4	0.0074	2

*For wet weight.
[†]Sources: 1, Rakusa-Suszczewski (1972); 2, Ozernyuk (1985); 3, Yarzhombek (1986); 4, Ryzhkov (1976); 5, Lindroth (1946); 6, Kamler (1976).
[‡]Upper lethal temperature for adult fish.
[§]Four-cell stage.

indiv^{-1} h^{-1}) were observed in three species of *Acipenser* (Korzhuev *et al.*, 1960), and a tenfold increase in *Clupea harengus pallasi* (Eldridge *et al.*, 1977) (see also Laurence; 1969; Davenport and Lønning, 1980; Kaushik *et al.*, 1982; and Fig. 4.15). This increase is mostly attributable to the activity associated with repayment of oxygen debt by larvae freed from egg capsules. A drop of respiratory quotient (RQ) below 0.6 and a depression of lactic acid content in carp at hatching were shown by Kamler *et al.* (1974). Turning to oxygen consumption of carp (Fig. 4.15(A)), the peak occurred at the onset of feeding ability and was followed by a high level of respiration. Depletion of reserves stored in the yolk sac and resorption of tissue in these unfed larvae are accompanied by a decrease of oxygen consumption. A similar course of oxygen consumption is usually observed in embryos and unfed yolk-sac larvae (Fig. 4.15(B) and (C); Belyaeva, 1959; Winnicki, 1968; Laurence, 1969; Kaushik *et al.*, 1982), whereas in fed yolk-sac larvae, oxygen consumption accelerates after a short-term depression during the transition from endogenous to exogenous feeding (Fig. 4.15(C) and (D); Ryzhkov, 1976; Eldridge *et al.*, 1982; De Silva *et al.*, 1986; further examples in Ozernyuk, 1985).

A second type of interpretation is to generalize the time course of the absolute respiration rate (per individual and per unit time) with a smooth curve. In *Misgurnus fossilis* (Nejfakh, 1960) and *Gadus morhua* (Davenport and Lønning, 1980), the increase of respiration with age from fertilization to hatch was close to linear. The exponential model, however, was found to describe the increase of respiration of *Tautoga onitis* (Laurence, 1973), *Cyprinus carpio* (Kamler, 1976), *Oncorhynchus mykiss* (Kamler and Kato, 1983) and *Brachydanio rerio* (Ozernyuk and Lelyanova, 1985); earlier applications of the exponential model to respiration of fish embryos were reviewed by Devillers (1965). Respiration during the endogenous feeding period was described by one exponential curve or else the curve was divided into segments.

Different approaches were also applied to the relative respiration rate (per unit weight and per unit time). During embryogenesis this rate can be computed per unit weight of the intact egg, or per unit weight of dissected tissue. The latter measure of relative oxygen consumption will be considered below.

The most analytical approach is represented by works relating abrupt, short-term increases of the relative respiratory rate to intensified differentiation during the transition from one developmental step to another. The increases are followed by depressed respiratory rates; both alternate periodically during embryonic (review: Trifonova, 1949) and postembryonic developmental steps (review: Ryzhkov, 1976). No such abrupt changes of the relative respiratory rate were found in other studies (Devillers, 1965; Winnicki, 1968; Zhukinskij, 1986). It should be recalled, however, that the revelation of short-term events on the respiration v. time curve depends on the frequency of measurements, and may be clouded by the inherent variability of both respiration and developmental rate among individuals, as well as by temperature. The latter effect is shown in Fig. 4.15(B) for offspring of one female *Oncorhynchus mykiss* which were incubated at different temperatures.

General trends in relative respiratory rate per unit weight of tissue during longer segments of development have also been studied. Conflicting results have been obtained for embryos of different species. Random fluctuations of oxygen consumption rate per gram of embryo tissue were found in *Salmo salar* by Hayes *et al.* (1951) and Ozernyuk and Zotin (1983); a stable rate at embryonic steps I–V followed by a threefold increase before hatching was reported by Ryzhkov (1976) for the same species. The respiratory rate of *Brachydanio rerio, Cyprinus carpio, Ctenopharyngodon idella* (Ozernyuk and Lelyanova, 1985) and *Oncorhynchus mykiss* (Ozernyuk and Lelyanova, 1987) increased during embryogenesis. A distinct increase of the relative respiratory rate after hatching has been reported for endogenously feeding larvae of many fish species (Shamardina, 1954; Davenport and Lønning, 1980; Ozernyuk and Lelyanova, 1985, 1987). The peak in the relative rate of oxygen consumption is usually found at the onset of exogenous feeding, but it can appear somewhat earlier (De Silva *et al.*, 1986) or later (Shamardina, 1954). In further ontogenesis the oxygen consumption rate per unit weight is inversely related to size. The above alterations in the relative respiratory rate result from the development of respiratory organs (pages 159–60), from changes in the concentration of mitochondria (Ozernyuk, 1985; Ozernyuk and Lelyanova, 1985, 1987), and from relations between the increased rate of respiration and the rate of body growth.

The relationship between the absolute rate of respiration of animals at rest (R_r) and weight (W) is known (Zeuthen, 1955, 1970; Hemmingsen, 1960; Kleiber, 1961) as:

$$R_r = aW^b \tag{4.27}$$

Winberg (1956) found the relationship:

$$R_r = 0.3W^{0.8} \tag{4.28}$$

where R is in $cm^3 O_2$ indiv^{-1} h^{-1} at 20 °C and W in g wet weight, to represent the best fit for 364 pairs of data for mostly postlarval fishes. Thus, because:

$$R/W = aW^{(b-1)} \tag{4.29}$$

in postlarval fishes the relative oxygen consumption rate (R/W) decreases with increasing body size:

$$R/W = 0.3W^{-0.2} \tag{4.30}$$

Considerable research has been done on the relationship between respiration and body size in postlarval fish, but few data exist for endogenously feeding fishes. The relative respiration rate of endogenously feeding *Salmo trutta* larvae (from hatching to swimming-up) was found by Raciborski (1987) to increase with increasing body size: $(b-1)>0$, $b=1.83\pm0.41$ ($\pm95\%$ confidence interval). In older, mixed and exogenously feeding larvae (from swimming-up to the time of 75% mortality of starved larvae) he found the slope b to be 1.02 ± 0.19, i.e. there was almost no change in the relative respiratory rate with increasing weight in these larvae, $(b-1)\approx0$. In *Clupea harengus* endogenously feeding larvae, $b=1.1$ (Holliday *et al.*, 1964; Blaxter and Hempel, 1966). Thus, the zones of increasing and decreasing R/W rates are separated by a zone in which the relative respiratory rate is independent of weight. This was also shown for *Brachydanio rerio* (Ozernyuk, 1985).

The total energy expended in respiration (R_c) is obtained by cumulation of the daily metabolism (R_i) from the beginning of the life cycle (Fe) to successive days of development. Obviously, the cumulative metabolism increases with age (Fig. 4.16, Table 4.9), with egg size (Table 4.9), and is higher in larvae that obtain external food before yolk resorption is complete

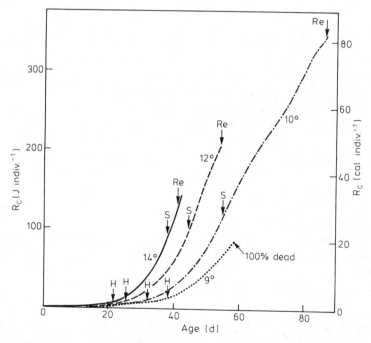

Fig. 4.16 Cumulative metabolism (R_c) of *Oncorhynchus mykiss* embryos and unfed larvae at four temperatures. Cumulation from fertilization (age 0) to successive days of life. Same experimental series as in Fig. 4.15 (B); H, S, Re as in Fig. 4.1. (Reproduced with permission from Kamler and Kato, 1983.)

Table 4.9 Cumulative metabolism (R_c, J indiv.$^{-1}$) from fertilization (Fe) to hatching (H), the onset of exogenous feeding (S) and the completion of yolk resorption (Re). Species are arranged in order of decreasing egg size

Species	Temp. (°C)	Cumulative metabolism (J indiv^{-1}) for the period:			Source†
		Fe–H	Fe–S	Fe–Re*	
Salmo salar m. sebago	‡	101.30	652.30	1527.87 f	1
S. trutta	3–11§	11.34	94.60	381.26 f	2
S. trutta	3–11§	11.34	94.60	331.28 u	2
Acipenser gueldenstaedti	20	1.81	39.84	–	3
Cyprinus carpio	23±2	0.38	1.26	3.14 u	4
C. carpio	21±2	0.33	1.05	2.51 u	5
Lepomis macrochirus	23.5	0.51	1.53	2.05 u	6

*f, Fed larvae; u, Unfed larvae.
†Recomputed from: 1, Ryzhkov (1976); 2, Raciborski (1987); 3, Korzhuev's data, cited in Ozernyuk (1985); 4, Kamler (1972a); 5, Kamler et al. (1974); 6, Toetz (1966).
‡Ambient temperatures: eggs incubated at 0.1–6.0 °C, larvae at temperatures >6°C.
§Ambient temperatures: 3–6 °C during egg incubation and then gradual increase to 11 °C.

(*Salmo trutta* in Table 4.9). The effects of temperature are apparently conflicting. The total amount of energy expended from fertilization to an identifiable event in early development decreased in *Oncorhynchus mykiss* with increasing temperature (Fig. 4.16) but increased in *Cyprinus carpio* (Table 4.9). Endogenously feeding larvae of *Oncorhynchus mykiss* between fertilization (day 0) and the 40th day of their life used a total of 16.7, 31.4, 62.0 and 118.5 J indiv^{-1} at 9, 10, 12 and 14 °C, respectively (Fig. 4.16), i.e. expenditure increased with temperature. Although the 'calendar age' of these larvae was the same, those incubated at higher temperature were developmentally more advanced, i.e. were 'physiologically older'. Cumulative metabolism, measured to any comparable event in early development (e.g. to eyeing, completion of yolk sac resorption etc.), decreased exponentially in *Oncorhynchus mykiss* with increasing temperature (Fig. 4.17), because development is more accelerated by temperature than is metabolism in that species (Table 4.6). Thus in *O. mykiss* an increase of temperature maximized tissue production within the early developmental stages (Table 4.6). The opposite (increase of cumulative metabolism and decrease of production at higher temperatures) was found in *Cyprinus carpio* (Table 4.9 and Fig. 4.11(D)) and in embryos of a gastropod, *Physa acuta* (Table 4.6).

Relationship between body growth and respiration

The data presented in the Sections 4.3 and 4.4 show the similarity between the time course of production and respiration in endogenously feeding fishes.

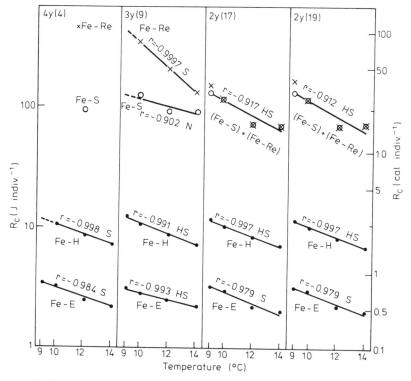

Fig. 4.17 Effect of temperature on cumulative metabolism of *Oncorhynchus mykiss*. Cumulation was done from fertilization (Fe) to eyeing (E), hatching (H), swimming-up (S) and completion of yolk resorption (Re). Larvae were unfed. The age and code number (in parentheses) of parental females are shown at top. N, not significant; S, $P<0.05$; HS, $P<0.01$. Semi-log scale. (Reproduced with permission from Kamler and Kato, 1983.)

A positive relationship between respiratory rate and growth rate in older embryos of *Salmo salar* at 10 °C was shown half a century ago by Privolnev (1938) and was confirmed by Hayes *et al.* (1951). Smith (1957) demonstrated a positive linear relationship between relative rate of metabolism (in cal per g dry embryo weight and h) and specific growth rate (g dry weight per g dry weight and day) in *Oncorhynchus mykiss*. He interpreted the intercept as indicating maintenance and the slope as related to growth. Engelmann (1966) showed a linear relationship between log production and log respiration (both in cal m^{-2} year $^{-1}$) in animal populations. McNeill and Lawton (1970) and Humphreys (1979) analysed relationships between log respiration and log production (both in cal m^{-2} year^{-1}); the latter author used 235 published energy budgets for wild animal populations. The slopes did not differ from 1.0 in 11 of 14 taxonomic groups into which the species were initially separated. The relationship for fish was: $\log R = 0.574 + 1.117 \log P (n = 9)$. The lowest production (P) was c. $100 \, \text{cal m}^{-2} \text{y}^{-1}$, which corresponds to $R = 634 \, \text{cal m}^{-2} \text{y}^{-1}$. In *Oncorhynchus mykiss* embryos and larvae prior to yolk sac resorption, the slope of the relationship between log cumulative respiration and log cumulative production was also close to 1 (see b in Fig. 4.18). Duncan and Klekowski (1975) demonstrated that the relationship between log cumulative respiration and log cumulative production for an individual throughout its life cycle is similar to Engelmann's (1966) relationship. If such comparisons are legitimate, a conclusion of c. 10 times higher metabolic costs per unit production in Humphrey's (1979) natural populations than in the hatchery-reared yolk-feeding *O. mykiss* can be tentatively drawn, a not surprising conclusion because the latter are endogenously feeding animals experiencing no feeding stimuli,

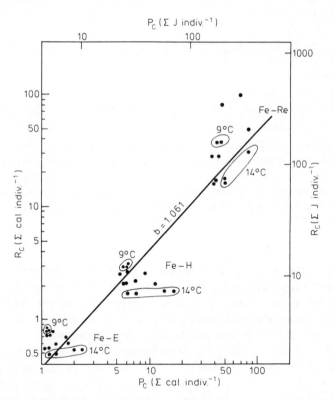

Fig. 4.18 Relationship between cumulative respiration (R_c) and cumulative production (P_c) in *Oncorhynchus mykiss*. Cumulations were done for periods from fertilization to eyeing (Fe–E), to hatching (Fe–H), and to final yolk resorption (Fe–Re). Offspring of four females was incubated and then reared without external food separately at 9, 10, 12 or 14 °C. Values for the extreme temperatures, 9 and 14 °C, are ringed. Log-log scale. (Modified from Kamler and Kato, 1983.)

paying no costs of food capture, meeting only low costs of food processing, and exhibiting limited motor activity.

Turning to environmental modifications of the relationship between respiration and production, McNeill and Lawton (1970) suggested that metabolic costs per unit production would increase in poikilotherm populations experiencing non-productive winter periods during which respiration continued. Similarly, in a laboratory population of a nematode, *Plectus palustris*, prolongation of the larval period due to food deficiency increased the metabolic expenses per unit production (Schiemer *et al.*, 1980). This effect can also be seen in Fig. 4.18, which illustrates higher metabolic values for a given level of production in *O. mykiss* at 9 than at 14 °C (see also Table 4.6). Therefore, high production efficiencies can be expected in yolk-feeding fishes, as can an enhancement of cumulative efficiency per developmental period at high temperatures in those species in which temperature accelerates development more than respiration. These expectations will be discussed further in Section 4.5.

Factors affecting metabolism during the endogenous feeding period

Temperature

Increasing temperature accelerates metabolic rate and depresses solubility of oxygen in ambient water. The effect of temperature on metabolic rate in aquatic poikilotherms has been the subject of numerous studies; Ivleva's (1981) review covers several hundred works. The increase of metabolic rate with increasing temperature can be described by the temperature coefficients Q_{10} or Q_1 (Section 4.1). According to Krogh's 'normal curve' (Ege and Krogh, 1914; Krogh, 1916), acceleration of the resting (standard) metabolic rates declines from low to high temperature (t) as follows:

$t(°C)$	0–5	5–10	10–15	15–20	20–25	25–30
Q_{10}	10.9	3.5	2.9	2.5	2.3	2.2

Winberg (1956) reviewed earlier works on the relationship between metabolic rate in postlarval fishes and temperature. He found that the Krogh's 'normal curve' approximated that relationship well, and he tabulated the correction factor q for converting respiratory rates to 20 °C according to Krogh's 'normal curve'. The factor $q = Q_{10}^{(20-t)/10}$ amounts to 2.67, 1.00 and 0.444 for $t = 10$, 20 and 30 °C, respectively (see also Winberg, 1971). Backiel (1977) described Krogh's 'normal curve' by the equation:

$$R_t = R_{20}(0.3\,e^{0.071t} - 0.24) \qquad (4.31)$$

Further reviews of the effect of temperature on metabolic rates in postlarval fish have been made by Fry (1957) and Brett and Groves (1979), to quote only two. Attention was paid to the ways in which the partitioning of metabolic costs between the subcomponents associated with activity and feeding depends on temperature.

Winberg (1983, 1987) has recently summarized the problem, taking into account Ivleva's (1981) findings. The relationship between the metabolic rate of various fully acclimated poikilothermic animals and temperature within their tolerated range is exponential; it can be expressed with a single temperature coefficient, $Q_{10} = 2.25$, $Q_1 = Q_{10}^{0.1} = 1.0845$. This coefficient is close to that ($Q_{10} = 2.3$) proposed for fish by Brett and Groves (1979).

The existence of a range of temperatures within which metabolism is less dependent upon temperature than in colder and warmer ranges was reported by Stroganov (1956) for fishes, and confirmed for other poikilotherms (e.g. Duncan and Klekowski, 1975; Galkovskaya and Sushchenya, 1978; review: Klekowski and Sazhina, 1985) as well as for larvae of three cyprinid fishes (Wieser and Forstner, 1986).

The relationship between metabolic rate and temperature in early ontogenesis was studied in *Esox lucius* by Lindroth (1942, 1946) and in *Clupea harengus* by Holliday *et al.* (1964). During embryonic development, the oxygen consumption rate of *Misgurnus fossilis* (Zotin and Ozernyuk, cited in Ozernyuk, 1985) and of *Huso huso* (Gershanovich, 1983) was related to temperature according to Krogh's 'normal curve'; similarly, in *Cyprinus carpio* embryos, the temperature dependence of the ammonia excretion rate was described by $Q_{10} = 2.36$ at 13–28 °C (Kaushik *et al.*, 1982).

Oxygen

The limiting effect of oxygen has a special significance for embryos of oviparous species; it is, besides predation, considered as the prevailing factor in fish embryonic development (Kryzhanovskij, 1949).

Blood circulation systems are poorly developed in fish embryos, with no haemoglobin in some species (Pliszka, 1953; Blaxter, 1969; De Silva *et al.*, 1986). In oviparous fishes, therefore, gas diffusion is expected to play a role during development inside the egg capsule. The partial pressure of oxygen in water is low in comparison to that of air (Fry, 1947). Oxygen diffusion through water is extremely slow (Prosser, 1961); its diffusion coefficient through animal muscle is even lower than that through water (0.096 and 0.204 mm³ O_2 per mm² for 1 mm of distance, 1 atm of partial pressure difference for oxygen and 1 h, respectively) (reconverted from Krogh, 1919). Thus, from physical considerations, partly anaerobic conditions are expected to prevail in eggs incubated in water.

A higher oxygen consumption and increased body size of *Salmo* embryos incubated in humid air, as compared to those incubated in water, was found by Winnicki (1968). The same was shown for various fish species in water with oxygen content above 100% air saturation (Gulidov, 1970). However, a high ambient oxygen level is not a common phenomenon in some natural spawning grounds, e.g, on the bottom of eutrophic water bodies or among plants at night. Moreover, the egg capsule (chorion) restricts the passage of oxygen. The diffusion coefficient for oxygen through a chorion is 0.076

(Hayes *et al.*, 1951, value reconverted to the above unit), a value lower than that for muscle. In consequence, oxygen consumption of *Acipenser stellatus* embryos artificially freed of chorion was higher than that of intact eggs (Skadovskij and Morosova, 1936); the same was found by Hayes *et al.* (1951) in *Salmo salar*, in which embryos removed from the chorion showed, moreover, a lower critical oxygen pressure, below which respiration becomes oxygen-dependent. Hence, the oxygen demand of a developing embryo is not always satisfied under natural conditions (Wickett, 1954) and can be greater than the maximum possible available from water (Winnicki, 1968; Alderdice, 1985).

Tolerance to low oxygen levels, to inhibition of cytochrome oxidase and to uncoupling oxidative phosphorylation varies both with fish species and with developmental stage. Anaerobic glycolysis is a less efficient process than aerobic metabolism, but it can be a supplementary source of energy in oocytes or mature non-activated eggs (this is illustrated for *Rutilus rutilus heckeli* and *Abramis brama* in Fig. 4.14) and during the cleavage phase of development. For example in intact embryos of *Misgurnus fossilis*, the rate of anaerobic glycolysis increased prior to gastrulation, during gastrulation the acceleration slowed down, and in early embryogenesis no anaerobic glycolysis was found (Milman and Yurowitzky, 1973). If removed from the chorion, embryos of *Acipenser stellatus* consumed 66%, 30% and 0% more oxygen than intact eggs at gastrulation, embryo formation and just before hatching, respectively (Skadovskij and Morosova, 1936). Developmental rate of early embryos remained almost unaffected by low oxygen level, whereas in older embryos the retarding effect increased progressively (*Salvelinus fontinalis* and *Oncorhynchus mykiss*, Garside, 1966, *Salmo salar*, Hamor and Garside, 1976). Thus, the embryo changes to a more aerobic organism as development proceeds (reviews: Smith, 1957; Devillers, 1965; Doudoroff and Shumway, 1970; Kamler, 1976; Gosh, 1985; Zhukinskij, 1986).

Many adaptations have been described which assist fish embryos in obtaining oxygen. The embryo relies on passive diffusion at the very beginning of development. Contractile movements of the blastoderm (cytoplasmic motor activity) and – after tail bud formation – neuromuscular movements of the embryo (Reznichenko *et al.*, 1967) are the respiratory movements which break down oxygen gradients in the perivitelline fluid of some species. The same role is played by the movements of pectoral fins observed in *Salmo ischchan* just before hatching (Ryzhkov, 1966). Chemical changes in water surrounding the egg which result from the embryo's respiration and excretion involve changes in water density which create convection currents. The latter mitigate oxygen gradients (Klyashtorin, 1982) which threaten fish embryos in stagnant water (Smith, 1957). In early embryos, gas exchange takes place through the body surface. In later embryos and early larvae, both embryonic respiratory organs and the body surface serve as gas–exchange surfaces. The time and extent of development of the embryonic respiratory organs (e.g. ducti Cuvieri, subintestinal vein, hepatic vein, and respiratory vessels in pectorals, in gill covers, in the ventral and/or dorsal fin fold) (Kryzhanovskij, 1949; Soin, 1968) differ in species living in well- or poorly oxygenated habitats and are one of the criteria of fish ecological classifications (Kryzhanovskij, 1949; Balon, 1975a). There are also adaptations of a chemical nature (carotenoids – review: Dąbrowski *et al.*, 1987) and behavioural ones (ventilation of eggs by parents guarding them) which improve the respiratory conditions of embryos within egg capsules.

A substantial change of oxygen conditions occurs at hatching. A dramatic increase of oxygen consumption (examples, pages 147, 151) and temporary depression of respiratory quotient and lactic acid content (Kamler *et al.*, 1974) suggest a repayment of oxygen debt. A unicellular layer of red muscle fibres (the 'red layer') exhibiting a strong cytochrome oxidase activity surrounds the body of newly hatched cyprinid larvae and serves as the temporary organ of gas exchange (El-Fiky *et al.*, 1987; El-Fiky and Wieser, 1988). Gills are rudimentary structures in newly hatched larvae (*Acanthopagrus schlegeli*, Iwai and Hughes, 1977; cyprinids, El-Fiky *et al.*, 1987, El-Fiky and Wieser, 1988). As shown by calculations made from *Cyprinus carpio* larvae by Osse (1989), gills cannot serve as the major respiratory organ. Early cyprinid larvae are powered entirely aerobically (El-Fiky *et al.*, 1987).

Egg size

Interspecific comparisons (Table 4.8) show that the relative oxygen consumption rate ($mg^{-1} h^{-1}$) at fertilization fluctuates haphazardly, whereas the oxygen consumption per egg increases with egg size. This is a result of a higher amount of cytoplasm and an elevated mass of mitochondria in larger eggs (Ozernyuk, 1985). The cumulative amount of oxygen consumed by embryos until hatching is positively related to egg size (Table 4.9); the same can be said about the cumulative respiration of endogenously feeding larvae hatched from eggs of various sizes.

Intraspecific studies on unfertilized eggs (Zhukinskij and Gosh, 1970, 1974; Gosh and Zhukinskij, 1979; Gosh, 1985) showed that small eggs derived from both young and old female fish respired less per 1 mg egg weight than large eggs produced by middle-aged females of the same species.

Metabolic criteria for optimization in fish early development

The problem has two aspects: how to select the best eggs for incubation, and what are the optimum conditions of incubation? Two tentative approaches to that problem will be presented here.

The fertilizability of eggs and viability of embryos at different developmental stages are positively related to oxygen consumption rate, cytochrome oxidase and ATPase activity, and anaerobic glycolysis rate; these were studied in *Rutilus rutilus heckeli*, *Abramis brama*, *Cyprinus carpio*, *Ctenopharyngodon idella* and *Hypophthalmichthys molitrix* (Zhukinskij and Gosh, 1970, 1974, 1988; Gosh, 1985). The numerous criteria for estimation of fish egg quality were reviewed by Zhukinskij and Gosh (1988). They grouped the criteria into four categories – first, visual–tactile, second biometric, third, cytomorphological, and fourth physiological–biochemical – and evaluated them from the points of view: degree of association with embryo viability and possibility of application as a easy quick test. Polarographic measurement of oxygen consumption was found to be the superior criterion for estimating, within 10–15 min, egg quality as 'high', 'medium', or 'low'. Zhukinskij and Gosh (1988) recommended this test for selective purposes.

The optimum temperatures for incubation of *Misgurnus fossilis*, *Oncorhynchus mykiss* and *Huso huso* were studied by Ozernyuk and co-authors (cited in Ozernyuk, 1985). Oxygen consumption was measured in embryos at 'equivalent' stages within a range of temperatures for which the duration of a single mitotic cycle (τ_m, min – Section 4.1) was also assessed. Then the oxygen consumption rates (in mm^3 O$_2$ per τ_m and individual) were computed. The resulting curves were U-shaped. At both the low and high ends of the temperature range, the metabolic costs of development were higher than those in the intermediate zone. For example, in *Misgurnus fossilis* at cleavage, oxygen consumptions of 0.04, 0.02 and 0.04 mm^3 O$_2$ τ_m^{-1} egg^{-1} were found at 4, 15–20, and 30 °C, respectively. The zone of low costs is considered by Ozernyuk (1985) to be the optimum temperature for development. The highest survival of embryos was found by Vovk (1974) at these temperatures.

4.5 BUDGETS OF ENERGY OR MATTER

A fish egg contains a predetermined amount of reserves, for which the conflicting demands of growth and metabolism compete. Thus the efficiency of energy or matter transfer from yolk to larval tissue contributes to the final size of a larva. Large larvae could be favoured in the first period because of their better ability to catch prey and avoid predators (Hunter, 1972), therefore size can be understood as fitness (Begon, 1984). There are two approaches to transformation efficiency in the endogenous feeding period. The first considers how the energy invested in an egg by a female parent is partitioned; this approach is specific to the endogenous feeding period. The alternative is to examine the fate of the yolk energy actually consumed by the developing organism; such a presentation is comparable to the energy budget for any animal.

Partitioning of egg energy or matter

The main components of energy contained in an egg at fertilization ($C.e._0$) are, in order of increasing energy content: germinal disc ($G.d.$), egg capsule (chorion, $e.c.$) and yolk (Y_0):

$$C.e._0 = G.d. + e.c. + Y_0 \qquad (4.32)$$

Few data exist about embryo size at the beginning of development. For example, in *Oncorhynchus mykiss* 0–2 h after fertilization it is 0.03 mg dry weight or $c.$ 0.1% of egg dry weight (Ryzhkov, 1976). Most studies therefore neglect this component. Chorion size is positively related to egg size (Table 4.10; Blaxter and Hempel, 1966); it contributes up to $c.$ 10% of egg dry weight. The chorion is partly digested by chorionase prior to hatching; probably part of the substances released can be used by the embryo (Smith, 1957). A low caloric value of egg capsules at hatching (15.1 J per mg dry matter) was found in *Oncorhynchus mykiss* (Kamler and Kato, 1983). Similar low values (14.1, 15.4 and 17.0 J per mg dry matter) have been reported for egg capsules of aquatic invertebrates (Kosiorek, 1979; Khmeleva *et al.*, 1981; Pandian, 1970). The caloric value of larval tissues is usually lower than that of the original egg (Table 4.11).

During the endogenous feeding period the yolk energy (Y) is allocated between production (P), metabolism (R) and excretion (U). The latter is measured as ammonia and urea; some metabolic excreta are retained in the egg (Smith, 1947; Blaxter, 1969) and are discharged with perivitelline fluid

Table 4.10 Dry weight of chorion related to egg size

Species and stage*	Egg dry wt ($mg\ egg^{-1}$)	Chorion dry wt ($mg\ capsule^{-1}$)	%	Source†
Limanda ferruginea, H	0.0127‡	0.0006‡	4.7	1
Scophthalmus maximus, Fe	0.0422‡	0.0051‡	1.2.0	2
Morone saxatilis, Fe	0.274	0.018	6.6	3
M. saxatilis, Fe	0.300	0.025‡	8.2	4
Clupea harengus, Fe	0.100	0.021–0.033	21–33	5
C. harengus, Fe	0.440	0.066	15	5
Cyprinus carpio, ?	–	–	1.2	6
Esox lucius, ?	–	–	1.4	6
Oncorhynchus mykiss, H	10.0–10.4	0.248–0.282	2.5–2.7	7
O. mykiss, H	24.4	0.482	2.0	7
O, mykiss, H	33.8	1.926	5.7	7

*Fe, dry weight was determined at fertilization; H, at hatching.
†Sources: 1, Howell (1980); 2, Quantz (1985); 3, Eldridge *et al.* (1982); 4, Rogers and Westin (1981); 5, Blaxter and Hempel (1966); 6, König and Grossfeld (1913); 7, Kamler and Kato (1983).
‡Ash-free.

Table 4.11 Caloric value (J per mg dry matter) of initial eggs, yolk and larval tissues at hatching (H, dissected) or at the completion of yolk resorption (Re) in seven fish species. Four aquatic invertebrates, are shown for comparison

Species	Egg	Yolk	Larval body	Source*
Fish				
Hemiramphus sajori	23.0	–	21.8 Re	1
Lepomis macrochirus	24.4	–	21.3 Re+3d	2
Cyprinus carpio	25.2	–	19.9 Re	3
Micropterus salmoides	25.1	–	20.8 Re	4
Oncorhynchus mykiss	29.5†	–	26.5 Re†	5
O. mykiss	27.6	27.9	24.3 H	6
Salmo trutta, 1983	–	27.2	23.4 H	7
S. trutta, 1984	–	27.2	20.9 H	7
Scophthalmus maximus	–	23.0‡	15.2‡ H–Re	8
Invertebrates				
Tubifex tubifex (Oligochaeta)	21.0	–	18.1	9
Streptocephalus torvicornis (Crustacea)	24.3	–	27.2	10
Crepidula fornicata (Gastropoda)	25.7	–	18.7	11
Homarus gammarus (Crustacea)	25.8	–	18.9	12

*Sources: 1, Kimata (1982); 2, Toetz (1966); 3, Kamler (1976); 4, Laurence (1969); 5, Suyama and Ogino (1958); 6, Kamler and Kato (1983); 7, Raciborski (1987); 8, Quantz (1985); 9, Kosiorek (1979); 10, Khmeleva *et al.* (1981); 11, Pandian (1969); 12, Pandian (1970).
†Recomputed from chemical composition.
‡Ash-free dry matter.

at hatching (Kamler and Kato, 1983). From Equations 4.12, 4.13 and 4.32, the balance of egg energy for the endogenous feeding period is summed as:

$$C.e._0 = G.d. + e.c. + P_c + R_c + U_c + Y.r. \tag{4.33}$$

The sum of energy retrieved in the germinal disc ($G.d.$), egg capsules ($e.c.$), and body tissues (P_c), and that metabolized (R_c) and excreted (U_c) until a given moment, plus the energy remaining in the yolk ($Y.r.$), should equal the initial energy of the egg ($C.e._0$). This is illustrated in Table 4.12 for key developmental events of *Salmo trutta* embryos and unfed larvae. The sum of energy retrieved using gravimetric, calorimetric, respirometric and chemical methods was close to the initial amount of energy in the egg. For *Oncorhynchus mykiss* at hatch the sum retrieved was 91–104% of the initial energy in 15 egg size–incubation temperature combinations (Kamler and Kato, 1983). Examples of the time course of egg resource partitioning are given by Blaxter and Hempel (1963), Marr (1966), Heming (1982), and Raciborski (1987); further examples are depicted in Fig. 4.19.

Table 4.12 Egg energy partitioning among different components of the energy balance (J indiv.$^{-1}$) for consecutive developmental stages of *Salmo trutta*. The energy contained in the initial egg ($C.e._0$) is compared with the sum retrieved. No food was given to larvae. Further explanations in the text. Recomputed from table VIA (series 1983) in Raciborski (1987)

Stage	$C.e._0$	$\approx$	$G.d.$	+	$e.c.$	+	P_c	+	R_c	+	U_c	+	$Y.r.$	=	Sum
Fe	712	$\approx$	2.0	+	15.1	+	0	+	0	+	0	+	715.0	=	732
E	712	$\approx$	2.0	+	15.1	+	11.0	+	0.4	+	0.2	+	674.7	=	703
H	712	$\approx$	2.0	+	15.1	+	76.0	+	11.8	+	2.1	+	598.6	=	706
S	712	$\approx$	2.0	+	15.1	+	213.0	+	68.7	+	8.8	+	395.3	=	703
Re	712	$\approx$	2.0	+	15.1	+	449.6	+	236.6	+	16.9	+	0	=	720

Mean sum retrieved   713
95% conf. lim. 698–728

Partitioning of consumed energy or matter

The basic equation for an energy budget (Equation 2.1, page 7), when adapted for application to the endogenous feeding period in which no defecation occurs, has the form (Equation 4.12) $C_Y = P + R + U$, where C_Y is energy consumed from yolk, and P, R and U are production, metabolism and nonfaecal excretion, respectively, all expressed in terms of energy. Assimilated energy, A (e.g. Ricker, 1968) (=metabolizable energy, e.g. Brett and Groves, 1979) is defined by:

$$A = P + R \qquad (4.34)$$

$$A = C_Y - U \qquad (4.35)$$

Both energy and matter budgets can be expressed in the instantaneous form (C_{Yi}, P_i, R_i, U_i and A_i, in J or mg per unit time) or in the cumulative form (C_{Yc}, P_c, R_c, U_c and A_c, in J or mg per life period from fertilization – see Equation 2.1). A cumulative energy budget for the endogenous feeding period is shown in Fig. 4.20. A comparison between the instantaneous and cumulative yolk consumption (C_{Yi} and C_{Yc}) and body production (P_i and P_c) is given in Fig. 4.21.

Transformation efficiencies

Egg energy or matter transformation efficiencies

The efficiency (in %) with which the original egg ($C.e._0$) is converted into larval body tissues (P_c) in the absence of external food is defined by:

$$K_e = P_c \times 100 / C.e._0 \qquad (4.36)$$

The components P_c and $C.e._0$ are expressed in terms of energy or a chemical

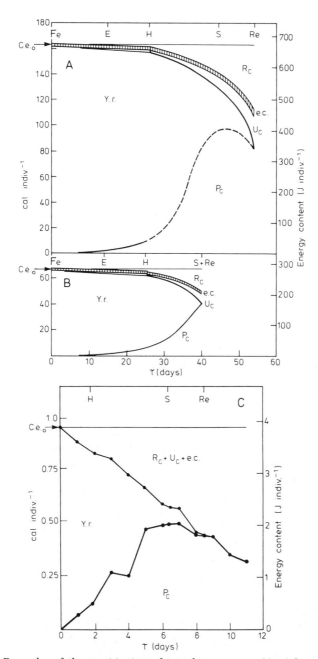

Fig. 4.19 Examples of the partitioning of initial egg energy ($C.e._0$) between tissue growth (P_c), metabolism (R_c), energy excreted (U_c), energy discarded with egg capsules (*e.c.*) and energy remaining in yolk (*Y.r.*) in endogenously feeding fish embryos and unfed larvae. (A) and (B) *Oncorhynchus mykiss* at 12 °C, eggs derived from (A) a three-year-old female and (B) a two-year-old female (reproduced with permission from Kamler and Kato, 1983). (C) *Lepomis macrochirus* at 23.5 °C (recomputed from Table 7 in Toetz 1966). Stages Fe to Re as in Fig. 4.1.

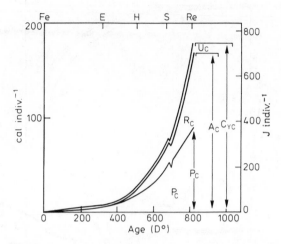

Fig. 4.20 Cumulative energy budget during the endogenous feeding period, calculated for embryos and unfed larvae of *Salmo trutta*. The cumulative consumption of yolk was computed as: $C_{Yc} = C.e._0 - e.c. - G.d. - Y.r.$ Energy in production (*P*), metabolism (*R*) and nitrogenous excreta (*U*) was cumulated from fertilization. For further explanation see Equations 4.12, 4.13 and 4.33; stages Fe to Re as in Fig. 4.1. (Reproduced with permission from Raciborski, 1987; data series 1984.)

compound (synthesis of one compound from another being neglected), or else in terms of dry weight. The last is only partly legitimate because the compositions of larval and egg dry matter are not equivalent (Table 4.11) and are variable (Chapter 3). The egg transformation efficiency can be used only for comparisons between endogenously feeding animals.

Oviparous fishes transform *c.* 44% of their initial egg energy into larval tissues (Tables 4.13 and 4.14). This efficiency is slightly lower (the difference is not statistically significant) than that for invertebrates (*c.* 50% of energy) because the latter save energy by passing all their yolk-dependent life inside egg capsules. A large body of information exists on oviparous fish K_e values computed in terms of dry matter (e.g. Heming, 1982; Mann and Mills, 1985; Docker *et al.*, 1986; review: Blaxter, 1969). The overall level is *c.* 60%, which is higher than that for energy because the weight-based efficiency does not allow for the lower energy value of the larval body. In viviparous fishes the dry weight of newborn fishes is higher and depends upon the composition of ovarian fluid; it amounts on average to 73% of the original egg (6 spp.), 259% (3 spp.) and 1728% (*Trygon violacea*) in fishes containing 1–3%, 5–9% and 13% organic matter in the fluid, respectively (recomputed from Grodziński, 1961). In a viviparous species, *Sebastes schlegeli*, the new-born fishes contained 93% of the energy initially stored in the egg. Ovarian fluid is absorbed in the embryonic hindgut; nutritional substances are provided from resorption of unfertilized ova (Boehlert *et al.*, 1986).

Eggs of fishes and aquatic invertebrates are non-cleidoic eggs (Needham,

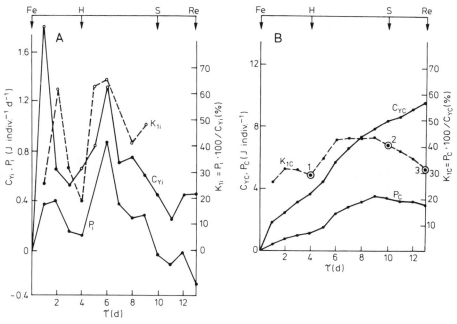

Fig. 4.21 Comparison between (A) instantaneous and (B) cumulative yolk consumption (C_{Yi} and C_{Yc}) and body production (P_i and P_c), respectively, in *Micropterus salmoides*. The instantaneous and cumulative gross efficiencies (K_{1i} and K_{1c}, respectively) are shown (scales on right). Note: K_{1i} cannot be computed for days 10–13 because of negative values of P_i. The cumulative efficiencies for intervals (1) Fe to H, (2) Fe to S, and (3) Fe to Re are ringed; cf. no. 10 in Table 4.15. (Recomputed from table 3 in Laurence, 1969.)

1950) which, unlike those of terrestrial insects, birds and reptiles, have to obtain at least water (Hayes, 1949; Dziekońska, 1958; Kamler, 1976; Lapin and Matsuk, 1979; Kimata, 1982; Section 3.3) and mineral salts (Table 4.14) from the environment or parent. In contrast, the amount of protein, lipids and carbohydrates per individual decreases during endogenous feeding (Fig. 4.12). Marine species retain more protein for growth (s in Table 4.14; see also Vetter *et al.*, 1983) than freshwater (f) species, because less protein is oxidized in the former for metabolic purposes. Two ecophysiological advantages for marine and – especially – terrestrial embryos in oxidizing lipids as the main energy source (Needham, 1950; Pandian, 1969, 1970) are that lipid oxidation yields a greater amount of energy and water, and that it does not produce nitrogen metabolites whose removal requires considerable quantities of water.

Consumed yolk energy or matter transformation efficiencies

The efficiencies of consumed yolk utilizaton are more universal: they can be used for comparison with both yolk-feeding and exogenously feeding ani-

Table 4.13 Efficiencies of initial egg transformation into larval tissue, $K_e = P_c \times 100 / C.e._0$ (%), from fertilization to final yolk absorption without external food. Efficiencies computed in terms of energy for five oviparous fish species are compared with efficiencies computed for five aquatic invertebrates

No.	Species	Variable	Temp. (°C)	K_e (%)	Source[*]
1.	*Tautoga onitis*	t, 16 °C	16	36.3	1
	T. onitis	t, 19 °C	19	25.5	1
	T. onitis	t, 22 °C	22	25.8	1
2.	*Cyprinus carpio*	t, 21 °C	21 ± 2	50.0	2
	C. carpio	t, 23 °C	23 ± 2	38.1	2
3.	*Micropterus salmoides*	–	18–21	44.8	3
4.	*Lepomis macrochirus*	–	23.5	45.4	4
5.	*Oncorhynchus mykiss*	–	10	52.3	5
6.	*Oncorhynchus mykiss*	$C.e._0$ 280 J egg^{-1}	12	60.5	6
	O. mykiss	$C.e._0$ 285 J egg^{-1}	12	58.0	6
	O. mykiss	$C.e._0$ 681 J egg^{-1}	12	50.0	6
	O. mykiss	$C.e._0$ 900 J egg^{-1}	12	32.0	6
	O. mykiss	4 egg sizes × 4 temp.		28.5–72.0	6
K_e energy, 5 fish spp., period Fe–Re				44 ± 7‡	1–6
K_e energy, 5 invertebrate spp., period Fe to (H=Re)†				50 ± 10‡	7–11

[*] Sources: 1, recomputed from Laurence (1973); 2, recomputed from Kamler (1976); 3, Laurence (1969); 4, recomputed from Toetz (1966); 5, recomputed from Smith (1957); 6, Kamler and Kato (1983); 7, Khmeleva *et al.* (1981); 8, Roshchin and Mazlev (1978); 9, recomputed from Kosiorek (1979); 10, Pandian (1969); 11, Pandian (1970).
† This period is appropriate because the invertebrates resorb their yolk before hatching.
‡ Mean ±95% confidence intervals.

mals. For energy they are expressed as percentages. The yolk assimilation efficiency is given by:

$$K_A = A \times 100 / C_Y = (P + R) \times 100 / (P + R + U) \qquad (4.37)$$

The efficiency of consumed energy utilization for growth (the gross conversion efficiency) is given by:

$$K_1 = P \times 100 / C_Y = P \times 100 / (P + R + U) \qquad (4.38)$$

The efficiency of assimilated energy utilization for growth (the net conversion efficiency) is given by:

$$K_2 = P \times 100 / A = P \times 100 / (P + R) \qquad (4.39)$$

Hence, any efficiency can be computed from the remaining two, e.g. $K_2 = K_1 \times 100 / K_A$. The instantaneous efficiencies (K_{Ai}, K_{1i}, K_{2i}) are computed from parameters measured for a moment in time; K_{1i} and K_{2i} cannot

Table 4.14 Comparison of egg energy and matter transformation into larval tissue, $K_e = P_C \times 100 / C.e._0$ (%), in unfed fishes and aquatic invertebrates developing in fresh waters (f) or in the sea (s)

No.	Species	Temp. (°C)	Energy	Protein	Lipids	Carbohy-drates	Ash	Source*
	Fishes, Fe–Re							
1.	*Cyprinus carpio*, f	23 ± 2	38.1	38.1	35.0	17.5	135.0	1†
2.	*Oncorhynchus mykiss*, f	–	44.2	43.1	45.9	–	102.5	2†
3.	*Lepomis macrochirus*, f	23.5	45.4	58.6	–	–	–	3†
4.	*Hemiramphus sajori*, s	19–21	55.9	66.0	41.2	34.9	111.6	4†
	Aquatic invertebrates, Fe to (H=Re)‡							
5.	*Tubifex tubifex*, f	24	46.8	55.9	54.5	16.8	455.0	5†
6.	*Crepidula fornicata*, s	14	61.0	85.1	50.7	63.6	966.7	6
7.	*Homarus gammarus*, s	16 ± 1	60.1	87.7	47.4	73.2	504.3	7

* Sources: 1, Kamler (1976); 2, Suyama and Ogino (1958); 3, Toetz (1966); 4, Kimata (1982); 5, Kosiorek (1979); 6, Pandian (1969); 7, Pandian (1970).
†Values are recomputed.
‡ This period is appropriate because the invertebrates resorb their yolk before hatching.

be computed for negative growth. Cumulative efficiencies (K_{Ac}, K_{1c}, K_{2c}) are computed from parameters cumulated for a period of life (Section 2.2). A comparison between K_{1i} and K_{1c} is given in Fig. 4.21. For a nitrogen budget, $R = U$ (discussion: Fischer, 1976).

Yolk assimilation efficiency, calculated in terms of cumulative energy (K_{Ac}), ranged between 62% and 99% in embryos of *Oncorhynchus mykiss* (stages Fe to H) at 9, 10, 12 and 14 °C (Kamler and Kato, 1983) and between 70% and 98% in embryos of *Salmo trutta* (stages Fe to H) at c. 4 °C (Raciborski, 1987). The presence of high efficiencies in embryos is not surprising because there is no obvious defecation prior to external feeding. However, embryonic assimilation deviates clearly from 100%, therefore the assumption that in embryos $A = C$ (Ryzhkov, 1976; Calow, 1977) oversimplifies the problem.

About half of the yolk energy input is converted to tissues during the yolk-feeding period (Table 4.15, Fig. 4.22). The maximum values of endogenous K_{1c} approach the best possible efficiencies (70–80%, Calow, 1977); the overall level is higher than that for exogenously feeding animals (Table 4.15). High gross efficiencies are typical of the endogenous feeding period, in which no energy is lost with faeces and active metabolism (which reduces efficiency) is low, especially during early development.

Fish embryos developing within egg capsules convert the assimilated energy into their bodies with a high efficiency, K_{2c} c. 84% (Table 4.16, Fig. 4.22). The net conversion efficiency computed for a complete endogenous

Table 4.15 Cumulative gross efficiencies computed in terms of energy as $K_{1c} = P_c \times 100/C_{Yc}$ (%) in yolk-feeding, unfed fishes developing in fresh waters (f) or in the sea (s), compared with exogenously feeding animals

No.	Interval	Species	Variable	Temp. (°C)	K_{1c} (%)	Source*
1.	Fe–Re	*Salmo salar*, f	–	10	41.2†	1
2.	H–S	*S. salar*, f	flat surface	6.3	33‡	2
	H–S	*S. salar*, f	astro-turf	6.3	56‡	2
3.	H–S	*Salmo trutta*, f	flat surface	7.9	38‡	3
	H–S	*S. trutta*, f	astro-turf	7.9	56‡	3
4.	Fe–H	*Salmo trutta*, f	–	c. 3–7	65	4
	Fe–S	*S. trutta*, f	–	c. 3–8	59–62	4
	Fe–Re	*S. trutta*, f	–	c. 3–11	55–62	4
5.	Fe–H	*Oncorhynchus mykiss*, f	–	ambient	43.4‡	5
	Fe–S	*O. mykiss*, f	–	ambient	52.9‡	5
6.	Fe–H	*Oncorhynchus mykiss*, f	4 egg sizes × 4 temp.	9, 10, 12, 14	42–77	6
7.	Fe–H	*Morone saxatilis*, f	–	18	53.9	7
	Fe–S	*M. saxatilis*, f	–	18	56.2	7
8.	Fe–H	*Morone saxatilis*, f	–	18	37.7	8
	Fe–S	*M. saxatilis*, f	–	18	43.8	8
9.	Fe–H	*Lepomis macrochirus*, f	–	25.3	52.2‡	9
	Fe–S	*L. macrochirus*, f	–	25.3	56.7‡	9
	Fe–Re	*L. macrochirus*, f	–	25.3	45.4‡	9
10.	Fe–H	*Micropterus salmoides*, f	–	18–21	29.1‡	10
	Fe–S	*M. salmoides*, f	–	18–21	40.7‡	10
	Fe–Re	*M. salmoides*, f.	–	18–21	31.6‡	10

11.	Fe–H	*Scophthalmus maximus*, s	salinity 17‰	15	39.6‡	11
	Fe–S	*S. maximus*, s	salinity 17‰	15	44.0‡	11
	Fe–Re	*S. maximus*, s	salinity 17‰	15	32.7‡	11
	Fe–H	*S. maximus*, s	salinity 32‰	15	45.4‡	11
	Fe–S	*S. maximus*, s	salinity 32‰	15	53.9‡	11
	Fe–Re	*S. maximus*, s	salinity 32‰	15	47.5‡	11
12.	Fe–H	*Limanda ferruginea*, s	temperature	8	54.4‡	12
	H–Re	*L. ferruginea*, s	temperature	8	32.4‡	12
	F–H	*L. ferruginea*, s	temperature	12	74.6‡	12
	H–Re	*L. ferruginea*, s	temperature	12	40.2‡	12
13.	Fe–H	*Clupea harengus pallasi*, s	–	–	74.4	13
	Fe–Re	*C. harengus pallasi*, s	–	–	43.9	13

Yolk-feeding fish, mean $\pm$ 95% conf. interval[a] — 48.3 $\pm$ 4.1

K_{Ii} for exogenously feeding animals:

Carnivore, young well-fed fish, generalized equation		20	14
Carnivore, young well-fed fish, generalized equation		29‡	15
10 fish species having different food habits,	10 g§	c. 22–60	16
10 fish species having different food habits,	200 g§	c. 12–18	16
Aquatic consumers		13–35	17

*Sources: 1, Hayes (1949); 2, Hansen and Møller (1985); 3, Hansen (1985); 4, Raciborski (1987); 5, Ryzhkov (1976); 6, Kamler and Kato (1983); 7, Eldridge *et al.* (1981b); 8, Eldridge *et al.* (1982); 9, Toetz (1966); 10, Laurence (1969); 11, Quantz (1985); 12, Howell (1980); 13, Eldridge *et al.* (1977); 14, Winberg (1956); 15, Brett and Groves (1979); 16, Miura *et al.* (1976); 17, Welch (1968).

†Computed as: $P_c \times 100/(P_c + R_c + U_c)(\%)$.

‡Recomputed.

§Mean weight.

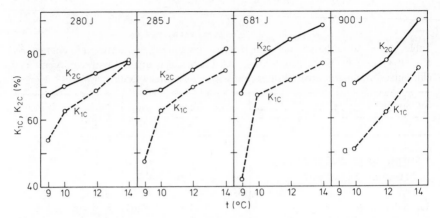

Fig. 4.22 Effect of temperature (t, °C) and initial egg size, ranging from 280 to 900 J egg^{-1}, on the cumulative gross and net energy efficiencies: $K_{1c} = P_c 100 / C_{Yc}$ (%) and $K_{2c} = P_c 100 / A_c$ (%) in *Oncorhynchus mykiss* embryonic development (stages Fe to H). a, mass mortality at 9 °C before hatching. (Reproduced with permission from Kamler and Kato, 1983.)

feeding period (Fe to Re) is lower (55%) because metabolic output increases after hatching. Invertebrates reduce their costs by remaining within egg capsules until resorption is complete, then invest the saved energy in growth: their net conversion efficiency (72%) exceeds that of fishes (55%, Fe to Re). The lower efficiencies reported for young carnivorous fishes in experiments, as compared with yolk-sac fishes, are attributable to feeding behaviour and possible higher costs of food processing (increased R_f component of metabolism, Equation 4.26) as well as to higher activity metabolism (R_a) in the carnivores. The lowest K_2 efficiency (9.8%) is reported for natural populations, where more energy is required to capture prey and avoid predators. Fig. 4.22 shows again that the equality $K_1 = K_2$ (i.e. $C = A$) cannot be applied to fish embryos.

No effect of egg size on efficiency can be shown at the interspecific level (Tables 4.13 to 4.16). However, intraspecific comparisons (*Clupea harengus*, Blaxter and Hempel, 1963; *Oncorhynchus mykiss*, Fig. 3.15(B) and Table 4.13) showed lower efficiencies in larvae from larger eggs. Larger larval size resulting from larger eggs (Section 3.10), their longer development (Mann and Mills, 1985), and values of b ⩾ 1 in Equation 4.27 all may contribute to this difference. However, egg size does not always determine efficiency of yolk utilization (Beacham *et al.*, 1985).

Because of the greater metabolic expenditures of larger embryos or larvae, the efficiencies fall as development proceeds (Tables 4.15 and 4.16; Gray, 1928a; Needham, 1931; Ivlev, 1939b; Smith, 1957; Lasker, 1962; Hansen and Møller, 1985). Salmonid larvae reared on an artificial, corrugated substratum converted yolk to body tissues more efficiently than the more active larvae reared on the usual flat surface (2 and 3 in Table 4.15). Marine

Table 4.16 Cumulative net efficiencies, computed in terms of energy as $K_{2c} = P_c \times 100 / A_c$ (%), in yolk-feeding unfed fishes as compared with embryonic net efficiencies in aquatic invertebrates and with exogenously feeding fishes

No. Species	Variable	Temp. (°C)	K_{2c} (%)	Mean (%)	Source*
Fishes, Fe to H					
1. *Micropterus salmoides*	–	18–21	94.0†		1
2. *Oncorhynchus mykiss*	4 egg sizes × 4 temp.	9–14	67–90	84‡	2
3. *Salmo trutta*	–	*c.* 3–7	85.0		3
Fishes, Fe to Re					
4. *Micropterus salmoides*	–	18–21	78.0†		1
5. *Oncorhynchus mykiss*	–	10	55.9†		4
6. *Salmo trutta*	–	*c.* 3–11	55–70		3
7. *Cyprinus carpio*	temperature	21 ± 2	62.6†,[a]	55§	5
C. carpio	temperature	23 ± 2	51.8†,[b]		5
8. *Tautoga onitis*	temperature	16	28.5†		6
T. onitis	temperature	22	35.9†		6
Yolk-feeding aquatic invertebrates, Fe to (H = Re)			*c.* 72[c]		7–11
K_{2i} for exogenously feeding animals:					
Carnivore, young well-fed fish, generalized equation			25		12
Carnivore, young well-fed fish, generalized equation			40†		13
Huso huso and *Acipenser ruthenus* juveniles, < 350 g			18–59		14
Fish and social insect natural populations			9.8 ± 1.8[d]		15

*Sources: 1, Laurence (1969); 2, Kamler and Kato (1983); 3, Raciborski (1987); 4, Smith (1957); 5, Kamler (1976); 6, Laurence (1973); 7, Pandian (1969); 8, Pandian (1970); 9, Galkovskaya *et al.* (1976); 10, Roshchin and Mazelev (1978); 11, Khmeleva *et al.* (1981); 12, Winberg (1956); 13, Brett and Groves (1979); 14, Gershanovich *et al.* (1987); 15, Humphreys (1979).
† Recomputed.
‡ Mean for nos 1–3.
§ Mean for nos 4–8.
[a] 66.6% if computed from $P_c \times 100 / (C.e_{,0} - e.c. - U_c)$.
[b] 50.0% if computed from the alternative equation.
[c] 10 original values from five species.
[d] Mean ± 95% conf. intervals.

and freshwater species do not differ in the general level of efficiency (Tables 4.13 to 4.16), but in brackish water the efficiency of yolk utilization by larvae of *Scophthalmus maximus* was lower than in seawater (11 in Table 4.15) in which they floated passively. The increase of metabolic rate with temperature and a reduction of developmental time are the factors contributing to changes in the efficiencies at different temperatures (Sections 4.1, 4.3 and 4.4). Efficiencies are seen either to increase with increasing temperature (12 in Table 4.15, 8 in Table 4.16, Fig. 4.22; Jones, 1972; Howell, 1980), or to decrease (1 and 2 in Table 4.13, 7 in Table 4.16; Heming, 1982). The highest efficiencies at the optimum temperatures were shown by Ryzhkov (1976) for salmonid larvae, by Elliott (1976) and Brett and Groves (1979) for postlarval fish, and by Winberg (1987) for poikilothermic animals. Winberg tentatively explained the decrease of K_2 from the optimum towards lower temperatures by different types of dependence of metabolic and developmental rates on temperature.

Hence, the variation in efficiencies can be attributed to competition between growth and metabolism for yolk energy. Increased metabolism will inevitably result in a relative decrease in growth, and hence in efficiency. Docker *et al.* (1986) found a significant negative correlation between metabolism and yolk efficiency in *Cottus bairdi* prior to external feeding. In order of decreasing sensitivity to changes of metabolic output, the efficiencies can be listed as follows: K_2, K_1, K_A and K_e. For example, the parallel decrease of both K_{2c} and K_{1c} in embryos of *Oncorhynchus mykiss* at temperatures decreasing from 14 to 10 °C (Fig. 4.22) can be explained by the increase of metabolism cumulated for the embryonic period (Fig. 4.17). Between 10 and 9 °C the net efficiency (K_{2c}) continued to decrease at a similar rate (Fig. 4.22), whereas the gross efficiency (K_{1c}) was strongly reduced at 9 °C. That, at least partly, resulted from an increase of cumulative excreted energy (U_c), which was 0.2–1.6% of the initial egg energy upon incubation at 14–10 °C ($n = 12$), but as much as 3.1–5.6% at 9 °C ($n = 3$). The lower efficiency at 9 °C was accompanied by decreased survival. Docker *et al.* (1986) reported a significant positive correlation between survival and efficiency of yolk utilization in *Cottus bairdi*.

Ehrlich and Muszynski (1981) described an energy-saving behavioural mechanism in *Leuresthes tenuis*. Yolk-sac larvae selected a temperature of 25 °C, which was near the upper limit of viable temperatures and well above that for maximum growth efficiency. This behaviour minimizes the duration of the most vulnerable phase. After yolk sac resorption, fed larvae selected lower temperatures, *c.* 23 °C, which were close to the upper limit of the maximum growth efficiency range. In contrast, starved larvae selected 18 °C, the coldest temperature within the maximum growth efficiency range.

In summary, Chapters 3 and 4 synthesize studies on fish egg properties and utilization. Such data may be of some assistance for aquaculture of commercial species as well as in population and recruitment studies. Aquaculturists desire to know not only what the optimum conditions for incubation are, but also why they are the best. For the latter group of problems, one example will be given. Fish of the 0 to 1 age class are the most productive and may contribute at least 50% of the total productivity of all age classes (Mahon *et al.*, 1979). The proper estimation of production for the first year requires a reliable determination of the initial number and biomass. The latter was in some studies calculated from egg diameter. The data presented in Chapters 3 and 4 suggest that that method is inappropriate.

Chapter five

Mixed feeding period

5.1 GENERAL REMARKS

A **mixed feeding period** can be defined as a time when yolk reserves are reduced and the larval fish must commence external feeding. Mixed feeding usually occurs between the initiation of external feeding and the complete resorption of the yolk sac.

Some species hatch in an advanced state with respect to feeding ability, with a functional mouth and the alimentary tract well developed. They start feeding shortly after hatching. For example in the marine species *Gadus morhua* and *Melanogrammus aeglefinus* (Laurence, 1978), *Ammodytes personatus* (Yamashita and Aoyama, 1985) and *Scophthalmus maximus* (Quantz, 1985), the period from hatching to first feeding is shorter than that from beginning feeding to final absorption of yolk and/or oil globules; the ratios for these four species are 1:6, 1:6, 1:3.5 and 1:2, respectively. In contrast, larvae of many species need more time to attain feeding ability. For the freshwater fishes *Oncorhynchus mykiss*, *Salmo trutta*, *Lepomis macrochirus* and *Micropterus salmoides* the ratio is *c.* 1:0.5, regardless of considerable differences in egg size (Table 4.3). The jaws of *Sardinops caerulea* are not functional until the final resorption of the yolk sac (Lasker, 1962).

Food was first found in the anterior part of the alimentary tract of carp larvae on the third day after hatching (Matlak and Matlak, 1976). The composition of this first food differed in the larvae from the three ponds studied; *Rotatoria*, *Cladocera*, or small chironomid larvae were the dominant food items, thus reflecting the composition of the food organisms in the particular pond.

In aquaculture the time at which external food is first given to larvae influences their subsequent survival and growth. It is recommended that food be offered to fish larvae when they attain feeding ability, i.e. when they begin to swim and search for food with their yolk sac still unresorbed (e.g. Goryczko, 1968; Jones, 1972; Elliott, 1984; Appelbaum, 1989). The commencement of external feeding requires time: for example, in the presence of *Brachionus plicatilis* it took 3 d during which no food was found in the alimentary tract of *Acanthopagrus cuvieri* larvae having a functional mouth (Hussain *et al.*, 1981). Food was offered to *Salmo trutta* larvae from swimming-up, but larvae did not start to consume it until at least 8 d later; this delay accounts for *c.* 30% of the time between free-swimming and final yolk resorption (Raciborski, 1987). Larvae of *Clupea harengus pallasi* required 4–5 d after first feeding to begin rapid growth (Eldridge *et al.*, 1977). Similar observations have been reported for other fishes by MacCrimmon and Twongo (1980), Heming *et al.* (1982) and Appelbaum (1989). Baranova (1974) administered the first live food to *Cyprinus carpio* larvae on the first and third days after hatching. Although both groups started to capture prey on the third day, larvae exposed earlier to food showed better growth from the fourth day. *Clupea harengus* larvae supplied with a food density of 200 nauplii dm^{-3} started to consume them a day earlier than those offered 30 nauplii dm^{-3} (Pedersen *et al.*, 1987). Hence, the adjustment of larvae to external food has at least two components, a behavioural and a physiological one; the larvae must learn to accept external food and to process it (further information, Section 6.2).

Starved larvae of *Morone saxatilis* retained oil globules for a longer time than fed ones; the rate of oil absorption increased progressively with an increase of food concentration (Eldridge *et al.*, 1981a, 1982). Feeding delayed final yolk absorption in *Salmo trutta* (Raciborski, 1987).

Lack of suitable food often occurs in the field when larvae attain feeding ability. Blaxter and Hempel (1963) put forward the concept of the **point of no return** (PNR), which is a point in time when starved larvae lose their feeding ability, approach neutral buoyancy and became less active. After final yolk resorption, larvae of *Clupea harengus* and *Pleuronectes platessa* reach PNR about half-way between commencement of starvation and death (Blaxter and Ehrlich, 1974). Extreme dependence on early feeding was reported for *Engraulis mordax*, in which PNR was 1.5 d after final yolk resorption (Lasker *et al.*, 1970). In contrast, PNR was not revealed for *Morone saxatilis* (Rogers and Westin, 1981; Eldridge *et al.*, 1981a) even if 40–60% of larvae had already died of starvation: once food was supplied, mortality was arrested and the survivors grew and developed normally. Unfed *Leuresthes tenuis* survivors were able to feed when 80% of larvae had died from starvation (May, 1971). Flexibility in the timing of first feeding gives larvae an advantage under conditions of low food supply. Larvae withstand starvation longer at lower temperatures (Blaxter and Hempel, 1963; Lasker *et al.*, 1970; Jones, 1972; Horoszewicz, 1974; Laurence, 1978; Rogers and Westin, 1981). On the basis of oxygen consumption and depletion of energy reserves measured during starvation of *Coregonus lavaretus* larvae from hatching to 50% mortality, Dąbrowski (1976) predicted 50% survival time at 8.5 °C and extrapolated this prediction for the temperatures 5–17 °C using Winberg's (1956) temperature correction factor. Larvae derived from larger eggs are better able to tolerate delayed feeding (Section 3.10). Studies by Dąbrowski (1981) on *Coregonus pollan* larvae have indicated an earlier PNR resulting from increased utilization of energy reserves prior to hatching. Similarly Rösch (1989) demonstrated an earlier acceptance of a dry diet in larvae of *Coregonus lavaretus* resulting from delayed hatching. Smaller prey organisms delay PNR (Dąbrowski, 1981). A shorter-than-PNR food depreviation period for the first-feeding larvae of *Oncorhynchus mykiss* resulted in a lower final weight, but the length of the starvation period did not influence the subsequent specific growth rate (Escaffre and Bergot, 1985). Coregonid larvae, however, exhibited recovery growth after delayed first feeding (Dąbrowski, *et al.*, 1986b). Older larvae of *Pleuronectes platessa* were able to withstand longer starvation than early ones without reaching their PNR (Wyatt, 1972). Laurence (1978) generalized the problem for first-feeding larvae: '...regardless of temperature and species, the delayed feeding time after yolk absorption which still allows larvae to start successful feeding, seems to be approximately equal to 25% of the time from hatching to complete yolk absorption...'.

The cumulative energy budgets for *Salmo trutta* larvae at final yolk resorption (series 1984 in Raciborski, 1987) were:

unfed larvae, age $821\,D°$:
$$625\,C_{Yc} \approx 367\,P_c + 331\,R_c + 17\,U_c + 0\,F_c$$

fed larvae, age $848.5\,D°$:
$$926\,C_{(Y+d)c} \approx 444\,P_c + 381\,R_c + 19\,U_c + 5\,F_c$$

where C_{Yc} is cumulating consumption of energy from yolk, $C_{(Y+d)c}$ is cumulative consumption of energy from yolk and diet; all parameters were determined directly and expressed in $J\,indiv^{-1}$. The complete cumulative energy budgets are shown in Fig. 4.20 and 5.1, respectively, for unfed and fed larvae. The increased amount of energy consumed by the fed group resulted in an increase of growth (see also Fig. 4.9), respiration (see also Fig. 4.15) and excretion; defecation began. In *Monrone saxatilis*, growth in body length of larvae between the first feeding and complete resorption of oil globules was directly correlated with food density (Eldridge *et al.*, 1981b); the non-assimilated fraction of the energy consumed was low during endogenous feeding and increased after initiation of exogenous feeding (Eldridge *et al.*, 1982). During the mixed feeding period, the small amounts of remaining yolk in the unfed group of *Salmo trutta* were more effectively assimilated (higher K_A) and utilized for growth (higher K_1) than was energy from yolk plus artificial diet in the fed group (Raciborski, 1987).

The duration of the mixed feeding period is species- and temperature-dependent. It is longer in species producing larger eggs (Table 4.3) and in larvae developing at lower temperatures (Fig. 4.2(A)).

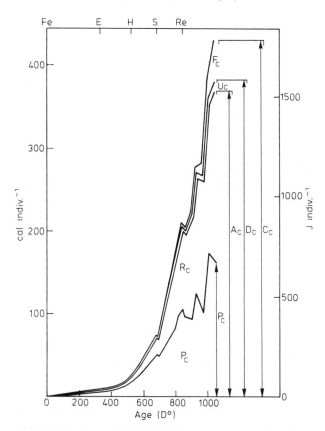

Fig. 5.1 Cumulative energy budget during fish endogenous, mixed and early exogenous feeding periods, illustrated for embryos and fed larvae of *Salmo trutta*. Cumulative consumption (C_c) is the sum of energy in the yolk absorbed and in the formulated diet eaten. Energy in production (P), metabolism (R) and nitrogenous excreta (U) was cumulated from fertilization; the sum of the above components forms digestible energy (D); in endogenous feeding $D = C_Y$ (Equation 4.12) but in exogenous feeding $D = C - F$. Energy lost in faeces (F) was cumulated from their appearance. For further explanation see Fig. 4.20. (Reproduced with permission from Raciborski, 1987; data series 1984.)

5.2 CRITICAL PERIODS IN FISH EARLY LIFE

The time of the transition from endogenous to external feeding is usually recognized as the **critical period**. The idea of critical periods is not supported by some investigators, however (e.g. Marr, 1956; Lindsey and Arnason, 1981).

The main factors which determine embryonic and larval mortality are as follows:

1. endogenous factors
2. exogenous factors
    (a) biotic factors: lack of food when larvae begin external feeding, predation, disease, parasitism
    (b) abiotic factors, e.g. low oxygen content, extremes of temperatures, pH or salinity, toxic substances

The alternative nature of growth and differentiation processes was assumed by Trifonova (1949), Olifan (1949), Privolnev (1953), Ryzhkov (1973, 1976, 1979) and Galkovskaya and Suchchenya (1978). An acceleration of growth occurs at the beginning of any developmental step (Section 4.3; example in Fig. 4.8). The transition from one developmental step to another is, according to these authors, accomplished within a short time in which differentiation is enhanced, growth rate decreases and oxygen consumption rate increases. (N.B.: the recent theory of saltatory ontogeny (Balon, 1986) also supports the concept of a fast transition from one stabilized developmental step to another via a rapid threshold.) The transition periods are those of high vulnerability to external factors and of increased mortality; they have been named the critical periods. The main critical periods in embryonic development are recognized as cleavage, early gastrulation and the beginning of organogenesis. Increased sensitivity and mortality of embryos of salmonids (Hayes, 1949), *Cyprinus carpio* (Tatarko, 1970) and *Coregonus albula* (Dąbrowski *et al.*, 1987) were found at gastrulation and before hatching. Lebedeva (1981) reported fertilization, cleavage, the beginning of gastrulation, tail bud formation, hatching and initiation of external feeding to be the critical periods in Coregonidae. According to Vladimirov (1970), Zhukinskij and Nedyalkov (1980) and Nedyalkov (1981), embryonic defects transmitted from parents or acquired during the sensitive early developmental stages may result in a failure of normal development and, finally, in mortality which manifests itself at critical periods. Leary *et al.* (1985) found a negative relationship between average heterozygosity at enzyme loci and the degree of bilateral asymmetry in *Oncorhynchus mykiss*; Danzmann *et al.* (1986) postulated that a faster developmental rate of heterozygotes would reduce the probability of accidents during a critical period.

The above works consider the critical periods connected primarily with endogenous factors. However, food shortage during mixed feeding has been recognized as the primary cause of mortality among fish larvae (Hjort, 1914; Nikolskij, 1974; Dąbrowski, 1975; Raciborski, 1987); it has been suggested that the year-class strength is then established. Dąbrowski (1975) calculated, using two methods, the minimal zooplankton density necessary to meet the maintenance expenditures of larvae of five freshwater species. He showed that the actual densities of zooplankton in water bodies are often much lower, indicating a key role for food availability during the mixed feeding period. After the mouth opens, considerable larval mortality can also be caused by ingestion of small air bubbles (Hussain *et al.*, 1981). Mass mortality of carp larvae during initiation of external feeding was caused by a pesticide (Kamler *et al.*, 1974) and by low pH (5.0–5.2), at which the swim bladder is prevented from filling with air and the larvae die from hunger (Korwin-Kossakowski, 1986, 1988). Thus, the mixed feeding period is the critical period, when considerable vulnerability to many factors is observed. However, sensitivity to low oxygen attains a maximum just before hatching (review: Doudoroff and Shumway, 1970), and sensitivity to sublethal salinity and ammonia peaks after hatching (Belyi, 1967 and Calamari *et al.*, 1981, respectively).

Toetz (1966) attempted a definition of the critical period generally applicable to all species. He evaluated numerous criteria – such as mortality, disappearance of the yolk sac, feeding and food-seeking movements, metabolism of starved larvae, tissue resorption and substrate catabolism – but no single criterion and no one combination of them proved to be sufficiently universal to define critical periods of all fish larvae.

Therefore, there are at least two controversies about the critical periods. Their existence is denied in some works. It seems, however, that early ontogeny of fishes can be considered as a series of vulnerable periods of which the most important is the initiation of external feeding. Critical periods are recognized in fishery practice (Leitritz and Lewis, 1976). Catastrophic events during critical periods may determine the strength of year classes in natural populations, especially in fishes in which reproduction is synchronized. There are also apparently conflicting opinions about the relative importance of endogenous and exogenous factors. The latter are dominant in natural populations. However, in aquaculture the external conditions are, in principle, controlled at an optimum level, and then the endogenous factors may acquire some importance.

Chapter six

Early exogenous
feeding period

6.1 DEVELOPMENT

General remarks

Balon (1986) proposed three life-history models for fishes: indirect, transitory and direct. **Indirect ontogeny** comprises five periods: embryo, larva, juvenile, adult and senescence. The **transitory** model has an alevin period as a vestige of the larval period. In the **direct** model the free embryo transforms directly into a juvenile. Balon (1986) showed that elimination of the larval period occurs in species producing few, but large, eggs; within such eggs (Section 4.1) incubation is prolonged. Hence, a shift along the r–K continuum (Pianka, 1974) in the K direction would be connected with an elimination of the larval period from fish life history.

As in the great majority of works on fish physiology and bioenergetics, we shall retain in this work the name **larva** for a hatched individual (Section 4.1). A larva differs greatly from an adult fish. A considerable conformity in many aspects can be observed in larvae of different species. Their head is well developed, whereas the rest of the body forms a thin vertical lamella. Gills are poorly developed, the alimentary tract is poorly differentiated and short at first feeding. The larval period ends with metamorphosis, when the axial skeleton is ossified and the undifferentiated median fin fold is no longer apparent. Anatomical and morphological characteristics have reached a state of development, similar to that in adults, except for the reproductive organs.

There is little information on the larval development of many fishes (Fuiman, 1984). Most studies concentrate on commercially important species. Descriptions for many freshwater species, done in a standardized way, were published in the 1950s and 1960s in *Trudy Inst. Morf. Zhivot.*

(*Trudy Instituta morfologii zhivotnykh im. A. N. Severtsova*); they continue to be in use. A few examples of these and other works are given below.

Salmonid larval development has been described by Ryzhkov (1966, 1976); that of *Salmo trutta* by Grudniewski (1961) and Raciborski (1987); *Coregonus pollan* by Dąbrowski (1981); *Cyprinus carpio* by Vasnetsov *et al.* (1957), Balon (1958a), Braginskaya (1960) and Peňáz *et al.* (1983) – a simplified example is given in Fig. 6.1; *Carassius* spp. by Dmitrieva (1957); *Rutilus rutilus* by Vasnetsov *et al.* (1957) and Lange (1960); *Abramis brama* by Vasnetsov *et al.* (1957); *Ctenopharyngodon idella* by Shireman and Smith (1983); *Stizostedion lucioperca* by Vasnetsov *et al.* (1957) and Dmitrieva (1960); *Scophthalmus maximus* and *S. rhombus* by Jones (1972); *Siganus fuscescens* by Kitajima *et al.* (1980); *Acanthopagrus cuvieri* by Hussain *et al.* (1981); and *Lepomis macrochirus* by Toetz (1966). See also Grodziński (1961) and Nikolskij (1974) for general descriptions of larval development.

Factors affecting developmental rate during exogenous feeding

The main factors contributing to the variability in developmental rate during early exogenous feeding are temperature and food.

Obviously, developmental rate increases with increasing temperature. The basic principles were presented in Section 4.1 for the endogenous feeding period. The relationship between larval developmental rate and temperature was shown for *Cyprinus carpio* by Tatarko (1966), Peňáz *et al.* (1983) (Fig. 6.2), Korwin–Kossakowski and Jezierska (1984) and Ivanov (1986). Korwin–Kossakowski and Jezierska (1984) report the relationship between larval development and temperature for *Tinca tinca*, *Coregonus lavaretus* and *C. albula* (review: Zhukinskij 1986). Fluctuating temperatures accelerated the developmental rate of *C. lavaretus* larvae beyond that shown under constant temperatures (Section 4.1); development rate of *Tinca tinca* larvae at the same temperature depended upon the 'thermal history' of spawners (Korwin–Kossakowski and Jezierska, 1984).

Studies of the effect of temperature on larval development are less advanced than those on embryogenesis. Larval development is a more difficult subject for such work, owing to difficulties in partitioning the effect of temperature from the influence of food, which is consumed more intensely at higher temperatures (Herzig and Winkler, 1985; Winberg, 1987; cf. Fig. 4.3(C) and 6.2). Three aspects of the influence of food on larval development will be considered in the following paragraphs, its absence (starvation), availability (concentration), and quality.

In *Coregonus lavaretus*, as shown by Korwin–Kossakowski and Jezierska (1984) and in *Cyprinus carpio* (Kamler *et al.*, 1990), delayed initial feeding prolonged the first larval steps; mass mortality was observed at steps B and C1, respectively, i.e. at the time when the well-fed larvae of both species commenced step D1 (for developmental steps see Fig. 6.1). Carp

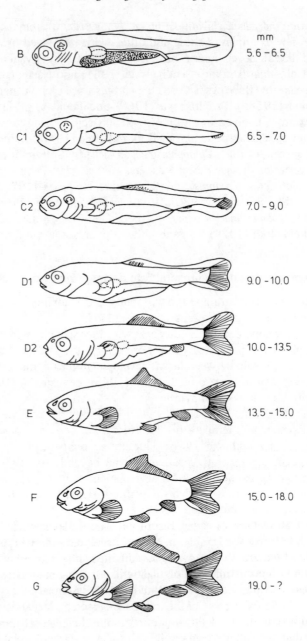

	mm
B	5.6 – 6.5
C1	6.5 - 7.0
C2	7.0 - 9.0
D1	9.0 - 10.0
D2	10.0 - 13.5
E	13.5 - 15.0
F	15.0 - 18.0
G	19.0 - ?

Fig. 6.1 Schematic representation of morphological changes of *Cyprinus carpio* larvae during mixed and exogenous feeding. (B) to (G) Developmental steps; step A, endogenous feeding, is omitted. (Simplified from Vasnetsov *et al.*, 1957 and Bragin- skaya, 1960.) Lengths (mm) are taken from Vasnetsov's original description. A more detailed description of carp development is given by Peňáz *et al.*, 1983.

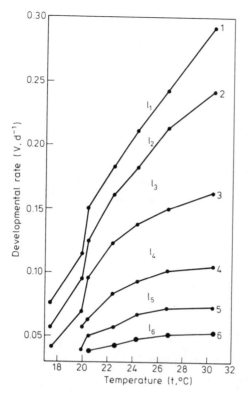

Fig. 6.2 Effect of temperature (t, °C) on the developmental rate $V = \tau^{-1}$, where τ is days from fertilization, in larvae of *Cyprinus carpio*. l_1 to l_6, larval development steps after Peňáz *et al.* (1983); 1–5 are boundaries between the larval steps. 6 is the boundary between the 6th larval step and the juvenile period. The same constant temperatures were maintained throughout the embryonic and larval development (recomputed from table 3 in Peňáz *et al.*, 1983).

larave deprived of food at step C1 showed no development prior to mass mortality (Kamler *et al.*, 1987). In *Morone saxatilis*, delayed initial feeding arrested development (Rogers and Westin, 1981). Morphological and histological changes occurring in the gut and liver during starvation of larvae of *Clupea harengus* and *Pleuronectes platessa* were described by Ehrlich *et al.* (1976).

Using a differential staining technique, Eldridge *et al.* (1981b) showed that the ossification rate in *Morone saxatilis* larvae was directly related to food concentration.

Carp larvae that were fed zooplankton were morphologically more advanced than larvae of the same age fed formulated diets (Fig. 6.3; Ivanov, 1986). Mass mortality of two-week-old larvae fed the inferior diet, St86, occurred at step D1, which had been reached by the zooplankton-fed group

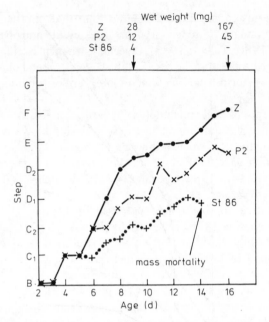

Fig. 6.3 Rate at which developmental steps (B–G, see Fig. 6.1) appear in *Cyprinus carpio* larvae at 26 °C fed natural food – zooplankton (Z) or two formulated diets (P2, St86) from the third day after hatching. Mean wet weights of larvae at the age of 9 and 16 d are shown (based on data in Kamler *et al.*, 1990).

a week earlier. Less suitable food depressed both developmental rate and growth rate (Fig. 6.3; Bergot *et al.*, 1989). The weight of early larvae of *Cyprinus carpio* at a given developmental step was diet-dependent (Kamler *et al.*, 1990). The composition of formulated diets modified the rate of formation of internal organs related to the alimentary tract (salmonid larvae, Ryzhkov, 1976) and influenced the relationship between organ size and body size (carp larvae, Szlamińska, 1984; *Coregonus shinzi palea*, Durante, 1986).

Thus, it is obvious that high temperature and suitable food accelerate differentiation of both morphological structures and internal organs of fish larvae, enabling them to pass the vulnerable period of life more rapidly. However, a general theory of the combined temperature-and-food effect on larval development remains to be developed.

6.2 FEEDING

General remarks

During the exogenous feeding period, food is orally ingested and digested in the alimentary tract. The exact measurement of amounts of food consumed by poikilotherms causes more difficulties than that of other components of the energy budget (Fischer, 1970a; Klekowski and Duncan, 1975b; Striganova, 1980; Talbot, 1985), especially when it is difficult to separate non-consumed food from faeces (Knights, 1985), metabolites, detritus etc. This is the case for fish larvae, which moreover consume food of very small particle size.

The daily rate of food consumption by postlarval fish was often estimated from their energy requirements or the relationship between food consumption and growth rate; since Winberg's (1956) monograph, this approach has been much developed (reviews: Majkowski and Waiwood, 1981; Majkowski and Hearn, 1984). Baranova (1974) applied Winberg's (1956) method to carp larvae.

An alternative approach (first proposed by Bajkov, 1935, modified by Elliott and Persson, 1978, reviewed by Talbot, 1985, and further reconsidered by Jobling, 1986a) is to evaluate the food intake from both the amount of food in the stomach (which is assessed in field samples or in laboratory-reared animals) and the time needed for gastric evacuation (Table 6.1). The original method was applied by Eldridge *et al.* (1981b) to larvae of *Morone saxatilis* fed *Artemia* nauplii. Non-assimilable markers, e.g. chromic oxide, powdered platinum, or metallic iron powder (combined with a radiographic method – Talbot and Higgins, 1983), or coloured food may

Table 6.1 Time required for food to pass through the gut of fish larvae

Species	Wet weight (mg)	Temp. (°C)	Food	Evacuation time (h)	Source*
Cyprinus carpio	1.45	24	Formulated diet	1[a]	1
C. carpio	4.22	24	Formulated diet	3–5[a]	1
C. carpio	20–30	25	Formulated diet	4.6[b]	2
C. carpio	2	25	Artemia nauplii	7[b]	2
Morone saxatilis	c. 0.2–20[c]	18	Artemia nauplii	2.3–4.5[d]	3
Ctenopharyngodon idella	9	28–33	Zooplankton	5.7–7.3[b]	2
Oncorhynchus kisutch	300	10	Formulated diet	50–60[b]	2
13 species (larvae)	–	–	–	8.6[e]	4
Postlarval carnivores	–	22–25	–	22.0[e]	4
Postlarval herbivores	–	22–25	–	6.0[e]	4

*Sources: 1, Szlamińska (1987c); 2, Yarzhombek (1986); 3, Eldridge *et al.* (1982); 4, Fänge and Grove (1979).
[a]50% evacuation.
[b]100% evacuation.
[c]Recomputed from dry weight; wet weight assumed to contain 15% dry weight.
[d]Time required for food to pass through the alimentary canal during continuous feeding.
[e]Mean gastric evacuation time.

be used in studies of food consumption, evacuation and digestion. The radioactive isotope ^{14}C was used by Panov and Sorokin (1967) to measure daily food consumption and other parameters of the energy budget (*A*, *P*, *R* and *FU*) in *Abramis brama* larvae of 15–49 mg wet weight fed different concentrations of *Bosmina*.

Another alternative is direct measurement of the difference between the amount of food offered and that remaining at the end of the feeding period. The advantage is that fish do not need to be starved or killed. Measuring the decline in number of prey is useful when studying early larvae feeding on rotifers which are too small to be separated from water or fish faeces. The difference method was used to estimate the number of *Brachionus plicatilis* consumed by *Clupea harengus pallasi* (Eldridge *et al.*, 1977) and to assess the consumption of mixed zooplankton by *Catostomus commersoni*, *Notropis cornutus* and *Poecilia reticulata* of 1–2 mg dry weight (*c.* 7–14 mg wet weight, Borgmann and Ralph, 1985). The difference between prey counts was also applied to *Cyprinus carpio* larvae (*c.* 2 mg wet weight) fed *Brachionus rubens* (Kamler *et al.*, 1986); corrections for energy losses from starved rotifers, owing to reproduction and metabolism, were proposed. Without these corrections the results would be over-estimated, to about twice the true values. Evaluation of the consumption of larger zooplankton (cladocerans) can be done by measurement of changes in prey dry weight; corrections have to be made for prey mortality (Kamler *et al.*, 1986). Daily consumption of formulated diet by fish larvae can be estimated by measuring changes in the dry weight of the diet. Corrections are made for losses of the diet in water (Section 7.2) (Fischer and Lipka, 1983). If no correction were made, diet consumption by carp larvae would be over-estimated by 60–70% (Kamler *et al.*, 1986).

Turning to gastric evacuation time (Table 6.1), numerous studies on postlarval fishes have shown that it is shortened by high temperature (review: Elliott and Persson, 1978). The relative gastric evacuation rate (mg g^{-1} h^{-1}) is significantly related to feeding rate by a power-law function ($r = 0.81$, 52 data points from 22 species, Pandian and Vivekanandan, 1985). The absolute rate of gastric evacuation (mg h^{-1}) is affected by diet composition and meal size (Jobling, 1986a). In carnivores, dry-pellet for-mulated feeds are emptied from the stomach more rapidly than natural prey organisms (Jobling, 1986b). Gastric retention time is shorter in post-larval herbivores than in carnivores (Table 6.1). The time needed for gastric evacuation is positively related to the consumer's body size (data for carp larvae: Kaushik and Dąbrowski, 1983; Szlamińska, 1987b). Fish larvae, most of which are predatory, empty their alimentary canals in a short time, similar to that of postlarval herbivores rather than carnivores (Table 6.1).

The alimentary tract is short and poorly differentiated in first-feeding larvae, especially in species producing small eggs. Its increase in relative length with age (Fig. 6.4; Stroband and Dąbrowski, 1981; Iwata, 1986) is paralleled by the development of intestinal foldings, mucosal volume and formation of the stomach in stomach-possessing species (Dąbrowski, 1986a; Segner *et al.*, 1989). Increase of gut length extends the duration of food exposure to proteases. Proteases of early larvae of *Cyprinus carpio* (step B) are able to digest only peptides, while high-molecular-weight soluble protein is not fully digested before step E (Ilina and Tureckij, 1986). The activity of digestive enzymes increases during exogenous feeding; it is more advanced in stomach-possessing larvae (Acipenseridae, Salmonidae), but in stomach-less larvae it remains retarded (Stroband and Dąbrowski, 1981; Dąbrowski, 1984a; Ilina and Tureckij, 1986; Buddington and Doroshov, 1986; Pedersen *et al.*, 1987; Segner *et al.*, 1989). In *Coregonus fera* the absorptive performance of the intestine and rectum develops during the first two weeks of exogenous feeding (Loewe and Eckmann, 1988). Feeding success (the ratio of successful bites to total bites) is low in first-feeding larvae (Blaxter, 1969; Barrahona-Fernandes and Girin, 1977; Drost and Boogaart, 1985), many of which have empty digestive tracts (Shireman *et al.*, 1988). Foraging efficiency, the ratio of energy gained from the prey to the total energy costs of food uptake, increases with experience (Drost and Boogaart, 1985; Townsend and Winfield, 1985; Osse and Drost, 1989).

Thus, the combination of the functional, anatomical, physiological and behavioural characteristics of early larvae restricts their ability to ingest and digest prey, but changes in these characteristics lead to a rapid improvement in feeding performance.

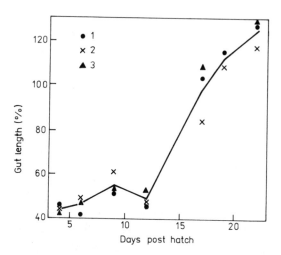

Fig. 6.4 Relative gut length, expressed as percentage of body length, in 987 *Cyprinus carpio* larvae from three ponds (1–3) during the first 3 weeks post hatch (computed from table I in Matlak and Matlak, 1976).

Factors affecting exogenous feeding

Fish feeding rate is affected by a number of interrelated factors; Brett (1979), Brett and Groves (1979), Fänge and Grove (1979) and Pandian and Vivekanandan (1985) have summarized the literature concerning postlarval fishes. The extensive literature on fish feeding includes relatively little work on larvae. Two factors determining the rate of live food consumption – fish size and food availability – will be considered here.

Body size

The two aspects of relations between fish feeding and their body size are prey size selectivity and the feeding rate–predator size relationship. Rich inter-specific data for marine fish larvae collected by Last (1980) demonstrate a dependence of the width of selected prey upon the predator fish's total length (Fig. 6.5). The small gape of early larvae in many species obliges them to consume very small food organisms (Blaxter, 1969). Studies by Hartmann (1986) performed on five fish species from Lake Constance throughout their life cycle showed that the upper limit of selected prey size range and its

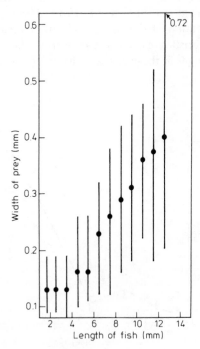

Fig. 6.5 Relationship between the width of prey and total length of 8 902 larvae of 20 marine fish species. Points, mean of all species; vertical lines, range between smallest and largest means of single species at any length group (from table 4 in Last, 1980).

median are strongly correlated with mouth size ($r^2 = 0.98$ for both). Thus, fish larvae are size selective. The percentage of *Gadus morhua* larvae having prey in their guts was related to gape (Knutsen and Tilseth, 1985). A 2.5-fold increase in width of copepod nauplii produces an order-of-magnitude increase in prey dry weight (Hunter, 1981). The general ontogenetic sequence in 0^+ class cyprinids was phytoplankton – rotifers – crustaceans – chironomid larvae; however, a high diversity of prey size was observed, especially in smaller larvae which, besides the dominant small food items, include in their diet the largest prey they can swallow (Mark *et al.*, 1987). Larvae of *Stizostedion lucioperca* and *Perca fluviatilis* select food items within a narrow spectrum; they are more susceptible to food deficiency than *Rutilus rutilus* and *Abramis brama* larvae, which exhibit more plasticity towards food size (Kudrinskaya *et al.*, 1976).

The dependence of relative feeding rate on body size in larvae, and, for comparison, small juveniles is illustrated in Fig. 6.6 for *Cyprinus carpio* fed

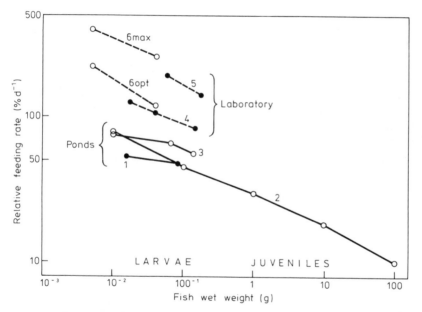

Fig. 6.6 Body size dependence of relative feeding rate (% body weight d^{-1}) in *Cyprinus carpio* larvae and juveniles fed live food. ○, Feeding rate expressed in terms of wet weight for food and fish; ●, feeding rate expressed in terms of dry weight (and, in (4) and (5), in terms of energy because no differences were found between the caloric value of zooplankton and fish body dry matter in these investigations). Pond data from: 1, Krivobok (1953); 2, Yarzhombek (1986); 3, Szarowski (unpublished data). Laboratory foods: 4, *Moina* monoculture at 26 °C (Kamler *et al.*, 1987); 5, mixed natural zooplankton at 26 °C (Kamler *et al.*, in press); 6, *Artemia* nauplii at 24 °C (Bryant and Matty, 1980), 6 opt, feeding rates for optimum growth and conversion efficiency, 6 max, maximum rates (recomputed mean values). Log–log scale.

live food. Feeding rate expressed as per cent of body weight is very high in small larvae and decreases as body size increases. Schiemer *et al.* (1989) report daily maximum feeding rates (dry weight per dry weight basis) of *c.* 77, 40 and 20%, respectively, in *Rutilus rutilus* larvae of 1, 10 and 100 mg dry weight, fed zooplankton at 20 °C. Another feature revealed by Fig. 6.6 is flexibility of carp feeding. The lower consumption values in ponds can be attributed to lower temperatures and food availability.

Food availability

Food availability is a product of the prey size spectrum, prey mobility, patchiness of prey distribution and prey density; these were considered for *Rutilus rutilus* larvae by Schiemer *et al.* (1989). Feeding rate increases asymptotically with increased prey density to reach a maximum level, the satiation level. Figure 6.7 illustrates this for larvae of *Abramis brama*; see also Eldridge *et al.* (1981b) for *Morone saxatilis* larvae and Schiemer *et al.* (1989) for *Rutilus rutilus* larvae. The relationship between feeding rate and prey density is modified by prey size: larger prey increase the relative feeding rate, and the satiation level is attained at lower density (Pandian and Vivekanan-dan, 1985). Shortly after first feeding, the intestine of *Clupea harengus* larvae had a higher content of trypsin at higher prey density (Pedersen *et al.*, 1987).

Scarcity or unsuitability of food are analogous in their effect to high levels of competition, hence they are expected to increase size variability (Koyama and Kira, 1956; Begon, 1984). The smallest individuals may be eaten by the largest ones; cannibalism related to larval size variability was observed by Taniguchi (1981) in *Cynoscion nebulosus*. Cannibalism was also reported for

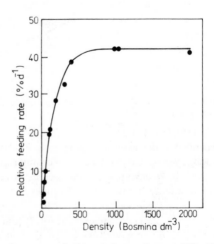

Fig. 6.7 Relative feeding rate of *Abramis brama* larvae (28–49 mg wet weight) as a function of prey (*Bosmina* sp.) density. The relative feeding rate is expressed in terms of carbon for prey and predator (from tables 2 and 3 in Panov and Sorokin, 1967).

Ctenopharyngodon idella larvae and juveniles fed plant food (Urban, 1984), for *Morone saxatilis* × *Morone chrysops* hybrid larvae longer than 25 mm fed fine zooplankton (Shireman *et al.*, 1988), for *Acanthopagrus cuvieri* larvae 10–15 mm (Hussain *et al.*, 1981) and for carp larvae fed a formulated diet of low quality (Lukowicz and Rutkowski, 1976), an insufficient amount (Charlon *et al.*, 1986) or starved from the age of 12 d (Kamler *et al.*, 1987).

6.3 BODY GROWTH

General remarks

Body weight (*W*) is a power function of length (*L*): $W = aL^b$ (Equation 4.21). The length exponent b was independent of temperature in juveniles of two tilapine fishes (Price *et al.*, 1985). The exponent b often shows values close to 3 in postlarval fishes, e.g. 3.06 in *Scophthalmus maximus* > 7 mm (Jones, 1972), 2.96 and 3.00 in *Cyprinus carpio* 5–100 g wet weight (Kamler, 1972b), 2.97 in *C. carpio* 0.36–2250 g (Oikawa and Itazawa, 1985), 3.02 in *Salmo trutta* 0.8–108.5 g (Mortensen, 1977), and 2.57–3.36 in 16 populations of 9 species of Pleuronectinae 17–143 cm in length (Fadeev, 1987). The slope b = 3 indicates isometric growth, in which body form remains unchanged. However, no simple mathematical relationship beween weight and length exists for the whole life cycle (Ricker, 1968). Non-isometric growth with b > 3 has been found in some fish larvae (Table 6.2; Hughes *et al.*, 1986; Opuszyński *et al.*, 1989). These larvae, which initially are thin, increase in height and/or width faster than in length (Matlak, 1966; Jones, 1972; Oikawa and Itazawa, 1985; Osse *et al.*, 1986). Osse *et al.* provide a hydrodynamic explanation for the shift in form as the size of *Cyprinus carpio* larvae increases. A reduction in viscosity effects, which are predominant in small larvae, minimizes the metabolic costs of locomotion. Hence, the shift in form in early life of fishes is of adaptive significance.

Recently Bergot *et al.* (1989) and Szlamińska *et al.* (1989) found that within the carp larval period, the fastest changes in the height/length ratio, in the condition factor ($CF = 100 \, W / L_s^3$, where L_s is standard length), and in the total length/body length ratio, occurred at larval steps D1 and D2, i.e. at body length 9–14 mm (for developmental steps see Fig. 6.1), and then levelled off (see also Section 6.4). The slope b (Equation 4.21) attained maximum values (4.5–5) at steps D1–D2 (9–14 mm); later it decreased, approaching 3 near metamorphosis.

Hence, the use of the condition factor to evaluate the condition (fatness) of fish larvae is misleading and the relative condition factor:

$$RCF = 100 \, W / L^b \qquad (6.1)$$

should be used (Ricker, 1968; Ehrlich *et al.*, 1976).

Table 6.2 Weight (W) to length (L) relationship in exogenously fed fish larvae, $W = aL^b$

Species*	Size†	$b \pm 95\%$ C.I	v‡	Source§
Cyprinus carpio f, w[ab]	1–200 mg W_w	4.68 ± 0.17	–	1
C. carpio f, w[ac]	1–600 mg W_w	4.59 ± 0.16	–	1
C. carpio l, w[a]	6.9–17.2 mm L_S	4.81 ± 0.22	4.82	2
C. carpio l, w[a]	1.6–6.2 mm L_S	4.25 ± 0.36	–	3
C. carpio l, w[a]	1.6–9.7 mm L_S	4.30 ± 0.33	–	3
C. carpio l, w[d]	1.6–2.6 mm L_S	3.84 ± 0.52	–	3
C. carpio l, w[d]	1.6–3.1 mm L_S	3.97 ± 0.52	–	3
C. carpio l, w[a]	< 20 mmL_S	4.47	–	4
C. carpio l, w[d]	5.9–21.3 mm L_S	4.87 ± 0.06	4.99	5
Salmo trutta f, w	80–930 mg W_w	3.82 ± 0.18	3.87	6
S. trutta l, w[e]	50–100 mg W_w	4.10 ± 1.21	4.37	7
S. trutta l, d[e]	8–20 mg W_d	4.66 ± 1.11	4.89	7
S. trutta l, w[f]	50–200 mg W_w	5.06 ± 1.25	5.44	7
S. trutta l, d[f]	8–30 mg W_d	5.63 ± 1.36	6.02	7
Scophthalmus maximus l, w[a]	< 7 mm	4.67 ± 0.52	5.00	8
Clupea harengus l, d[a]	1–100 mg W_d	4.57	–	9
Pleuronectes platessa l, d[a]	1–20 mg W_d	3.91	–	9
Pagrus major l, w	< 6.4 mm L_t	4.38	–	10
Siganus fuscescens l, w[a]	< 15 mm L_t	3.27	–	11

*Origin: f, larvae collected from the field; l, laboratory-reared larvae. Type of equation: d, equation is for dry weight; w, equation is for wet weight.
†Size of larvae used, measured as: L_s, standard length; L_t, total length; W_d, dry weight; W_w, wet weight.
‡GM functional coefficient, computed according to Ricker (1973).
§Sources: 1, Kamler (1972b); 2, Oikawa and Itazawa (1985); 3, Szlamińska and Przybył (1986); 4, Osse *et al.* (1986); 5, Szlamińska *et al.* (1989); 6, Mortensen (1977); 7, Raciborski (1987); 8, Jones (1972); 9, Ehrlich *et al.*, (1976); 10, Kitajima *et al.* (1976); 11, Kitajima *et al.* (1980).
[a] Fed zooplankton.
[b] Golysz carp family (strain) no. 4.
[c] Family no. 5.
[d] Fed formulated diet.
[e] Series 1983.
[f] Series 1984.

During fish starvation, 'negative growth' occurs. A slight decrease of length in starved early larvae (shrinkage) was reported by Ehrlich *et al.* (1976) for *Clupea harengus* and by Eldridge *et al.* (1981b) for *Morone saxatilis*, although no changes in length, or else a slight increase, have been reported as well. There is a decrease in fat content, followed by a decrease in protein content and an increase in water and ash content (review: Love, 1970).

Hence length is not a sensitive index of trophic state; energy changes during starvation or feeding are usually more marked than changes in wet weight (Section 4.3; Dąbrowski, 1976). Specific growth rates in terms of length, wet weight, dry weight, and energy are shown in Table 6.3 for carp larvae in different nutritive conditions. Most earlier growth data in larval aquaculture are expressed in wet weight, but energy or dry weight are recommended as more reliable measures of growth.

Factors affecting body growth during early exogenous feeding

Growth rate is particularly susceptible to the effects of temperature, body size, food quality and amount of food. Much has been achieved in that field as applied to postlarval salmonids (reviews: Brett, 1979; Brett and Groves, 1979; Elliott, 1982).

Temperature

Temperature exerted little effect on larval growth of three cyprinids between 20 and 25 °C (Q_{10} 1.29–1.67), but this effect was strong (Q_{10} 3.9) at lower temperatures (15–20 °C) in *Rutilus rutilus* larvae (Wieser *et al.*, 1988b). The general relationships between growth and temperature were presented in Section 4.3, but the discussion was confined to constant temperatures, which are unusual under natural conditions. The rate of energy accumulation by carp larvae and juveniles reared at fluctuating temperatures was higher than that obtained at constant temperatures (Konstantinov and

Table 6.3 Specific growth rate (G, % d^{-1}) for length (L), wet weight (W_w), dry weight (W_d) and energy content ($C.e.$) computed from Equation 4.16 for different feeding groups of *Cyprinus carpio* larvae at 26 °C

Feeding group	Age (d)	G_L	G_{W_w}	G_{W_d}	$G_{C.e.}$	Source[*]
Starved	13–17[a]	–	−7.2	−11.5	−11.5	1
Diet St P	13–24	2.6[b]	11.5	12.7	13.2	
Zooplankton						
(*Moina* sp.)	13–24	7.0[b]	27.2	30.2	31.0	
Starved	2–7[c]	0[d]	−6.0	−7.3	−9.5	2
Diet P2	2–16	3.9[d]	20.9	23.3	24.2	
Zooplankton						
(natural)	2–16	8.1[d]	36.0	37.1	36.6	

[*]Sources: 1, Kamler *et al.* (1987); 2, Kamler *et al.* (in press).
[a]Cannibalism.
[b]Expressed in terms of total length (L_t).
[c]Mass mortality.
[d]Expressed in terms of standard length (L_s).

Zdanovich, 1985). A clear increase of growth rate and of efficiency of utilization of assimilated energy for growth at fluctuating temperatures as compared with constant ones was shown by Galkovskaya and Sushchenya (1978) for aquatic invertebrates. The basic principles of the combined effect of temperature and ration size on postlarval fish growth were elaborated by Brett *et al.* (1969) and Elliott (1976). They showed with salmonid juveniles that the optimum temperature for growth decreases with a decrease of ration level.

Body size

A power-law equation is usually the best model for the inverse relationship between specific growth rate and body size; for summarizing reviews see Zaika (1972) for aquatic invertebrates and Brett (1979) for postlarval fish. An example for carp larvae is shown in Fig. 6.8. In general, specific growth rate decreases from roughly 50% d^{-1} in fish larvae weighing 1–10 mg to 1–2% d^{-1} in fish juveniles of 100 g (Hogendoorn, 1980; Dąbrowski, 1986a). Aguirre (1986) showed a high rate of protein synthesis in endogenously feeding larvae of *Coregonous schinzi palea* and a decrease in this rate as a function of body size of exogenously feeding larvae.

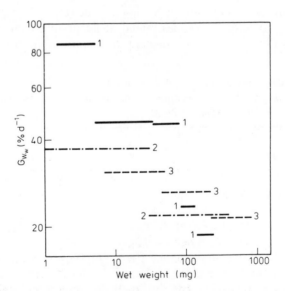

Fig. 6.8 Relationship between the specific growth rate for wet weight at 26 °C and wet weight of *Cyprinus carpio* larvae in ponds (1, Skácelová and Matěna, 1981) and in the laboratory: 2, fed mixed live zooplankton (Kainz and Gollmann, 1980); 3, fed a formulated diet based on yeast and beef liver (Bergot *et al.*, 1989). The growth rates were converted from the original temperatures to 26 °C using Winberg's (1956) conversion factor.

Food composition

The effect of food composition on growth was reported by Taniguchi (1981) for *Cynoscion nebulosus*, whose larvae grew larger when fed copepods than when fed rotifers. A sudden acceleration of growth of *Rutilus rutilus* larvae in nature paralleled changes of food composition: a disappearance of phytoplankton and an increase in the proportion of crustaceans ingested (Wieser et al., 1988b). Figure 6.8 and Table 6.4 illustrate the dependence of growth on food quality for larvae of *Cyprinus carpio*, which, having no stomach, are most susceptible to dietary deficiencies (Dąbrowski, 1948a,b). In Table 6.4, groups A–F are arranged in order of decreasing growth rate. In general, larvae that were fed live foods grew better than those fed artificial diets. Remarkable progress has been observed in formulated diets since yeast was introduced (Appelbaum, 1977; Appelbaum and Dor, 1978; further discussion, Section 7.2).

Consumption rate

A decreased rate of protein synthesis with fasting was reported by Fauconneau (1984) for *Cyprinus carpio* larvae weighing 50 mg. With increasing consumption rate (C), production (P) increases: from 'negative growth' (energy loss) at starvation, it crosses the $P=0$ level at the maintenance consumption and continues to rise. The optimum consumption (C_{opt}) is the value at which the gross conversion efficiency $K_1 = P/C$ reaches a maximum. For example, the maintenance rations at $20\,°C$ for c. 1–2 mg ash-free dry weight *Catostomus commersoni* larvae, *Notropis cornutus* juveniles and newborn *Poecilia reticulata* were 0.19, 0.13 and 0.08 mg food $mg^{-1}d^{-1}$, respectively (Borgmann and Ralph, 1985). The C_{opt} values in terms of wet weight for food and fish in *Cyprinus carpio* larvae fed on *Artemia* nauplii at $24 \pm 0.5\,°C$ were 200–250% d^{-1} and 100–120% d^{-1} for larvae 5 and 10 d old, respectively (Bryant and Matty, 1980).

Food concentration

The dependence of larval growth on food concentration has attracted considerable attention (Panov and Sorokin, 1967; Wyatt, 1972; Baranova, 1974; Taniguchi, 1981; Eldridge et al., 1981b; Borgmann and Ralph, 1985; McMullen and Middaugh, 1985). A typical curve is shown in Fig. 6.9. The threshold food concentration is the one below which growth is negative. It was (in terms of ash-free dry weight for food and fish) 0.18 mg $mg^{-1} d^{-1}$ for *Notropis cornutus* at $20\,°C$ and 0.37 mg $mg^{-1} d^{-1}$ for *Catostomus commersoni*, which requires a higher food concentration (Borgmann and Ralph, 1985). The threshold food concentration for *Abramis brama* larvae 15–49 mg wet weight was 45–60 *Bosmina* dm^{-3} (Panov and Sorokin, 1967); for *Morone saxatilis* larvae (L_s 6 mm) at $18\,°C$ it was 100 *Artemia* nauplii dm^{-3} (Eldridge et al., 1981b). In *Menidia peninsulae*, larval growth increased as food concentration increased from 500 to 5000 food organisms dm^{-3} at 30

Table 6.4 Comparison of the specific growth rate $(G, \% \, d^{-1})$ for wet weight in *Cyprinus carpio* larvae. G_t values were converted to $G_{26°}$ values using Winberg's (1956) temperature conversion factor

Group*	t (°C)	Wet weight range (mg)	G_t (% d^{-1})	$G_{26°}$ (% d^{-1})	Source†
A	22–30	1.5–8.4	86.0‡	86.0	1
	22–30	8.4–33.5	46.2‡	46.2	1
	22–30	33.5–83.8	45.8‡	45.8	1
	22–30	83.8–134.7	23.7‡	23.7	1
	22–30	134.7–236.8	18.8‡	18.8	1
B	21±1.0	1–31	24.5‡	37.0	2
	21±1.0	31–400	14.2‡	21.4	2
	25	1.5–46.3	24.5‡	26.5	3
	24±0.5	1.6–9.7	25.7§	30.3	4
	20±2	2.4–255	23.3‡	38.2	5
	30±1	2.4–360	25.1‡	18.3	5
	24	1.5–8.5	24.8§	29.3	6
	24	8.5–18.5	11.2§	13.2	6
	26	1.7–263	36.0§	36.0	7
C	24.9	1.5–35.0	31.5‡	34.0	8^a
	21	1.5–134	15.0‡	22.7	9
	24±0.5	1.1–1.5	6.6‡	7.8	10^b
	24±0.5	1–14	50.0‡	59.0	10^c
	24±0.5	15–16.9	2.3‡	2.7	10^b
	24±0.5	11–58	33.0‡	38.9	10^c
	25	24.7–187.9	10.1§	10.9	11
	26	1.5–8.3	28.3§	28.3	12
	26	12.7–253.0	27.3§	27.3	12
D	21	25–131.0	11.6‡	17.5	9
	23–25	1.5–1500.0	13.8‡	16.3	13
	25	4.3–28.3	18.9§	20.4	3
	25	18.3–36.2	11.4§	12.3	3
	24±0.5	1.6–12.7	29.6§	34.9	4
	20±2	2.3–103.0	19.0‡	31.2	5
	30±1	2.4–191.0	21.9‡	16.0	5
	26	12.7–45.0	11.5§	11.5	12
E	22±1	1.2–43.4	19.8‡	27.5	14
	28–32	1.9–107.4	36.7‡	26.8	15^d
	25	7.8–19.7	4.6‡	5.0	11
	25	14.5–141.4	11.4‡	12.3	11
	25	1.5–13.8	15.8§	17.1	3
	24±0.5	1.6–3.1	9.6§	11.3	4
	24	7.5–46.7	26.1‡	30.8	16
	24	46.7–221.9	22.3‡	26.2	16

Table 6.4—contd.

Group*	t (°C)	Wet weight range (mg)	G_t (% d^{-1})	$G_{26°}$ (% d^{-1})	Source†
	24	221.9–787.4	18.1‡	21.3	16
	26	1.7–4.0	20.9§	20.9	7
F	21	2.5–49	8.8‡	13.3	9
	26	1.8–2.2	10.3§	10.3	12

* Arranged in order of decreasing growth rate. A, natural food in ponds; B–F, larvae reared under controlled conditions: B, mixed live zooplankton, C, monocultures of live food; D, mixed foods (live foods and formulated diets); E, diet based on yeast; F, other diets (modified from Kamler et al., 1987).

† Sources: 1, Skácelová and Matěna (1981); 2, Kainz and Gollmann (1980); 3, Dąbrowski (1984b); 4, Szlamińska and Przybył (1986); 5, Ilina (1986); 6, Szlamińska (1987a); 7, Kamler et al. (in press b); 8, Baranova (1974); 9, Lukowicz and Rutkowski (1976); 10, Bryant and Matty (1980); 11, Dąbrowski (1982b); 12, Kamler et al. (1987); 13, Huisman (1979); 14, Appelbaum (1977); 15, Ostroumova et al. (1980); 16, Bergot et al. (1989).

‡ G_t values computed from the authors' data.

§ Original G_t values.

ᵃ Maximum value used.

ᵇ Shortage of food.

ᶜ Excess of food.

ᵈ Feed 'Equizo' (undisclosed composition): assignment to group E uncertain.

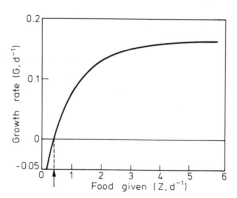

Fig. 6.9 Specific growth rate (G) for *Catostomus commersoni* at 20 °C as a function of food supply (Z). Growth rate and food offered are expressed in mg of ash-free dry weight per ash-free initial weight of fish per day. The curve was calculated from the model given by Borgmann and Ralph (1985):

$$G = (G_m + G_0)(1 - \exp(-kZ)) - G_0$$

where G_m is maximum specific growth rate, $0.164\,d^{-1}$, G_0 is weight loss at no food given, $0.071\,d^{-1}$, and k is 0.970. Arrow, threshold food concentration, below which growth is negative.

and 25 °C, whereas at 20 °C growth was retarded and no influence of food concentration was found (McMullen and Middaugh, 1985). From the relationship between growth and food concentration, Baranova (1974) recommended 300–350% body weight d^{-1} as the optimum amount of live food at 25 °C for the first week of active feeding of *Cyprinus carpio* larvae.

The mean food concentrations in water bodies is often below the threshold concentration for fish larvae, so a patchy distribution of prey is important (Panov and Sorokin, 1967). Food availability in non-eutrophic water bodies is a decisive factor for growth of cyprinid larvae, whereas in eutrophic water bodies year-to-year differences of growth can be attributed to temperature conditions (Schiemer *et al.*, 1989).

6.4 METABOLISM

During larval development the temporary superficial respiratory organ – the red layer of muscle fibres – decreases gradually (El-Fiky *et al.*, 1987). The number of gill filaments, gill lamellae and filament length increase (cyprinids: El-Fiky *et al.*, 1987; El-Fiky and Wieser, 1988; Osse, 1989). Gill area increases; the slopes of the power-law relationship of gill area to body mass are 7.066 and 1.222 for carp larvae 1.6–2.8 mg and 2.8–330 mg weight, respectively (Oikawa and Itazawa, 1985). In freshly hatched larvae of *Anabas testudineus*, skin is the major site of gaseous exchange, gills become functional later; the development of scales in the skin hinders cutaneous respiration and coincides with the beginning of air-breathing of these fish, which inhibit warm and poorly oxygenated tropical swamps (Hughes *et al.*, 1986; review: Hughes and Al-Kadhomyi, 1988). Cyprinid larvae, which are at first mostly aerobically fuelled (El-Fiky *et al.*, 1987; El-Fiky and Wieser, 1988; Hinterleitner *et al.*, 1989), later develop anaerobic enzymes (Hinterleitner *et al.*, 1989).

Fish, unlike homeotherms, do not expend large amounts of energy in thermoregulation. Hence the three main subcomponents of respiratory metabolism (R) in fish are: R_r resting (standard) metabolism; R_a, activity metabolism; and R_f – feeding metabolism: $R = R_r + R_a + R_f$ (Equation 4.26, page 144).

Resting (standard) metabolism

The effect of size on resting (standard) metabolism is shown by $R = aW^b$ (Equation 4.27, page 152). N.B.: this relationship is usually determined for routine metabolism, i.e. in fishes showing a normal spontaneous activity; the effect of size on the maximum activity metabolism is different.

The majority of the values of the weight exponent b found in the animal kingdom (Hemmingsen, 1960; Kleiber, 1961), including postlarval fish (Winberg, 1956; Brett and Groves, 1979; Elliott, 1982; Ozernyuk, 1985) are within the range of 0.70–0.86. However, Zeuthen's (1955, 1970) inter-specific comparisons showed that poikilothermic metazoans with body weight below 0.1 g exhibit b values of 0.9–1.0, whereas in large poikilo-therms and homeotherms, b is 0.7–0.8. It is interesting that a similar change of the R on W relationship occurs during ontogenesis of some poi-kilotherm species (Zeuthen, 1955, 1970; Dąbrowski, 1986c). This is shown for fish in Table 6.5 and Fig. 6.10(A). Thus, the zone in which respiratory rate per unit tissue weight increases with increasing body size in en-dogenously feeding fishes (b>1, Section 4.4) and the zone of decreasing respiratory rate in subadult and adult fishes (b=0.70–0.86) are separated by a transitional zone in which a size-independent respiratory rate is observed (b≈1). The last has been reported for endogenously feeding or mixed-feeding larvae (Section 4.4), for exogenously feeding larvae (Fig. 6.10(A); Table 6.5; Wieser *et al.*, 1988a), and for juveniles (Table 6.6) of some species. Thus, the transitional zone is not associated with developmen-

Table 6.5 Ontogenetic changes in the effect of size on metabolism, $R = aW^b$

Species	Temp. (°C)	Body wet weight (g)	b ± 95% C.I.	Source*
		Oxygen consumption		
Cyprinus carpio (larvae)[a]	20	0.002–3.8[a]	0.98	1
C. carpio (larvae)	26	0.002–0.25	0.98 ± 0.05	2
C. carpio (larvae)	20	0.001–0.6	0.98 ± 0.05	3
C. carpio (juveniles)	20	1.2–45.7	0.80 ± 0.05	3
C. carpio (juveniles)	20	2.5–3 487.0	0.85 ± 0.02	4
C. carpio (juveniles)	20	3.8–15.7	0.82	5
Misgurnus fossilis (young)	–	–	0.92	6
M. fossilis (adults)	–	–	0.81	6
Ctenopharyngodon idella (larvae)	21	0.001–0.2[bc]	1.05	7
C. idella (juveniles)	22	40.0–110.0[c]	0.76	8
		Ammonia excretion by starved fish		
Carassius auratus cuvieri (larvae)	20	0.001–0.01[b]	1.0–1.2	9
C. auratus cuvieri (larvae + subadults)	20	0.030–3.0[b]	0.7–0.8	9

*Sources: 1, Winberg and Khartova (1953); 2, Kamler *et al.* (1987); 3, Kamler (1972b); 4, Winberg (1956); 5, Kausch (1968); 6, Ozernyuk (1985); 7, Urban (1982); 8, Fischer (1970b); 9, Iwata (1986).
[a] Mostly larvae: in 115 of 123 measurements, body weight was markedly below 1 g.
[b] Size read from the graph.
[c] Animal food.

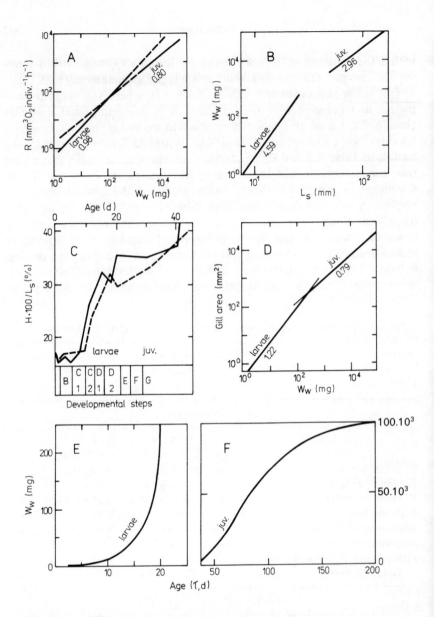

Fig. 6.10 Ontogenetic changes of respiration and growth types in *Cyprinus carpio*. Numbers under regression lines indicate the slopes. (A) Respiration (*R*) in relation to wet weight (*W*$_w$) (data from Kamler, 1972b; broken curves are extrapolations); (B) wet weight in relation to standard length (*L*$_s$) (data from Kamler, 1972b for the Gołysz carp family (strain) no. 5); (C) age course of the ratio of height (*H*) to standard length in carp family no. 4 (solid curve) and no. 5 (broken curve) (data from Matlak, 1966; for developmental steps see Fig. 6.1); (D) gill area in relation to body wet weight (data from Oikawa and Itazawa, 1985); (E) and (F) growth curves (data from Kamler, 1976): (E), an exponential curve for larvae; (F), an S-shaped curve for juveniles. Log–log scale in (A), (B), and (D).

Table 6.6 Size dependence of oxygen consumption at different temperatures. General equation: $R = aW^b$, where R is oxygen consumption (mm^3 O$_2$ indiv^{-1} h^{-1}) and W is body wet weight (g)

Species	Temp. (°C)	Body wet weight (g)	a(95% C.L. for a)	b±95% C.I.	Source*
Salmo trutta	3.5	1.87–39.6	98(91–106)	0.939±0.040	1
S. trutta	8.0	0.55–16.0	141(128–154)	0.911±0.056	1
S. trutta	12.0	0.25–20.0	226(216–237)	0.934±0.034	1
Huso huso	12	0.07–8.6	275 –	0.959	2
H. huso	16	0.09–9.5	358 –	0.949	2
H. huso	20	0.07–9.1	387 –	0.907	2
H. huso	24	0.09–8.4	532 –	0.914	2

*Sources: 1, Mortensen (1985); 2, Gershanovich (1983).

tal advancement (cf. fig. 3 in Wieser and Forstner, 1986). In larvae of many species, b values of 0.70–0.86 have been observed (e.g. Kudrinskaya, 1969; Melnichuk, 1969; Laurence, 1978; Eldridge et al., 1982; Wieser et al., 1988a; Schiemer et al., 1989). The ontogenetic changes in the respiration-to-body-size relationship in *Cyprinus carpio* (Fig. 6.10(A)) are associated with a shift in body shape (Fig. 6.10 (B) and (C)); see also Section 6.3) and are closely related to the development of the fish–water interface area (gills, Fig. 6.10 (D)). Similar changes in slope of the regression of gill area against body weight occur in several fish within the wet weight range 0.05–1.0 g and are related to metamorphosis (review: Hughes and Al-Kadhomyi, 1988). The decrease of weight-specific metabolic rate with increasing body size at later steps of individual ontogeny is related to a decrease in the concentration of mitochondria (review: Ozernyuk, 1985). Bertalanffy (1964) showed that particular metabolic types (size-dependencies of respiration) are associated with growth types. If the growth curve is exponential, then the respiration is proportional to weight ($b \approx 1$ in Equation 4.27); this is observed in exogenously feeding carp larvae (Fig. 6.10 (E) and (A)). On the other hand, S-shaped growth curves are, according to the Bertalanffy (1964) theory, associated with respiration proportional to a value intermediate between the surface area and weight ($0.66 < b < 1$); this is the case with juvenile carp (Fig. 6.10(F) and (A); N.B.: the slow growth rate at the end of the period depicted in Fig. 6.10(F) results from wintering, as day 190 was 18 December). The change of carp body shape and metabolic type is such that a 10 g carp juvenile respires less than would a carp larva weighing 10 g, and conversely, the energy expenditures of a larva are lower than those of a hypothetical juvenile of the same size (broken curves in Fig. 6.10(A); see also Calow, 1984). In summary, then, the change of metabolic type and growth type in early ontogeny of carp and similar species can be considered as an energy-saving adaptation.

Resting (standard) metabolism is significantly affected by temperature. The basic relationships are given in Section 4.4. The intercept a in Equation 4.27 noticeably increases with increasing temperature (example in Table 6.6). The weight exponent b in *Salmo trutta* was not affected by temperature (Elliott, 1976; Table 6.6).

In fish the participation of oxidative metabolism of amino acids in total metabolism is higher than in mammals. Ammonia – the main end product of protein catabolism in fish – is easily excreted in fresh waters (Brett and Groves, 1979; Pandian and Vivekanandan, 1985; Aguirre, 1986; Section 4.5.). The mean $N_{ammonia}$ excretion in starved carp larvae amounted to 79.5% of total $N_{ammonia} + N_{urea}$ excreted (Kaushik *et al.*, 1982). Protein is an important energy source for fish. Buddington and Doroshov (1986) showed with larvae of *Acipenser transmontanus*, and Dąbrowski (1986c) with *Coregonus shinzi palea* weighing 20–700 mg, that nitrogen retention is lower in fish larvae than in older fish. The reduction of relative rate of metabolic nitrogen excretion $N_{ammonia} + N_{urea}$ with age is illustrated in Table 6.7 for carp larvae. Both low protein level in the diet and high temperature reduce nitrogen excretion by juvenile fish (Knights, 1985).

Activity metabolism

Fry (1957) and Brett and Groves (1979) summarized earlier data on energy expenditures for motor activity (R_a) in postlarval fishes. In general, as much as 25% of energy consumed with food can be invested in active metabolism of carnivorous fish (see also Priede, 1985; Brafield, 1985). The mean resting (standard) metabolic rate is $1.21 \, J \, g^{-1} h^{-1}$ (range 0.33–3.10) in temperate zone fishes weighing 50–100 g and fully acclimated to temperature (Brett and Groves, 1979). The maximum metabolic rate (the energy expended at both maximum temperature and activity) ranged from 2.09 to 13.81

Table 6.7 Relative excretion of different forms of nitrogen ($\mu g \, N \, mg^{-1} d^{-1}$) at 26 °C by *Cyprinus carpio* larvae aged 4–5 and 9–10 d fed a formulated diet (St 82) or zooplankton monoculture (recomputed from Urban-Jezierska *et al.*, 1989)

Age (d)	Dev. step*	Diet	N_{rv} =	$N_{ammonia}$ +	N_{urea} +	$N_{protein}$	$N_{protein}$ (%)
4–5	C1	St82	7.75	3.82	0.57	3.36	43
4–5	C1	Zoopl.	8.13	4.45	0.80	2.88	35
9–10	C1	St82	4.93	1.92	0.48	2.53	51
9–10	C2	Zoopl.	2.23	1.94	0.29	0	0

* Steps as in Fig. 6.1.

$J g^{-1} h^{-1}$, thus, metabolic rate can rise to values 2–3 times, maximally up to 10 times, the resting rate. In fish larvae, swimming activity was found to increase oxygen consumption up to 2.5 times above the resting level in *Salmo trutta* (Carrick, 1981), up to 3.5 times in *Sardinops caerulea* at 14 °C (Lasker and Theilacker, 1962), 2–3 times (Dąbrowski, 1986b) or 1.5–15 times (Dąbrowski et al., 1986a) in *Coregonus shinzi palea* at c. 14 °C, and 2.4–2.9 times above routine level in three cyprinids at 12–24 °C (Wieser and Forstner, 1986). Dąbrowski (1986b) developed a circular chamber in which oxygen consumption and ammonia excretion were measured in fish larvae and early juveniles forced to swim by optomotor reaction. Both oxygen consumption rate (R/W) and ammonia nitrogen excretion rate $(N_{ammonia}/W)$ were related exponentially to swimming speed (s):

$$R/W \text{ or } N_{ammonia}/W = ae^{bs} \tag{6.2}$$

In fish larvae, protein is an energy source to cover the increasing energy costs at high activity level: the slope b in Equation 6.2 was higher for ammonia excretion than for oxygen consumption at every body size (Dąbrowski, 1986a,b,c,d; Dąbrowski et al., 1986a). The participation of protein catabolism in total energy used by *Coregonus shinzi palea* of 112 mg increased exponentially with increasing swimming speed from 21.8% at rest (standard metabolism) to 41.5% at a velocity of 4 body lengths s^{-1} (Dabrowski, 1986b).

A power-law negative relationship exists between the relative energy expenditure for locomotion $(J g^{-1} km^{-1})$ and the body wet weight of aquatic organisms (Schmidt-Nielsen, 1972; Dąbrowski, 1986c) over a large range of 14 orders of magnitude, from sea urchin sperm to individuals of *Thymallus* (Pavlova, 1987). The negative slope is greater for fish larvae and juveniles under 1 g wet weight than for larger fish. The costs of locomotion calculated for fish weighing 0.01, 0.1 and 1 g were respectively 988, 109 and $12 J g^{-1} km^{-1}$ (Dąbrowski, 1986c).

In summary, then, locomotor activity is expensive for early fish larvae, in which energy-saving mechanisms and, in particular, protein-saving mechanisms are poorly developed. Wieser et al. (1988a) concluded from their study on cyprinid larvae that: 'the animals probably have to use some kind of 'switching strategy' since high rates of growth and high levels of swimming activity appear to be mutually exclusive...'.

Feeding metabolism

Priede (1985) considered the conflicting needs of energy for activity metabolism (R_a) and for food conversion (specific dynamic action, SDA) in young postlarval fishes. He found that these needs compete for metabolic scope; performing vital functions beyond the limits of metabolic scope decreases the chance of survival.

Fish larvae capture prey by suction (Drost and Boogaart, 1985, 1986;

Osse *et al.*, 1986; Osse and Drost, 1989). The total energy costs of one feeding act (swimming forward during attack and suction) are only a fraction of 1% of the energy gained with a prey. The cost/benefit ratio increases when smaller prey are captured, when they are mobile, and when they occur at low densities (Wyatt, 1972; Kudrinskaya *et al.*, 1976). Frequent missed attacks also increase the cost/benefit ratio; aiming accuracy is low in early larvae and improves with age (Drost *et al.*, 1988). Hence, the costs of prey capture by fish larvae are low but may be of some importance at the beginning of exogenous feeding.

More energy seems to be expended to cover the costs of food conversion. From their review of postlarval fishes, Brett and Groves (1979) concluded that SDA can constitute *c.* 14% of the energy consumed in food. The ratio of total feeding metabolism (R_f) to resting (standard) metabolism (R_r) can be < 1 – this is the case when food ration falls below the maintenance level. In young, well-fed fish the ratio can amount to 5.8; an average value is 3.7. An increase of nitrogen excretion rate after feeding was described in carp larvae by Kaushik and Dąbrowski (1983); the amplitude and duration of the increase was related both to fish size and to amount of nitrogen ingested. Dąbrowski *et al.* (1986a) found an effect of feeding history on metabolism of *Coregonus shinzi palea* juveniles 190–390 mg wet weight that were unfed for 24 h prior to measurement. The proportion of energy obtained from protein was lowest in fish that had been fed a lipid-rich diet and highest on a lipid-deficient diet.

A substantial part of R_f may reflect the metabolic costs of growth (Jobling, 1985). This was considered for endogenously feeding fishes in Section 4.4. Figure 6.11 illustrates this problem for exogenously feeding carp larvae. The close relationship between the costs of growing and body growth, in terms of energy per individual per day, is demonstrated for six feeding groups of larvae receiving live food, four formulated diets, or starved. In carp larvae the mass-specific rates of oxygen consumption (RW^{-1}) reflected the specific growth rates induced by diets (Kamler *et al.*, 1987), whereas in roach larvae they were independent of the rates of growth induced by feeding levels (Wieser *et al.*, 1988a).

6.5 BUDGETS OF ENERGY OR MATTER

General remarks

An energy budget for a well-fed, growing fish larva is calculated from the 'classical' Equation 2.1 (page 7 and Table 6.8, row 1). All the efficiencies: the assimilation efficiency $K_A = A \times 100 / C$, the gross conversion efficiency $K_1 = P \times 100 / C$, and the net conversion efficiency $K_2 = P \times 100 / A$, are applicable. However, low levels of energy input are often observed for fish larvae in the field or for larvae which are fed with deficient diets in culture.

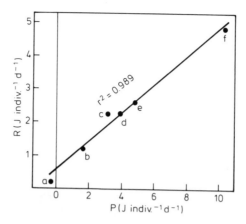

Fig. 6.11 Relationship between oxygen consumption (R) and growth (P) in six feeding groups of *Cyprinus carpio* larvae: a, starved; b–e, fed formulated diets; f, fed live zooplankton. A two-week feeding test was performed at 26 °C. The time course of both respiration and growth (daily measurements, expressed in terms of energy) was defined with exponential equations, separately for each feeding group ($P<0.001$). The R and P values for groups a to f were calculated for days 4–5 of feeding (days 6–7 post hatch) – one day before mass mortality of the starved group. $R=0.568+0.421P$, $r^2=0.989$, $P<0.001$ (raw numerical data from Kamler *et al.*, in press).

Other types of energy budgets then occur (Klekowski *et al.*, 1967), in which the efficiencies K_1 and K_2 are not applicable (Table 6.8). At the maintenance energy intake (row 2) an individual's energy resources remain constant, i.e. $P=0$, $A=R$, $K_A=R/C$. At an energy intake below the maintenance level (row 3), negative growth occurs. In starved larvae (row 4), the losses of body reserves (the negative growth) equal the sum of metabolic output, and none of the efficiencies K_A, K_1 and K_2 are applicable.

Table 6.8 Energy budgets at different feeding levels*

No.	Feeding state	Energy of			Energy budget equation
		Food ration	Growth	Faeces	
1.	Fed	$C>0, C>C_{main}$	$P>0$	$F>0$	$C=P+R+F+U$
2.	Fed	$C>0, C=C_{main}$	$P=0$	$F>0$	$C=R+F+U$
3.	Fed	$C>0, C<C_{main}$	$P<0$	$F>0$	$C+P=R+F+U$
4.	Starved	$C=0$	$P<0$	$F=0$	$P=R+U$

*Symbols: C, consumption; F, faeces; P, production; R, metabolism; U, nonfaecal excretion. C_{main} is the maintenance consumption, the energy necessary to maintain an individual's resources at a constant value.

Factors affecting transformation efficiencies

Age

The low feeding performance of early larvae (Section 6.2) and their poorly developed energy- and protein-saving mechanisms (Section 6.4) would lead us to expect their transformation efficiencies to be low initially and to improve with age. This is shown in Fig. 6.12 for carp larvae in which the zooplankton-fed group (1B) attained a high level of gross efficiency K_1,

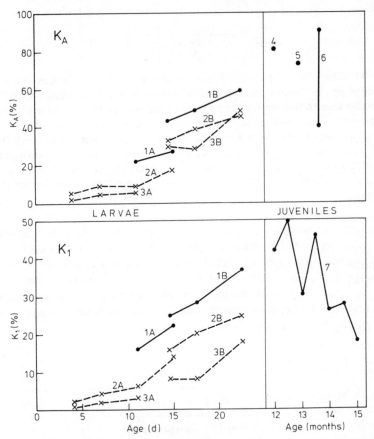

Fig. 6.12 Age course of assimilation efficiency $K_A = (P+R)100/C$ and gross conversion efficiency $K_1 = P \times 100/C$, both in terms of energy for P, R and C in *Cyprinus carpio* larvae reared at 26 °C and in juveniles. Carp larvae diets (A, Kamler *et al.*, in press; B, Kamler *et al.*, 1987) were: 1A, mixed zooplankton; 2A, formulated diet P2; 3A, diet St 86; 1B, zooplankton monoculture; 2B, diet St P; 3B, diet St 82E. Juveniles, fed live foods: 4, from Winberg's (1956) generalized equation for well-fed carnivorous fish; 5, as in 4, Brett and Groves (1979); 6, Krasnoper (1985) – range; 7, time course for juvenile carp in ponds (Ivlev's data in Winberg, 1956).

exceeding 35% at the end of the larval period (developmental step F, day 22–23 after hatching). An increase of K_1 for energy was also reported by Bryant and Matty (1980) for carp larvae aged 5–10 d as compared with those aged 1–5 d (see also Macháček *et al.*, 1986). The net efficiency K_2 for energy increased slightly in *Rutilus rutilus* larvae, from 68.6 to 71.9% and from 72.3 to 75.5% at 15 and 20 °C, respectively, during the first few weeks after hatching (Wieser *et al.*, 1988b). Iwata (1986) has demonstrated an increase of carbon assimilation efficiency during early exogenous feeding of *Carassius auratus cuvieri*, *C. auratus grandoculis* and *Hypophthalmichthys molitrix*. Fish larvae are unusual in this respect; efficiencies decrease progressively with size and age of fish juveniles (Curve 7 in Fig. 6.12; Staples and Nomura, 1976, for *Oncorhynchus mykiss*; Ryzhkov *et al.*, 1982, for *O. kisutch* and *Salvelinus alpinus*; review: Miura *et al.*, 1976) and other poikilotherms (reviews: Klekowski and Fischer, 1975; Calow, 1977). The increase of K_A and K_1 during larval development and their decrease in juveniles was shown by Urban (1982) in *Ctenopharyngodon idella* fed animal food.

It seems, then, that two peaks of efficiency occur in fish ontogenesis: the first in early embryogenesis (Section 4.5), and the second at metamorphosis; they are separated by depressed efficiencies during the mixed feeding period.

Diet type

Formulated diets are less effectively assimilated and converted into larval body tissue than is live food; this difference is conspicuous in stomachless fish larvae (Fig. 6.12). Depressed digestibility (e.g. Jobling, 1986b), reduced assimilability (e.g. Urban-Jezierska *et al.*, 1989) and lower gross conversion efficiency in terms of energy and protein (Lukowicz, 1979; Urban, 1982; Urban-Jezierska *et al.*, 1989) in young fishes fed compound diets, as compared with those fed live foods, is attributed by Jobling (1986b) to rapid evacuation of formulated diets from the alimentary tract (Section 6.2). Particles of dry diet remained almost unchanged microscopically during passage through the gut of *Coregonus lavaretus* larvae, while planktonic organisms were broken down within the anterior part of the alimentary tract (Rösch and Segner, 1990). Formulated diets were consumed by carp larvae in large surplus, especially by early larvae; the feed convertion ratios in terms of energy, i.e. the ratios of energy ingested daily to the energy contained in the larval body, amounted to: 9.5–15.7 at the fourth day of life, 3.4–7.7 (value for zooplankton, 2) at the 11th day, and 1.5–2.1 (zooplankton, 1.4) at the 15th day (Kamler *et al.*, in press). Brett (1971) reported for six feeding groups of *Oncorhynchus nerka* receiving different diets that the maximum food intake increased, and the conversion efficiency decreased, in the sequence of diets from the most to the least effective for growth; the same has been shown for carp larvae (Kamler *et al.*, 1987; Kamler *et al.*, in press). Early larvae aged 4–5 days, especially those receiving the formulated diet, egested large amounts of protein (Table 6.7). Older (9–10 d) larvae fed the diet continued to lose a considerable amount of protein, while the zooplankton-fed larvae reached the next developmental step (C2) and no more protein was found in their faeces + urine (*FU*). The same was found by Urban (1984) in *Ctenopharyngodon idella* fed live food: the high protein excretion shown by early larvae declined in further development. Low retention of energy and matter from formulated diets seems to be a specific property of fish larvae. Elvers and eel fingerlings of *c.* 2.5 g reached high efficiencies (K_A 94–95%, K_1 46–66%, Knights, 1985) on diets specifically designed for them, containing lower proportions of indigestible material than live foods would.

Ration size

Gross conversion efficiency (K_1) varies with ration size. It increases steeply with increasing food consumption at low ration levels, and slows down at higher food intakes. This is shown in Fig. 6.13 for *Abramis brama* larvae; the same was reported for juveniles of *Trachurus japonicus* by Azeta and Kimura (1971). Further investigations showed that at higher feeding levels, K_1 often decreases in fishes (Brett *et al.*, 1969; Elliott, 1976; Bryant and Matty, 1980; reviews: Brett and Groves, 1979; Elliott, 1982) and in other poikilotherms (review: Calow, 1977). This decrease is attributed to the enhanced metabolic expenditures for food processing (SDA) at high consumption levels. Hence, the above general relationship seems to be one more factor contributing to low growth efficiencies exhibited by early fish larvae reared indoors under conditions of excess food.

In this Chapter I demonstrated that fish larvae are different from older fish in many respects. This creates specific problems for larval-fish culturists. Questions relating to feeding of fish larvae in aquaculture will be summarized briefly in Chapter 7.

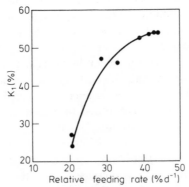

Fig. 6.13 Relationship between gross conversion efficiency $K_1 = P \times 100 / C$ and the relative feeding rate, both expressed in terms of carbon for fish and food, in *Abramis brama* larvae 28–49 mg wet weight. Negative growth was observed at feeding rates below 10% d^{-1} (from tables 2 and 3 in Panov and Sorokin, 1967).

Chapter seven

Feeding of fish larvae in aquaculture

7.1 EVALUATION OF EFFECTS

The main objectives of fish culture are to maximize survival and growth, which accordingly are measured to evaluate the effects of rearing technologies. Survival and growth are usually computed from samples taken at the beginning and at the end of the experimental period. These determinations are open to errors which may reduce the reliability of comparisons.

Estimation of survival from the initial and final numbers of fish larvae assumes a constant rate of mortality during the intervening period. In fact, the survival curve may also be concave (indicating early mortality) or convex (late mortality).

The growth of young carp larvae shorter than 10 mm reported from ponds (Fig. 6.8, Table 6.4, see also Ostroumova *et al.*, 1980 – 50–70% d^{-1}) is unexpectedly high as compared with that reported from indoor tests with natural food. A possible explanation is that metabolites produced by larvae cultured at high densities and substances lost from starters (Mukhina, 1963; Urban-Jezierska *et al.*, 1984) may – in spite of water exchange – be partly responsible for hampering larval growth. An alternative explanation invokes a stimulating effect of fluctuating temperatures (Galkovskaya and Sushchenya, 1978; Konstantinov and Zdanovich, 1985) on larval growth in the wild, an effect which would be absent under constant temperatures in controlled conditions. Another alternative is that pond data are more biased than culture data, though both in the direction of being too high. That is, greater mortality was experienced by the smaller members of a brood in laboratory experiments on *Salmo trutta* larvae (Hansen, 1985), and stronger size-selective mortality of larvae can be expected in natural water bodies where slower-growing, smaller individuals are more susceptible to predation. The growth rate computed from field data would therefore be based to a greater extent on the faster-growing survivors, and consequently, growth in ponds would be more strongly over-estimated. The opposite (under-estimation of growth) is shown by Ricker (1979) for postlarval fishes in which natural and fishery factors cause the faster-growing (larger) individuals to die earlier. The relationship between the population growth rate (i.e. that observed in survivors) and the true average growth rate of individual larvae, in connection with size variability and selective mortality, remains unsolved in the evaluation of effects of rearing technologies. Another possible source of error is larval cannibalism (Section 6.2; Bergot *et al.*, 1986).

7.2 LIVE FOODS V. FORMULATED DIETS

The principal causes of mortality of fish larvae in the wild are food deficiency, predation, unfavourable temperature and pathogens. In principle, these are eliminated in indoor cultures under controlled conditions with sufficient exogenous food supplied. However, it is not always possible to supply natural food – live zooplankton organisms taken from the field – of adequate size, at appropriate times and in sufficient amounts. The mass production of living feeds has been extensively used in Japan; their nutritive value to fish larvae depends upon the chemical properties of their own food (Watanabe *et al.*, 1978; Fujita *et al.*, 1980; review: Watanabe *et al.*, 1983). Zooplankton monocultures take up much space and skilled labour; *Artemia* nauplii are expensive. Another alternative is formulated diets.

During the past decade, studies on application of formulated diets to cyprinid and coregonid larvae have greatly advanced (Ostroumova *et al.*, 1980; Dąbrowski, 1982b; Charlon and Bergot, 1984; Rösch and Appelbaum, 1985; Charlon *et al.*, 1985, 1986; Rösch and Dąbrowski, 1986; Bergot *et al.*, 1986, 1989; Segner *et al.*, 1988; Kamler *et al.*, in press; Table 6.4). The turning point was the introduction of yeast (single-cell protein, SCP) as one of the major ingredients of formulated diets (Appelbaum, 1977; Appelbaum and Dor, 1978). Nevertheless, the development of the larvae fed formulated diets can be retarded (Section 6.1), their growth can be impeded (Section 6.3), and the energy and matter transformation efficiencies are often depressed (Section 6.5).

The inferior results of rearing fish larvae on formulated diets have been explained in many ways. The morphological, functional and physiological properties of the larval alimentary tract have been considered, as has the nature of the diets.

A short and poorly developed alimentary tract, rapid evacuation of food from it, and low production of digestive enzymes all constrain larval food digestion, especially of formulated diets (Section 6.2). However, important interspecific differences exist in that respect. Dąbrowski (1984a) assigned larval fish to three categories according to the properties of their alimentary tract. Salmonids have a functional stomach at first feeding. They, and other stomach-possessing larvae, utilize dry diets efficiently from the very beginning of exogenous feeding. For example, the instantaneous gross efficiencies K_{1i} for *Salmo trutta* of c. 200 mg wet weight fed a dry diet were c. 38% for energy and c. 45% for protein (Raciborski, 1987), values which are within the range for older fishes 5 to 250 times as heavy.

The second group are species in which the stomach develops later in ontogenesis. A delayed first intake of formulated diets, in comparison with that of live food, was shown by Rösch and Appelbaum (1985), Rösch and Dąbrowski (1986), Segner *et al.* (1988) and Rösch (1989) for *Coregonus lavaretus*, which belongs to the second group. Increased mortality during the first three weeks of feeding was attributed to the death of larvae that did not accept dry food (Segner *et al.*, 1988). However, dry food can be used as an initial food for *C. lavaretus* larvae (Rösch and Appelbaum, 1985; Segner *et al.*, 1988) despite some retardation of their growth.

Cyprinid fishes, which have no stomach throughout ontogenesis, form the third group. Eleven artificial diets were provided to *Coregonus* sp. larvae and six diets to *Rutilus rutilus* larvae during the first 40 d of active feeding (Köck and Hofer, 1989). Better results were reported for *Coregonus* sp. than for *R. rutilus*, in which growth was negligible on several diets. High food consumption (in the case of formulated diets) and accelerated gut evacuation rate leads to dilution of digestive enzymes, failure of their reabsorption in the hindgut, losses of proteins (Table 6.7), decreased growth, and increased mortality of *R. rutilus* (Köck and Hofer, 1989). *Cyprinus carpio* is a 'difficult' species to which to feed formulated diets from the very beginning of exogenous feeding (Dąbrowski, 1984a); various aspects are shown in Figs. 6.3, 6.4, 6.8, 6.11 and 6.12, and in Tables 6.3, 6.4 and 6.7.

Turning now to the nature of diets, inert particles stimulate the foraging activity of predator larvae poorly. Low effects of formulated diets have also been attributed to a deficit of essential substances. These may be highly unsaturated fatty acids, amino acids, vitamins, and minerals, singly or in combination. Lack of some substances in compound diets depresses survival and impairs growth (Bergot *et al.*, 1986, and many others) and prevents metamorphosis in fish larvae (Flüchter, 1982). However, fish larvae, with their rapid changes of nutrient requirements, are a difficult subject for routine nutritional studies (Dąbrowski, 1984a).

A lack of exogenous enzymes is another property distinguishing dry diets from live food. The proteolytic activity of zooplankton enzymes is high (Ostroumova *et al.*, 1980), and these enzymes may provide a substantial contribution to the total enzymic activity of the larval alimentary tract (Lauff and Hofer, 1984), but Ilina and Tureckij (1986) did not confirm the importance of dietary enzymes to early carp larvae at developmental steps B and C1. Attempts were made to imitate the conditions created in the larval alimentary tract by prey proteolytic enzymes. However, pretreated (hydrolysed) diets or diets with added digestive enzymes had little effect on larval growth (Aoe *et al.*, 1974; Dąbrowski *et al.*, 1979; Kamler *et al.*, 1987). Moreover, the hydrolysed starters were less stable in water than nonhydrolysed ones (Urban–Jezierska *et al.*, 1984).

In general, dry diets for larvae are susceptible to leaching because of their small particle size, *c.* 0.05–0.70 mm in diameter. Losses of nutrients from starter particles begin immediately after contact with water. Dissolved substances and particles smaller than 20 μm were defined as inaccessible to cyprinid larvae, i.e. lost. Two compounds of the losses were supposed to exist (Urban–Jezierska *et al.*, 1984): first, fine 'dust' surrounding the proper particles of diet – this is lost immediately after immersion of the starter in water and remains suspended in the water; second, substances slowly leached from the particles – these comprise vitamins, minerals, carbohydrates and free amino acids which remain dissolved in the water. The losses of dry matter, total nitrogen and protein from our earlier starters were *c.* 50% after 10 s immersion (Urban–Jezierska *et al.*, 1984). According to Littak *et al.* (1980), carp larvae consume at most one-half of the amount of diet offered. In our later diets (Kamler *et al.*, 1989), the losses of dry matter were reduced to *c.* 30–40%; the control diet Ewos C-20 was the most stable, with losses of 33.8% (range 31.4–36.3), 32.7% (30.9–34.4), and 28.7% (27.4–30.0), respectively, for the fractions 00, 0 and 1. An adverse effect on fish larvae of pollution caused by both fine suspended matter and dissolved substances is presumed to exist.

Other advantages of zooplankton over formulated diets are the ability of the former to remain suspended in water and their plasticity of shape (Van der Wind, 1979; Rösch and Dąbrowski, 1986), which permits fish larvae to ingest them more easily.

It seems, then, that the successful adaptation of fish larvae to formulated diets will not be achieved by a simple manipulation of diet composition, but that the improvement of feeding techniques is a vital and complex problem. Recommendations have been made by Dąbrowski (1984a), Szlamińska (1988) and Appelbaum (1989); here a few aspects will be listed.

Diet particle micro-encapsulation seems to be a solution to the loss of substances from diets. Prolonged suspension of food particles in the water column ensures their greater availability to fish larvae. So does frequent feeding; a rearing system which gave satisfactory growth of carp larvae fed on diets was described by Charlon and Bergot (1984), although Knights (1985) reported high energy transformation efficiency due to low energy expenditure on feeding metabolism (R_f) in young eels fed once a day. Slow water motion ensures the mobility of food particles, which attracts the larvae's attention. The size of food particles administered has to increase as larval size increases (Fig. 6.5). The need to keep tanks clean is a specific problem with dry feeds (Rösch and Appelbaum, 1985; Rösch and Dabrowski, 1986); it can be partly resolved by a special design of the rearing system (Charlon and Bergot, 1984). A tendency towards rearing fish larvae at high temperatures – approaching the upper limit for growth – is observed (Lirski, 1985), but the question of how they will adapt to life in the natural environment still remains open. Stress and other disturbances should be avoided, because of the possible depression of feeding and non-optimal partitioning of assimilated energy, leading to increased activity metabolism (R_a) and consequently a possible reduction in growth. The optimization of the amount of food given in relation to size and temperature still remains unsolved for many commercial fish larvae.

Although in some laboratory experiments, satisfactory growth of carp larvae was achieved exclusively on dry diets (Ostroumova *et al.*, 1980; Dąbrowski, 1982b; Charlon and Bergot, 1984; Charlon *et al.*, 1986; Bergot *et al.*, 1989; Kamler *et al.*, in press), dry diets should not be recommended as the only food for first feeding of cyprinid larvae in intensive cultures. The diets can be used after previous feeding with zooplankton, and at the beginning they should be complemented with live food (Imam and Habashy, 1972; Lukowicz and Rutkowski, 1976; Littak *et al.*, 1980; Bryant and Matty, 1981; Dąbrowski, 1984a,b; Okoniewska *et al.*, 1986; Kouřil and Hamáčkova, 1989; Table 6.4).

References

Abramova, N. B. and Vasileva, M. N. (1973) Some properties of embryonal mitochondria in *Misgurnus fossilis*. *Ontogenez*, **4**, 288–93. (In Russian)

Aguirre, P. (1986) Métabolisme protéique chez des larves de Corégones (*Coregonus schinzi palea*, Cuv. & Val.) et chez les larves d'Esturgeons (*Acipenser baeri* Brandt). Influence d'alimentation. Diplôme d'Etudes, Bordeaux, Univ. Bordeaux. (42 pp.)

Airaksinen, K. J. (1968) Preliminary note on the winter spawning vendace (*Coregonus albula* L.) in some Finnish lakes. *Annls zool. fenn.*, **5**, 312–14.

Albaret, J. J. (1982) Reproduction et fécondité des poissons d'eau douce de Côte d'Ivoire. *Revue Hydrobiol. trop.*, **15**, 347–71.

Alderdice, D. F. (1985) A pragmatic view of early life history studies of fishes. *Trans. Am. Fish. Soc.*, **114**, 445–51.

Alderdice, D. F. and Forrester, C. R. (1971) Effects of salinity and temperature on embryonic development of the petrale sole (*Eopsetta jordani*). *J. Fish. Res. Bd Can.*, **28**, 727–44.

Alderdice, D. F. and Forrester, C. R. (1974) Early development and distribution of the flathead sole (*Hippoglossoides elassodon*). *J. Fish. Res. Bd Can.*, **31**, 1899–918.

Alderdice, D. F. and Velsen, F. P. J. (1978) Relation between temperature and incubation time for eggs of chinook salmon (*Oncorhynchus tshawytscha*). *J. Fish. Res. Bd Can.*, **35**, 69–75.

Alderdice, D. F., Jensen, J. O. T. and Velsen, F. P. J. (1988) Preliminary trials on incubation of sablefish eggs (*Anoploma fimbria*). *Aquáculture*, **69**, 271–90.

Alderdice, D. F., Wickett, W. F. and Brett, J. R. (1958) Some effects of temporary exposure to low dissolved oxygen levels on Pacific salmon eggs. *J. Fish. Res. Bd Can.*, **15**, 229–49.

Allen, K. R. (1951) The Horokiwi stream. A study of a trout population. *Fish Bull. N. Z.*, **10**, 1–278.

Anokhina, L. E. (1960) Relationships between egg size variability and lipid contents in *Clupea harengus pallasi maris-albi* Berg. *Dokl. Akad. Nauk SSSR*, **133**, 960–3. (In Russian)

Anthony, J. S. and Millot, J. (1972) Première capture d'une femelle de Coelacanthe en état de maturité sexuelle. *C.r. hebd. Séanc. Acad. Sci., Paris*, D, **274**, 1925–6.

Anukhina, A. M. (1968) On the quality of eggs of the White Sea navaga in relation to the abundance of offspring. *Vop. Ikhtiol.*, **8**, 1004–8. (In Russian)

Aoe, H., Ikeda, K. and Saito, T. (1974) Nutrition of protein in young carp. II.

Nutritive value of protein hydrolyzates. *Bull. Jap. Soc. scient. Fish.*, **40**, 375–9.

Appelbaum, S. (1977) Geeigneter Ersatz für Lebendnahrung von Karpfenbrut? *Arch. FischWiss.*, **28**, 31–43.

Appelbaum, S. (1989) Can inert diets be used more successfully for feeding larval fish? (Thoughts based on indoor feeding behavior observations). *Pol. Arch. Hydrobiol.*, **36**, 435–7.

Appelbaum, S. and Dor, U. (1978) Ten day experimental nursing of carp (*Cyprinus carpio* L.) larvae and dry feed. *Bamidgeh*, **30**(3), 85–8.

Arutyunova, N. V. and Lizenko, E. I. (1985) Relation of lipid component contents to fertilization per cent in eggs of *Acipenser stellatus*, in, *Tezisy dokl. VI vsesoyuzn. konf. po ekologicheskoj fiziologii ryb*, *Vilnius, 1985* (eds Yu.B. Virbitskas, I. A. Barannikova, A. F. Karpevich, *et al.*) Akademiya Nauk SSSR, Vilnius, pp. 284–5. (In Russian)

Azeta, M. (1981) Some considerations on the high mortality during the larval stage of fish with special reference to the fluctuations of population—Based on the population dynamics investigations of early life history of anchovy. Report of Fisheries Resources Investigations by the Scientists of the Fisheries Agency, Japanese Government, No. 22, 7–28 (In Japanese, English summ.)

Azeta, M. and Kimura, S. (1971) Experimental estimation of food requirement of young jack mackerel, *Trachurus japonicus*. *Bull. Seikai Reg. Fish. Res. Lab.*, **39**, 15–31. (In Japanese, English summ.)

Backiel, T. (1952) Rozwój gruczołów płciowych sielawy w cyklu rocznym [Structural changes in the gonads of *Coregonus albula* during the annual cycle] *Roczn. Nauk roln.*, **64**, 271–95. (In Polish, English summ.)

Backiel, T. (1977) An equation for temperature dependent metabolism or for the 'normal curve' of Krogh. *Pol. Arch. Hydrobiol.*, **24**, 305–9.

Backiel, T. and Zawisza, J. (1988) Variations of fecundity of roach (*Rutilus rutilus*) and perch (*Perca fluviatilis*) in Polish lakes. *Pol. Arch. Hydrobiol.*, **35**, 205–25.

Bagenal, T. B. (1969) The relationship between food supply and fecundity in brown trout, *Salmo trutta* L. *J. Fish Biol.*, **1**, 167–82.

Bagenal, T. B. (1971) The interrelation of the size of fish eggs, the date of spawning and the production cycle. *J. Fish Biol.*, **3**, 207–19.

Bagenal, T. B. (1973) Fish fecundity and its relations with stock and recruitment. *Rapp. P.-v. Réun. Cons. perm. int. Explor. Mer*, **164**, 186–98.

Bagirova, Sh. M. (1976) Lipid composition of fish tissues. *Tezisy dokladov, 23 Sezd Vsesoyuzn. Gidrobiol. Obshch., Riga. 11–15 May 1976*, Zinatne, Riga, Vol. 2, pp. 107–8. (In Russian)

Bajkov, A. D. (1935) How to estimate the daily food consumption of fish under natural conditions. *Trans. Am. Fish. Soc.*, **65**, 288–9.

Ballard, W. W. (1973) Normal embryonic stages for salmonid fishes, based on *Salmo gairdneri* Richardson and *Salvelinus fontinalis* (Mitchill). *J. exp. Zool.*, **184**, 7–26.

Balon, E. (1958a) Vývoj dunajskeho kapra (*Cyprinus carpio carpio* L.) v priebehu predlárvalnej fazy a larválnej periódy. [Die Entwicklung des Donauwildkarpfens (*Cyprinus carpio carpio* L.) im Laufe der praelarvalen Phase und der larvalen Periode]. *Biol. Práce*, **4**, 5–54. (In Slovakian, German summ.)

Balon, E. (1958b) O okresowości w życiu ryb. [Über Entwicklungsstufen im Leben der Fische]. *Biul. Zakł. Biol. Stawów PAN*, **7**, 5–15. (In Polish, German summ.)

Balon, E. K. (1975a) Reproductive guilds of fishes: a proposal and definition. *J. Fish.*

Res. Bd Can., **32**, 821–64.

Balon, E. K. (1975b) Terminology of intervals in fish development. *J. Fish. Res. Bd Can.*, **32**, 1663–70.

Balon, E. K. (1977) Early ontogeny of *Labeotropheus* Ahl, 1927 (Mbuna, Cichlidae, Lake Malawi), with a discussion on advanced protective styles in fish reproduction and development. *Env. Biol. Fishes*, **2**, 147–76.

Balon, E. K. (1986) Saltatory ontogeny and evolution. *Riv. Biol. – Biol. Forum*, **79**, 151–90.

Bams, R. A. (1970) Evaluation of a revised hatchery method tested on pink and chum salmon fry. *J. Fish. Res. Bd Can.*, **27**, 1429–52.

Baranova, V. P. (1974) Dependence of the carp larvae growth rate on conditions of rearing. *Izv. Gos. Nauchno–Issled. Inst. Ozern. Rechn. Rybn. Khoz.*, **92**, 66–78 (In Russian, English summ.)

Barrahona–Fernandes, M. H. and Girin, M. (1977) Effect of different food levels on the growth and survival of laboratory–reared sea–bass larvae (*Dicentrarchus labrax* (L.)), in, 3rd Meeting of the I.C.E.S. Working Group on Mariculture, Brest, France, May 10–13, 1977, *Actes de Colloques du C.N.E.X.O.*, **4**, 69–84.

Bartel, R. (1971) Czynniki decydujące o wielkości jaj pstrąga tęczowego (*Salmo gairdneri* Richardson) [Key factors affecting egg size in rainbow trout (*Salmo gairdneri* Richardson)]. *Roczn. Nauk roln., Ser. H*, **93**, 7–34. (In Polish, English summ.)

Beacham, T. D. and Murray, C. B. (1985) Effect of female size, egg size, and water temperature on developmental biology of chum salmon (*Oncorhynchus keta*) from the Nitinat River, British Columbia. *Can. J. Fish. aquat. Sci.*, **42**, 1755–65.

Beacham, T. D., Withler, F. C. and Morley, R. B. (1985) Effect of egg size on incubation time and alevin and fry size in chum salmon (*Oncorhynchus keta*) and coho salmon (*Oncorhynchus kisutch*). *Can. J. Zool.*, **63**, 847–50.

Begon, M. (1984) Density and individual fitness: asymmetric competition, in, *Evolutionary Ecology* (23rd Symp. Br. Ecological Soc., Leeds, 1982) (ed. B. Shorrocks), Blackwell, Oxford, pp. 175–94.

Belyaeva, V. N. (1959) Respiration of sazan, *Cyprinus carpio* L. at early developmental stages. *Dokl. Akad. Nauk SSSR*, **125**, 443–5. (In Russian)

Belyi, N. D. (1967) Development of eggs of *Lucioperca lucioperca* (L.) and *Abramis brama* (L.) in the lower Dnieper waters of different salinity. *Vopr. Ikhtiol.*, **7**, 187–91. (In Russian)

Berg, K., Lumbye, J. and Ockelmann, J. (1958) Seasonal and experimental variations of the oxygen consumption of the limpet *Ancylus fluviatilis* (O. F. Müller). *J. exp. Biol.*, **35**, 43–73.

Bergot, P., Charlon, N. and Durante, H. (1986) The effect of compound diets feeding on growth and survival of coregonid larvae. *Arch. Hydrobiol., Beih Ergebn. Limnol.*, **22**, 265–72.

Bergot, P., Szlamińska, M., Escaffre, A.-M. and Charlon, N. (1989) Length–weight relationship in common carp (*Cyprinus carpio* L.) larvae fed artificial diets. *Pol. Arch. Hydrobiol.*, **36**, 503–10.

Bernatowicz, S., Dembiński, W. and Radziej, J. (1975) *Sielawa* [*Vendace*], Państwowe Wyd. Rolnicze i Leśne, Warsaw. (In Polish)

Bertalanffy, L. von (1964) Basic concepts in quantitative biology of metabolism. *Helgoländer wiss. Meeresunters.*, **9**, 5–37.

Blaxter, J. H. S. (1969) Development: eggs and larvae, in, *Fish Physiology*, Vol. 3, *Reproduction and Growth* (eds W. S. Hoar and D. J. Randall), Academic Press, NY, pp. 177–252.

Blaxter, J. H. S. and Ehrlich, K. F. (1974) Changes in behaviour during starvation of herring and plaice larvae, in *The Early Life History of Fish* (ed. J. H. S. Blaxter) Springer, Heidelberg, pp. 575–88.

Blaxter, J. H. S. and Hempel, G. (1963) The influence of egg size on herring larvae (*Clupea harengus* L.). *J. Cons. perm. int. Explor. Mer.*, **28**, 211–40.

Blaxter, J. H. S. and Hempel, G. (1966) Utilization of yolk by herring larvae. *J. mar. biol. Ass. U.K.*, **46**, 219–34.

Boehlert, G. W., Kusakari, M., Shimizu, M. and Yamada, J. (1986) Energetics during development in kurosoi, *Sebastes schlegeli* Hilgendorf. *J. exp. Mar. Biol. Ecol.*, **101**, 239–56.

Boëtius, I. and Boëtius, J. (1980) Experimental maturation of female silver eels, *Anguilla anguilla*. Estimates of fecundity and energy reserves for migration and spawning. *Dana*, **1**, 1–28.

Bogucki, M. (1928) Badania nad przepuszczalnością błon oraz ciśnieniem osmotycznym jaj ryb łososiowatych [Recherches sur la perméabilité des membranes et sur la pression osmotique des oeufs des Salmonides]. *Acta Biol. Exp. Vars.* 2(2), 19–46. (In Polish, French summ.)

Bogucki, M. (1930) Recherches sur la perméabilité des membranes et sur la pression osmotique des oeufs des Salmonides. *Protoplasma*, **9**, 345–69.

Bogucki, M. and Trzesiński, P. (1950) Fluctuations in the water and fat content of the cod. *J. Cons. perm. Int. Explor. Mer*, **16**, 208–10.

Borgmann, U. and Ralph, K. M. (1985) Feeding, growth and particle-size-conversion efficiency in white sucker larvae and young common shiners. *Env. Biol. Fishes*, **14**, 269–79.

Bottrell, H. H. (1975) The relationship between temperature and duration of egg development in some epiphytic Cladocera and Copepoda from the River Thames, Reading, with a discussion of temperature functions. *Oecologia*, **18**, 63–84.

Boulekbache, H. (1981) Energy metabolism in fish development. *Am. Zool.*, **21**, 377–89.

Boyd, M. (1928) A comparison of the oxygen consumption of unfertilized and fertilized eggs of *Fundulus heteroclitus*. *Biol. Bull. mar. biol. Lab., Woods Hole*, **55**, 92–100.

Brafield, A. E. (1985) Laboratory studies of energy budgets, in, *Fish Energetics: New Perspectives* (eds P. Tytler and P. Calow) Croom Helm, London, pp. 257–81.

Brafield, A. E. and Llewellyn, M. J. (1982) *Animal energetics*, Blackie, Glasgow.

Braginskaya, R. Ya. (1960) Developmental stages of cultured carp. *Trudy Inst. Morf. Zhivot.*, **28**, 129–49. (In Russian)

Braum, E. (1964) Experimentelle Untersuchungen zur ersten Nahrungsaufnahme und Biologie an Jungfishen von Blaufelchen (*Coregonus wartmanni* Bloch), Weissfelchen (*Coregonus fera* Jurine) und Hechten (*Esox lucius* L.). *Arch. Hydrobiol.*, suppl. **28**, v, 183–244.

Brett, J. R. (1962) Some considerations in the study of respiratory metabolism in fish, particularly salmon. *J. Fish. Res. Bd Can.*, **19**, 1025–38.

Brett, J. R. (1971) Growth responses of young sockeye salmon (*Oncorhynchus nerka*) to different diets and planes of nutrition. *J. Fish. Res. Bd Can.*, **28**, 1635–43.

Brett, J. R. (1979) Environmental factors and growth, in *Fish Physiology*, Vol. 8 (eds W. S. Hoar, D. J. Randall and J. R. Brett), Academic Press, NY, pp. 599–675.

Brett, J. R. and Groves, T. D. D. (1979) Physiological Energetics, in, *Fish Physiology*, Vol. 8 (eds W. S. Hoar, D. J. Randall and J. R. Brett), Academic Press, NY, pp. 279–352.

Brett, J. R., Shelbourn, J. E. and Shoop, C. T. (1969) Growth rate and body composition of fingerling sockeye salmon, *Oncorhynchus nerka*, in relation to temperature and ration size. *J. Fish. Res. Bd Can.*, **26**, 2363–94.

Brittain, J. E., Lillehammer, A. and Saltveit, S. J. (1984) The effect of temperature on intraspecific variation in egg biology and nymphal size in the stonefly, *Capnia atra* (Plecoptera). *J. anim. Ecol.*, **53**, 161–9.

Brown, M. E. (1957) Experimental studies on growth, in, *Physiology of Fishes*, Vol. I – *Metabolism* (ed. M. E. Brown), Academic Press, NY, pp. 361–400.

Bryant, P. L. and Matty, A. J. (1980) Optimisation of *Artemia* feeding rate for carp larvae (*Cyprinus carpio* (L.)). *Aquaculture*, 21, 203–12.

Bryant, P. L. and Matty, A. J. (1981) Adaptation of carp (*Cyprinus carpio*) larvae to artificial diets. I. Optimum feeding rate and adaptation age for a commercial diet. *Aquaculture*, **23**, 275–86.

Buddington, R. K. and Doroshov, S. I. (1986) Development of digestive secretions in white sturgeon juveniles (*Acipenser transmontanus*). *Comp. Biochem. Physiol.*, **83A**, 233–8.

Buznikov, G. A. (1955) Materials on developmental physiology and biochemistry of teleostean fishes. *Vop. Ikhtiol.*, 3, 104–25. (In Russian).

Calamari, D., Marchetti, R. and Vilati, G. (1981) Effect of long-term exposure to ammonia on the developmental stages of rainbow trout (*Salmo gairdneri* Richardson). *Rapp. P.-v. Réun. Cons. perm. int. Explor. Mer*, **178**, 81–6.

Calow, P. (1977) Conversion efficiencies in heterotrophic organisms. *Biol. Rev.*, **52**, 385–409.

Calow, P. (1978) *Life Cycles: an evolutionary approach to the physiology of growth, reproduction and ageing.* Chapman and Hall, London.

Calow, P. (1979) The cost of reproduction – a physiological approach. *Biol. Rev.*, **54**, 23–40.

Calow, P. (1984) Economics of ontogeny – adaptational aspects, in, *Evolutionary Ecology* (23rd Symp. Br. Ecological Soc. Leeds, 1982) (ed. B. Shorrocks), Blackwell Sci. Publ., Oxford, pp. 81–104.

Calow, P. C. (1985) Adaptive aspects of energy allocation, in *Fish Energetics: New Perspectives* (eds P. Tytler and P. Calow), Croom Helm, London, pp. 13–31.

Carlson, A. R. and Siefert, R. E. (1974) Effect of reduced oxygen on embryos and larvae of lake trout (*Salvelinus namaycush*) and largemouth bass (*Micropterus salmoides*). *J. Fish. Res. Bd Can.*, **31**, 1393–6.

Carrick, T. R. (1981) Oxygen consumption in the fry of brown trout (*Salmo trutta* L.) related to pH of water. *J. Fish Biol.*, **18**, 73–80.

Charlon, N. and Bergot, P. (1984) Rearing system for feeding fish larvae on dry diets. *Aquaculture*, **41**, 1–9.

Charlon, N., Durante, H., Escaffre, A. M. and Bergot, P. (1985) Alimentation artificielle des larves de carpe (*Cyprinus carpio* L.). Posters summary, in, Symp. Aquaculture Carp and Related Species, Evry, 2–5 September 1985, INRA, ITAVI, p. 31.

Charlon, N., Durante, H., Escaffre, A. M. and Bergot, P. (1986) Alimentation artificielle des larves de carpe (*Cyprinus carpio* L.). *Aquaculture*, **54**, 83–8.

Chechenkov, A. V. (1973) Vendace physiological state as related to age, sex and maturation, in *Ekologicheskaya Fiziologiya Ryb. Tezisy Dokl. Vsesoyuzn. Konf. po Ekologicheskoj Fiziologii Ryb, Moskva, 1973*) (ed. N. S. Stroganov), Minist. Rybnogo Khoz. SSSR, Moscow, pp. 112–14. (In Russian)

Chełkowski, Z., Domagała, J. and Trzebiatowski, R. (1986) Zależność rozmiarów ikry od wielkości troci (*Salmo trutta* L.) rzeki Regi pod kątem selekcji stada [Application of the relationship between egg size and female size in the selection of the Rega River trout (*Salmo trutta* L.) for reproduction]. Abstract, 13th Congr. Polish Hydrobiological Soc., Szczecin, 16–19 September 1986, Pol. Tow. Hydrobiol. and Akademia Rolnicza, Szczecin, pp. 27–8. (In Polish)

Chełkowski, Z., Domagała, J., Trzebiatowski, R. and Woźniak, Z. (1985) The effect of the size of sea trout (*Salmo trutta* L.) females from Rega River upon the size of eggs. *Acta Ichthyol. Piscatoria*, **15**, 37–71.

Chernyaev, Zh. A. (1981) Effect of temperature and light factors on embryonic development of coregonid fishes in the Baikal Lake, in, *Tezisy dokl. vtorogo vsesoyuzn. soveshch. po biologii i biotekhnike razvedeniya sigovykh ryb, Petrozavodsk, 1981* (ed. Yu. S. Reshetnikov), Minist. Rybnogo Khoz. SSSR, Akademiya Nauk SSSR, Petrozavodsk, pp. 22–5. (In Russian)

Ciechomski, J. de (1966) Development of the larvae and variations in the size of the eggs of the Argentine anchovy, *Engraulis anchoita* Hubbs and Marini. *J. Cons. perm. int. Explor. Mer*, **30**, 281–90.

Ciechomski, J. D. de (1973) The size of the eggs of the Argentine anchovy, *Engraulis anchoita* (Hubbs & Marini) in relation to the season of the year and to the area of spawning. *J. Fish Biol.*, **5**, 393–8.

Ciepielewski, W. (1971) Tworzenie się pierścienia rocznego na łuskach sielawy (*Coregonus albula* L.) [Annual ring formation on the scales of *Coregonus albula* L.]. *Roczn. Nauk Roln., Ser. H.*, **93**, 25–34. (In Polish, English summ.)

Coburn, M. M. (1986) Egg diameter variation in eastern North American minnows (Pisces: Cyprinidae): correlation with vertebral number, habitat and spawning behaviour. *Ohio J. Sci.*, **86**, 110–20.

Cole, L. C. (1954) The population consequences of life history phenomena. *Q. Rev. Biol.*, **29**, 103–37.

Constantz, G. D. (1979) Life history patterns of a livebearing fish in contrasting environments. *Oecologia*, **40**, 189–201.

Cooney, J. D. and Gehrs, C. W. (1980) The relationship between egg size and naupliar size in the calanoid copepod *Diaptomus clavipes* Schacht. *Limnol. Oceanogr.*, **25**, 549–52.

Cowx, I. G. (1990) The reproductive tactics of roach, *Rutilus rutilus* (L.) and dace *Leuciscus leuciscus* (L.) populations in the rivers Exe and Culm, England. *Pol. Arch. Hydrobiol.*, **37**, 193–208.

Craig, J. F. (1977) The body composition of adult perch, *Perca fluviatilis*, in Windermere with reference to seasonal changes and reproduction. *J. Anim. Ecol.*, **46**, 617–32.

Craig, J. F. (1978) A note on ageing in fish with special reference to the perch, *Perca fluviatilis* L. *Verh. Internat. Verein. Limnol.*, **20**, 2060–4.

Craig, J. F. (1980) Growth and production of the 1955 to 1972 cohorts of perch,

Perca fluviatilis L. in Windermere. *J. Anim. Ecol.*, **49**, 291–315.

Craig, J. F. (1982) Population dynamics of Windermere perch. *Rep. Freshwat. Biol. Ass.*, **50**, 49–59.

Craig, J. F. (1985) Aging in fish. *Can. J. Zool.*, **63**, 1–8.

Craig, J. F. and Kipling, C. (1983) Reproduction effort versus the environment; case histories of Windermere perch, *Perca fluviatilis* L., and pike, *Esox lucius* L. *J. Fish Biol.*, **22**, 713–27.

Craig, J. F., Kenley, M. J. and Talling, J. F. (1978) Comparative estimations of the energy content of fish tissue from bomb calorimetry, wet oxidation and proximate analysis. *Freshwat. Biol.*, **8**, 585–90.

Cummins, K. W. and Wuycheck, J. C. (1971) Caloric equivalents for investigations in ecological energetics. *Mitt. int. Verein. theor. angew. Limnol.*, **18**, 1–158.

Dąbrowska, H., Grudniewski, C. and Dąbrowski, K. (1979) Artificial diets for common carp: effect of addition of enzyme extracts. *ProgveFish Cult.*, **41**(4), 196–200.

Dąbrowski, K. (1975) Okres krytyczny w życiu wylęgu ryb. Próba energetycznego określenia minimum pokarmowego. [Point of no return in the early life of fishes. An energetic attempt to define the food minimum]. *Wiad. ekol.*, **21**, 277–93. (In Polish, English summ.)

Dąbrowski, K. (1976) An attempt to determine the survival time for starving fish larvae. *Aquaculture*, **8**, 189–93.

Dąbrowski, K. R. (1981) The spawning and early life history of the pollan (*Coregonus pollan* Thompson) in Lough Neagh, Northern Ireland. *Int. Revue ges. Hydrobiol.*, **66**, 299–326.

Dąbrowski, K. (1982a) Seasonal changes in the chemical composition of fish body and nutritional value of the muscle of the pollan (*Coregonus pollan* Thompson) from Lough Neagh, Northern Ireland. *Hydrobiologia*, **87**, 121–41.

Dąbrowski, K. (1982b) Further study on diet formulation for common carp larvae. *Riv. It. Piscic. Ittiop.*, **27**, 11–29.

Dąbrowski, K. (1983) A note on the energy transformation in body and gonad of coregonid fish. *Arch. Hydrobiol.*, **97**, 406–14.

Dąbrowski, K. (1984a) The feeding of fish larvae: present 'state of the art' and perspectives. *Reprod. Nutr. Dévelop.*, **24**, 807–33.

Dąbrowski, K. (1984b) Influence of initial weight during the change from live to compound feed on the survival and growth of four cyprinids. *Aquaculture*, **40**, 29–40.

Dąbrowski, K. R. (1985) Energy budget of coregonid (*Coregonus* spp.) fish growth, metabolism and reproduction. *Oikos*, **45**, 358–64.

Dąbrowski, K. (1986a) Ontogenetical aspects of nutritional requirements in fish. *Comp. Biochem. Physiol.*, **85A**, 639–55.

Dąbrowski, K. R. (1986b) A new type of metabolism chamber for the determination of active and postprandial metabolism of fish, and consideration of results for coregonid and salmon juveniles. *J. Fish Biol.*, **28**, 105–17.

Dąbrowski, K. (1986c) Energy utilization during swimming and cost of locomotion in larval and juvenile fish. *J. appl. Ichthyol.*, **3**, 110–17.

Dąbrowski, K. (1986d) Active metabolism in larval and juvenile fish: ontogenetic changes, effect of water temperature and fasting. *Fish Physiol. Biochem.*, **1**, 125–44.

Dąbrowski, K., Kaushik, S. J. and Bergot, P. (1986a) Active metabolism in young coregonids: a new approach involving oxygen consumption and ammonia excretion. *Arch. Hydrobiol. Beih Ergebn. Limnol.*, **22**, 151–9.

Dabrowski, K., Łuczyński, M., Czeczuga, B. and Falkowski, S. (1987) Relationships among fish reproductive effort, carotenoid content in eggs and survival of embryos. *Arch. Hydrobiol. Suppl.* 79, **1**, 29–48.

Dąbrowski, K., Takashima, F. and Strüssmann, C. (1986b) Does recovery growth occur in larval fish? *Bull. Jap. Soc. scient. Fish.*, **52**, 1869.

Dahlgren, B. T. (1985) Relationships between population density and reproductive success in the female guppy, *Poecilia reticulata* (Peters). *Ekol. pol.*, **33**, 677–703.

Danzmann, R. G., Ferguson, M. M. and Allendorf, F. W. (1987) Heterozygosity and oxygen-consumption rate as predictors of growth and developmental rate of rainbow trout. *Physiol. Zool.*, **60**, 211–20.

Danzmann, R. G., Ferguson, M. M., Allendorf, F. W. and Knudsen, K. L. (1986) Heterozygosity and developmental rate in a strain of rainbow trout (*Salmo gairdneri*). *Evolution*, **40**, 86–93.

Daoulas, Ch. and Economou, A. N. (1986) Seasonal variation of egg size in the sardine, *Sardina pilchardus* Walb., of the Saronikos Gulf: causes and a probable explanation. *J. Fish Biol.*, **28**, 449–57.

Dave, G. (1985) The influence of pH on the toxicity of aluminium, cadmium, and iron to eggs and larvae of the zebrafish, *Brachydanio rerio. Ecotoxicol. Environ. Safety*, **10**, 253–67.

Davenport, J. and Lønning, S. (1980) Oxygen uptake in developing eggs and larvae of the cod *Gadus morhua* L. *J. Fish Biol.*, **16**, 249–56.

DeMartini, E. E. and Fountain, R. K. (1981), Ovarian cycling frequency and batch fecundity in the queenfish, *Seriphus politus*: attributes representative of serial spawning fishes. *Fishery Bull.*, **79**, 547–60.

De Silva, C.D., Premawansa, S. and Keembiyahetty, C. N. (1986) Oxygen consumption in *Oreochromis niloticus* (L.) in relation to development, salinity, temperature and time of day. *J. Fish Biol.*, **29**, 267–77.

Detlaf, T. A. and Detlaf, A. A. (1960) On dimensionless characteristics of the duration of development in embryology. *Dokl. Akad. Nauk. SSSR*, **134**, 199–202. (In Russian)

Devillers, C. (1965) Respiration et morphogenèse dans l'oeuf des Téléostéens. *Année biol.*, **4**, 157–86.

Devillers, C., Thomapoulos, A. and Colas, J. (1953) Différenciation bipolaire et formation de l'espace perivitellin dans l'oeuf de *Salmo irideus. Bull. Soc. zool. Fr.*, **78**, 462–70.

DeVlaming, V., Grossman, G. and Chapman, F. (1982) On the use of the gonosomatic index. *Comp. Biochem. Physiol.*, **73A**, 31–9.

Diana, J. S. and Mackay, W. C. (1979) Timing and magnitude of energy deposition and loss in the body, liver and gonads of northern pike (*Esox lucius*). *J. Fish. Res. Bd Can.*, **36**, 481–7.

DiMichele, L. and Taylor, M. H. (1980) The environmental control of hatching in *Fundulus heteroclitus. J. exp. Zool.*, **214**, 181–7.

DiMichele, L., Powers, D. A. and DiMichele, J. A. (1986) Developmental and physiological consequences of genetic variation at enzyme synthesizing loci in *Fundulus heteroclitus. Am. Zool.*, **26**, 201–8.

Dmitrieva, E. N. (1957) Morpho-ecological analysis of two species of *Carassius*. *Trudy Inst. Morf. Zhivot.*, **16**, 102–70. (In Russian)

Dmitrieva, E. N. (1960) Some data on the development of *Lucioperca lucioperca* in the Rybinskoe Dam Reservoir. *Trudy Inst. Morf. Zhivot.*, **28**, 107–28. (In Russian)

Docker, M. F., Medland, T. E. and Beamish, F. W. H. (1986) Energy requirements and survival in embryo mottled sculpin (*Cottus bairdi*). *Can. J. Zool.*, **64**, 1104–9.

Domurat, J. (1956) Rozwój embrionalny troci (*Salmo trutta* L.), szczupaka (*Esox lucius* L.) i płoci (*Rutilus rutilus* L.) w środowisku bezwodnym [Embryonic development of trout (*Salmo trutta* L.), pike (*Esox lucius* L.) and roach (*Rutilus rutilus* L.) in the environment deprived of water]. *Pol. Arch. Hydrobiol.*, **3**, 167–73. (In Polish, English summ.)

Domurat, J. (1958) Rozwój embrionalny szczupaka (*Esox lucius* L.) w oleju parafinowym [Development embryonnaire du brochet dans l'huile de paraffine]. *Pol. Arch. Hydrobiol.*, **5**, 19–23. (In Polish, French summ.)

Donaldson, L. R. and Olson, P. (1955) Development of rainbow trout brood stock by selective breeding. *Trans. Am. Fish. Soc.*, **15**, 93–101.

Doudoroff, P. and Shumway, D. L. (1970) Dissolved oxygen requirements of freshwater fishes. *F.A.O. Fish. Tech. Pap.* No. 86, F.A.O., Rome, 291 pp.

Dowgiałło, A. (1975) Chemical composition of an animal's body and of its food, in, *Methods for Ecological Bioenergetics* (IBP Handbook No. 24) (eds W. Grodziński, R. Z. Klekowski and A. Duncan) Blackwell Sci. Publ., Oxford, pp. 160–85.

Drost, M. R. and Boogaart, J. G. M. van den (1985) Food uptake mechanism of larval carp. Posters summary, in, *Symp. Aquaculture Carp and Related Species, Evry, 2–5 September 1985*, INRA, ITAVI, p. 32.

Drost, M. R. and Boogaart, J. G. M. van den (1986) The energetics of feeding strikes in larval carp, *Cyprinus carpio*. *J. Fish Biol.*, **29**, 371–9.

Drost, M. R., Osse, J. W. M. and Muller, M. (1988) Prey capture by fish larvae, water flow patterns and the effect of escape movements of prey. *Neth. J. Zool.*, **38**, 23–45.

Duncan, A. (1983) The influence of temperature upon the duration of embryonic development of tropical *Brachionus* species (Rotifera), in, *Limnology of Parakrama Samudra – Sri Lanka*, (ed. F. Schiemer), Dr W. Junk, The Hague, pp. 107–15.

Duncan, A. and Klekowski, R. Z. (1975) Parameters of an energy budget, in, *Methods for Ecological Bioenergetics* (IBP Handbook No. 24) (eds W. Grodzinski, R. Z. Klekowski and A. Duncan), Blackwell Sci. Publ., Oxford, pp. 97–147.

Durante, H. (1986) Influence d'alimentation sur la croissance des larves de la carpe (*Cyprinus carpio*) et de coregone (*Coregonus shinzi palea*). Aspects morphologiques. Thèse de 3 Cycle, Univ. Bordeaux, Bordeaux, France. (173 pp).

Dziekońska, J. (1956) Studies on embryonic development of bream in the Vistula Lagoon. *Pol. Arch. Hydrobiol.*, **3**, 291–305. (In Polish, English summ.)

Dziekońska, J. (1958) Studies on early development stages of fish. II. The influence of some environment conditions on the embryonic development of bream (*Abramis brama* L.) in the Vistula Delta. *Pol. Arch. Hydrobiol.*, **4**, 193–206. (In Polish, English summ.)

Eckmann, R. (1987) A comparative study on the temperature dependence of embryogenesis in three coregonids (*Coregonus* spp.) from Lake Constance. *Schweiz. Z. Hydrol.*, **49**, 353–62.

Ege, E. and Krogh, A. (1914) On the relation between the temperature and the respiratory exchange in fishes. *Int. Rev. ges. Hydrobiol. Hydrog.*, **7**, 48–55.

Ehrlich, K. F. and Muszynski, G. (1981) The relationship between temperature-specific yolk utilization and temperature selection of larval grunion. *Rapp. P.-v. Réun. Cons. perm. Int. Explor. Mer*, **178**, 312–13.

Ehrlich, K. F., Blaxter, J. H. S. and Pemberton, R. (1976) Morphological and histological changes during the growth and starvation of herring and plaice larvae. *Mar. Biol.*, **35**, 105–18.

Eldridge, M. B., Echeverria, T. and Whipple, J. A. (1977) Energetics of Pacific herring (*Clupea harengus pallasi*) embryos and larvae exposed to low concentrations of benzene, a monoaromatic component of crude oil. *Trans. Am. Fish. Soc.*, **106**, 452–61.

Eldridge, M. B., Whipple, J. A. and Bowers, M. J. (1982) Bioenergetics and growth of striped bass, *Morone saxatilis*, embryos and larvae. *Fishery Bull.*, **80**, 461–74.

Eldridge, M. B., Whipple, J. and Eng. D. (1981a). Exogenous energy sources as factors affecting mortality and development in striped bass (*Morone saxatilis*) eggs and larvae. *Rapp. P.-v. Réun. Cons. perm. int. Explor. Mer*, **178**, 568–70.

Eldridge, M. B., Whipple, J. A., Eng. D., Bowers, M. J. and Jarvis, B. M. (1981b) Effects of food and feeding factors on laboratory-reared striped bass larvae. *Trans. Am. Fish. Soc.*, **110**, 111–20.

El-Fiky, N. and Wieser, W. (1988) Life styles and patterns of development of gills and muscles in larval cyprinids (Cyprinidae; Teleostei). *J. Fish Biol.*, **33**, 135–45.

El-Fiky, N., Hinterleitner, S. and Wieser, W. (1987) Differentiation of swimming muscles and gills, and development of anaerobic power in the larvae of cyprinid fish (Pisces, Teleostei). *Zoomorphology*, **107**, 126–32.

Elliott, J. M. (1976) The energetics of feeding, metabolism and growth of brown trout (*Salmo trutta* L.) in relation to body weight, water temperature and ration size. *J. Anim. Ecol.*, **45**, 923–48.

Elliott, J. M. (1979) Energetics of freshwater teleosts. *Symp. zool. Soc. Lond.*, **44**, 29–61.

Elliott, J. M. (1982) The effects of temperature and ration size on the growth and energetics of salmonids in captivity. *Comp. Biochem. Physiol.*, **73B**, 81–91.

Elliott, J. M. (1984) Numerical changes and population regulation in young migratory trout, *Salmo trutta*, in a Lake District stream, 1966–1983. *J. Anim. Ecol.*, **53**, 327–50.

Elliott, J. M. and Davison, W. (1975) Energy equivalents of oxygen consumption in animal energetics. *Oecologia* (*Berl.*), **19**, 195–201.

Elliott, J. M. and Persson, L. (1978) The estimation of daily rates of food consumption for fish. *J. Anim. Ecol.*, **47**, 977–91.

Embody, G. C. (1934) Relation of temperature to the incubation periods of eggs of four species of trout. *Trans. Am. Fish. Soc.*, **64**, 281–92.

Engelmann, M. D. (1966) Energetics, terrestrial field studies and animal productivity. *Adv. Ecol. Res.*, **3**, 73–115.

Epler, P., Bieniarz, K. and Horoszewicz, L. (1981a), Effect of different thermal regimes on reproductive cycles of tench (*Tinca tinca* L.). Part III. Histological characteristics of ovaries. *Pol. Arch. Hydrobiol.*, **28**, 197–205.

Epler, P., Kamler, E. and Bieniarz, K. (1981b) Investigations on artificial maturation of females of European eel (*Anguilla anguilla* L.); quality of eggs. *Pol. Arch. Hydrobiol.*, **28**, 135–46.

Erickson, D. L., Hightower, J. E. and Grossman, G. D. (1985) The relative gonadal index: an alternative index for quantification of reproductive condition. *Comp. Biochem. Physiol.*, **81A**, 117–20.

Escaffre, A.-M. and Bergot, P. (1984) Utilization of the yolk in rainbow trout alevins (*Salmo gairdneri* Richardson): effect of egg size. *Reprod. Nutr. Dévelop.*, **24**, 449–60.

Escaffre, A.-M. and Bergot, P. (1985) Effet d'une alimentation précoce ou retardée sur la croissance d'alevins de truite arc-en-ciel (*Salmo gairdneri*) issus d'oeufs de tailles différentes. *Bull. Fr. Pêche Piscic.*, **296**, 17–28.

Fadeev, N. S. (1987) *North Pacific Pleuronectinae*, Agropromizdat, Moscow.

Fänge, R. and Grove, D. (1979) Digestion, in, *Fish Physiology*, Vol. 8 (eds W. S. Hoar, D. J. Randall and J. R. Brett), Academic Press, NY, pp. 162–260.

Farris, D. A. (1959) A change in the early growth rates of four larval marine fishes. *Limnol. Oceanogr.*, **4**, 29–36.

Fauconneau, B. (1984) The measurement of whole body protein synthesis in larval and juvenile carp (*Cyprinus carpio* L.). *Comp. Biochem. Physiol.*, **78B**, 845–50.

Faustov, V. S. and Zotin, A. I. (1967) Energetical characteristics of fish eggs belonging to different ecological groups. *Vop. Ikhtiol.*, **7**, 81–7. (In Russian)

Fenerich–Verani, N., Godinho, H. M. and Narahara, M. Y. (1984) The size composition of the eggs of curimbatá, *Prochilodus scrofa* Steindachner 1881, induced to spawn with human chorionic gonadotropin (HCG). *Acquaculture*, **42**, 37–41.

Ferguson, M. M., Danzmann, R. G. and Allendorf, F. W. (1985a) Developmental divergence among hatchery strains of rainbow trout (*Salmo gairdneri*). I. Pure strains. *Can. J. Genet. Cytol.*, **27**, 289–97.

Ferguson, M. M., Danzmann, R. G. and Allendorf, F. W. (1985b) Developmental divergence among hatchery strains of rainbow trout (*Salmo gairdneri*). II. Hybrids. *Can. J. Genet. Cytol.*, **27**, 298–307.

Fischer, Z. (1970a) Some remarks about food ration. *Pol. Arch. Hydrobiol.*, **17**, 177–82.

Fischer, Z. (1970b) Bilans energii i materii białego amura (*Ctenopharyngodon idella* Val.) żywionego pokarmem zwierzęcym lub roślinnym [Energy and matter budgets of grass carp (*Ctenopharyngodon idella* Val.) fed on animal or plant food]. Typescript., Inst. Biol. Doświadczalnej PAN, Warszawa, 89 pp. (In Polish)

Fischer, Z. (1976) Nitrogen conversion in carp (*Cyprinus carpio* L.). *Pol. Arch. Hydrobiol.*, **23**, 309–26.

Fischer, Z. (1983) Changes in nitrogen compounds in fish and their ecological consequences, in, *Nitrogen as an Ecological Factor* (22nd Symp. Br. Ecol. Soc., Oxford, 1981) (eds J. A. Lee, S. McNeill and I. H. Rorison) Blackwell Sci. Publ., Oxford, pp. 369–79.

Fischer, Z. and Lipka, J. (1983) The role of amino acids in nitrogen metabolism of artificially fed carp. *Pol. Arch. Hydrobiol.*, **30**, 363–79.

Fisher, R. A. (1930) *The Genetical Theory of Natural Selection*, Clarendon Press (O.U.P.), Oxford.

Florez, F. (1972) The effect of temperature on incubation time, growth and lethality of embryos, larvae and juveniles of the ide, *Idus idus* (L.). *Rep. Inst. Freshwat. Res. Drottningholm*, **52**, 50–64.

Flüchter, J. (1980) Review of the present knowledge of rearing whitefish (Coregonidae) larvae. *Aquaculture*, **19**, 191–208.

Flüchter, J. (1982) Substances essential for metamorphosis of fish larvae extracted from *Artemia. Aquaculture,* **27**, 83–5.

Fry, F. E. J. (1947) Effects of the environment on animal activity (Univ. Toronto Studies Biol. Ser. No. 55) *Publ. Ontario Fisheries Res. Lab.,* **68**, 62 pp.

Fry, F. E. (1957) The aquatic respiration of fish, in, *The Physiology of Fishes,* Vol. 1, *Metabolism* (ed. M. E. Brown), Academic Press, NY, pp. 1–63.

Fuiman, L. A. (1984) Ostariophysi: development and relationships, in *Ontogeny and Systematics of Fishes.* (An international symposium dedicated to the memory of Elbert Halvor Ahlstrom, La Jolla, 15–18 August 1983). Am. Soc. Ichthyol. and Herpetol., Spec. Publ. No. 1, pp. 126–37.

Fujita, S. and Yogata, T. (1984) Induction of ovarian maturation, embryonic development and larvae and juveniles of the amberjack, *Seriola aureovittata. Jap. J. Ichthyol.,* **30**, 426–34. (In Japanese, English summ.)

Fujita, S., Watanabe, T. and Kitajima, C. (1980) Nutritional quality of *Artemia* from different localities as a living feed for marine fish from the viewpoint of essential fatty acids, in *The Brine Shrimp Artemia.* Vol. 3. *Ecology, culturing, use in Aquaculture* (eds G. Persoone, P. Sorgeloos, O. Roels and E. Jaspers), Universa Press, Vetteren, pp. 277–90.

Galicka, W. (1984) Body composition of young rainbow trout (*Salmo gairdneri* Richardson) bred in cages in Głębokie Lake. *Pol. Arch. Hydrobiol.,* **31**, 371–85.

Galkina, Z. I. (1967) Quality of rainbow trout females and their offspring, in, *Obmen Veshchestv i Biokhimiya Ryb* (ed. G S. Karzinkin), Nauka, Moscow, pp. 75–9. (In Russian)

Galkina, Z. I. (1970) The dependence of egg size on the size and age of female salmon (*Salmo salar* L. and *Salmo irideus* Gib.). *Vop. Ikhtiol.,* **10**, 827–37. (In Russian)

Galkovskaya, G. A. and Sushchenya, L. M. (1978) *Growth of Aquatic Animals at Fluctuating Temperatures.* Nauka i Tekhnika, Minsk. (In Russian)

Galkovskaya, G. A., Semenyuk, G. A., Mityanina, I. F. and Eremova, N. G. (1976) Age changes of growth efficiency in some planktonic crustaceans. *Tezisy dokl. III Sezd. vses. gidrobiol. obshch. 3,* Zinatne, Riga, pp. 216–19. (In Russian)

Gall, G. A. E. (1974) Influence of size of eggs and age of female on hatchability and growth in rainbow trout. *Calif. Fish Game,* **60**, 26–35.

Garcia, A. and Braña, F. (1988) Reproductive biology of brown trout (*Salmo trutta* L.) in the Aller River (Asturias, Northern Spain). *Pol. Arch. Hydrobiol.,* **35**, 361–73.

Garside, E. T. (1966) Effects of oxygen in relation to temperature on the development of embryos of brook trout and rainbow trout. *J. Fish. Res. Bd Can.,* **23**, 1121–34.

Gerasimova, T. D. (1973) Metabolic indices in mature sexual products and in body of developing embryos of mirror carp, in, *Ekologicheskaya Fiziologiya Ryb* (Tezisy dokl. vsesoyuzn. konf. po ekologicheskoj fiziologii ryb, Moskva, 1973) (ed. N. S. Stroganov), Minis. Rybnogo Khoz. SSSR, Moscow, pp. 97–8. (In Russian)

Gershanovich, A. D. (1983) Influence of temperature on the metabolism, growth and food requirements of *Huso huso* and *Acipenser nudiventris* early developmental stages. *Vop. Ikhtiol.,* **23**, 238–43. (In Russian, English summ.)

Gershanovich, A. D., Pegasov, V. A. and Shatunovskij, M. I. (1987) *Ecology and Physiology of Juvenile Sturgeons,* Agropromizdat, Moscow. (In Russian)

Giese, A. (1967) Some methods for study of the biochemical constitution of marine invertebrates. *Oceanogr. mar. Biol. Ann. Rev.,* **5**, 159–86.

Gnaiger, E. (1980) Alters- und Phasenverteilung in der Populationsdynamik: Der additive Effekt schneller Temperaturänderung auf die Eienentwicklung von *Cyclops abyssorum tatricus*. *Jber. Abt. Limnol. Innsbruck*, **6**, 99–115.

Gnaiger, E. and Bitterlich, G. (1984) Proximate biochemical composition and caloric content calculated from elemental CHN analysis: a stoichimetric concept. *Oecologia (Berlin)*, **62**, 289–98.

Goryczko, K. (1968) Żywienie ryb łososiowatych paszą granulowaną [Feeding of salmonid fish with granulated food]. *Opracowanie Zakładu Gospodarki Rybackiej w Rzekach i Zbiornikach Zaporowych. Inst. Rybactwa Śródlądowego*, No. 26, 15 pp. (In Polish)

Goryczko, K. (1972) Inkubacja jaj, pochodzących z jesiennego tarła pstrąga tęczowego (*Salmo gairdneri* Richardson) [Incubation of eggs from the autumn spawning of rainbow trout (*Salmo gairdneri* Richardson)]. *Roczn. Nauk roln., Ser. H.*, **94**, 69–80. (In Polish, English summ.)

Gosh, R. I. (1985) *Metabolism of Fish Sexual Products and Embryos*, Naukova Dumka, Kiev. (In Russian)

Gosh, R. I. and Zhuhinskij, V. N. (1979) Efficiency of energy metabolism in *Rutilus rutilus heckeli* and *Abramis brama* eggs as related to female age. *Gidrobiol. Zhurnal*, **15**(3), 59–63. (In Russian, English summ.)

Gray, J. (1926), The growth of fish. 1. The relationship between embryo and yolk in *Salmo fario*. *Br. J. exp. Biol.*, **4**, 215–25.

Gray, J. (1928a) The growth of fish. 2. The growth rate of the embryo of *Salmo fario*. *Br. J. exp. Biol.*, **6**, 110–24.

Gray, J. (1928b), The growth of fish. 3. The effect of temperature on the development of the eggs of *Salmo fario*. *Br. J. exp. Biol.*, **6**, 125–30.

Grodziński, Z. (1961) *Anatomia i Embriologia Ryb* [*Anatomy and Embryology of Fishes*], Państwowe Wyd. Rolnicze i Leśne, Warsaw. (In Polish)

Grudniewski, C. (1961) Rozwój niektórych morfologicznych ocen okresu larwalnego troci jeziorowej (*Salmo trutta morpha lacustris* L.) z jeziora Wdzydze [Development of some morphological characters of the larval period in lake trout (*Salmo trutta morpha lacustris* L.) from Wdzydze Lake]. *Roczn. nauk Roln., Ser. D*, **93**, 568–622. (In Polish)

Gulidov, M. V. (1970) Morpho-physiological peculiarities of fish development in different oxygen conditions. Summary of PhD Thesis No. 097–Zoology, Inst. Morfologii i Ekologii Zhivotnykh im. A. N. Severtsova, Moscow, pp. 1–16. (In Russian)

Gunnes, K. (1979) Survival and development of Atlantic salmon eggs and fry at three different temperatures. *Aquaculture*, **16**, 211–19.

Hamor, T. and Garside, E. T. (1976) Developmental rates of embryos of Atlantic salmon, *Salmo salar* L., in response to various levels of temperature, dissolved oxygen, and water exchange. *Can. J. Zool.*, **54**, 1912–17.

Hamor, T. and Garside, E. T. (1977) Size relations and yolk utilization in embryonated ova and alevins of Atlantic salmon *Salmo salar* L. in various combinations of temperature and dissolved oxygen. *Can. J. Zool.*, **55**, 1892–8.

Hamor, T. and Garside, E. T. (1979) Hourly and total oxygen consumption by ova of Atlantic salmon *Salmo salar* L. during embryogenesis, at two temperatures and three levels of dissolved oxygen. *Can. J. Zool.*, **57**, 1196–200.

Hansen, T. (1985) Artificial hatching substrate: effect on yolk absorption, mortality and growth during first feeding of sea trout (*Salmo trutta*). *Aquaculture*, **46**, 275–85.

Hansen, T. J. and Møller, D. (1985) Yolk absorption, yolk sac constrictions, mortality, and growth during first feeding of Atlantic salmon (*Salmo salar*) incubated on astro-turf. *Can. J. Fish. aquat. Sci.*, **42**, 1073–8.

Hartmann, J. (1986) Interspecific predictors of selected prey of young fishes. *Arch. Hydrobiol.*, **22**, 373–86.

Hartree, E. F. (1972) Determination of protein: a modification of the Lowry method that gives a linear photometric response. *Analyt. Biochem.*, **48**, 422–8.

Hayes, F. R. (1949) The growth, general chemistry, and temperature relations of salmonid eggs. *Q. Rev. Biol.*, **24**, 281–308.

Hayes, F. R. and Armstrong, F. H. (1942) Physical changes in the constituent parts of developing salmon eggs. *Can. J. Res.*, D, **20**, 99–114.

Hayes, F. R. and Armstrong, F. H. (1943) Growth of the salmon embryo. *Can. J. Res.*, D, **21**, 19–33.

Hayes, F. R., Wilmot, J. R. and Livingstone, D. A. (1951) The oxygen consumption of the salmon egg in relation to development and activity. *J. exp. Zool.*, **116**, 377–95.

Heming, T. A. (1982) Effects of temperature on utilization of yolk by chinook salmon (*Oncorhynchus tshawytscha*) eggs and alevins. *Can. J. Fish. aquat. Sci.*, **39**, 184–90.

Heming, T. A., McInerney, J. E. and Alderdice, D. F. (1982) Effect of temperature on initial feeding in alevins of chinook salmon (*Oncorhynchus tshawytscha*). *Can. J. Fish. aquat. Sci.*, **39**, 1554–62.

Hemmingsen, A. M. (1960) Energy metabolism as related to body size and respiratory surfaces, and its evolution. *Rep. Steno mem. Hosp.*, **9**, 1–110.

Hempel, G. and Blaxter, J. H. S. (1967) Egg weight in Atlantic herring (*Clupea harrengus* L.) *J. Cons. perm. int. Explor. Mer*, **31**, 170–95.

Henderson, M. A. and Ward, F. J. (1978) Changes in the chemical composition, calorific and water content of yellow perch fry, *Perca fluviatilis flavescens*. *Verh. Internat. Verein. Limnol.*, **20**, 2025–30.

Herzig, A. and Winkler, H. (1985) Der Einfluss der Temperatur auf die embryonale Entwicklung der Cypriniden. *Öst. Fisch.* **38**, 182–96.

Herzig, A. and Winkler, H. (1986) The influence of temperature on the embryonic development of three cyprinid fishes, *Abramis brama*, *Chalcalburnus chalcoides mento* and *Vimba vimba*. *J. Fish Biol.*, **28**, 171–81.

Hinterleitner, S., Thurner-Flür, J., Wieser, W. and El-Fiky, N. (1989) Profiles of enzyme activity in larvae of two cyprinid species with contrasting life styles (Cyprinidae, Teleostei). *J. Fish Biol.*, **35**, 709–18.

Hirshfield, M. F. (1980) An experimental analysis of reproductive effort and cost in the Japanese medaka, *Oryzias latipes*. *Ecology*, **61**, 282–92.

Hjort, J. (1914) Fluctuations in the great fisheries of Northern Europe viewed in light of biological research. *Rapp. P.-v. Réun. Cons. perm. int. Explor. Mer*, **20**, 1–228.

Hoar, W. S. (1957) The gonads and reproduction, in, *The Physiology of Fishes*, Vol. 1. *Metabolism* (ed. M. E. Brown) Academic Press, NY, pp. 287–321.

Hogendoorn, H. (1980) Controlled propagation of the African catfish, *Clarias lazera* (C. & V.). III. Feeding and growth of fry. *Aquaculture*, **21**, 233–41.

Holliday, F. G. T., Blaxter, J. H. S. and Lasker, R. (1964) Oxygen uptake of developing eggs and larvae of the herring (*Clupea harengus*). *J. mar. biol. Ass. U.K.*, **44**, 711–23.

Horoszewicz, L. (1974) Przeżywalność karpia głodującego w wodzie o różnej temperaturze [Survival of carp fry starving in water of various temperature]. *Rocz.*

Nauk Roln., Ser. H, **96**, 45–56. (In Polish, English summ.)

Horoszewicz, L. (1981a) Effect of different thermal regimes on reproductive cycle of tench *Tinca tinca* (L.). Part I. Design for the experimental and seasonal changes in condition factors of spawners. *Pol. Arch. Hydrobiol.*, **28**, 169–86.

Horoszewicz, L. (1981b) Effect of different thermal regimes on reproductive cycles of tench *Tinca tinca* (L.). Part VIII. Towards understanding of reproduction mechanisms and requirements of controlled spawning. *Pol. Arch. Hydrobiol.*, **28**, 257–62.

Howell, H. (1980) Temperature effects on growth and yolk utilization in yellowtail flounder, *Limanda ferruginea*, yolk-sac larvae. *Fishery Bull.*, **78**, 731–9.

Hughes, G. M. and Al-Kadhomyi, N. K. (1988) Changes in scaling of respiratory systems during the development of fishes. *J. mar. biol. Ass. U.K.*, **68**, 489–98.

Hughes, G. M., Munshi, J. S. D. and Ojha, J. (1986) Post-embryonic development of water and air-breathing organs of *Anabas testudineus* (Bloch). *J. Fish Biol.*, **29**, 443–50.

Huisman, E. A. (1979), Integration of hatchery, cage and pond culture of common carp (*Cyprinus carpio* L.) and grass carp (*Ctenopharyngodon idella* Val.) in the Netherlands, in, *Bio-engineering Symposium, Traverse City, Michigan, USA, 15–18 October 1979*, pp. 1–20.

Humpesch, U. (1985) Inter- and intra-specific variation in hatching success and embryonic development of five species of salmonids and *Thymallus thymallus*. *Arch. Hydrobiol.*, **104**, 129–44.

Humpesch, U. H. and Elliott, J. M. (1980) Effect of temperature on the hatching time of eggs of three *Rhitrogena* spp. (Ephemeroptera) from Austrian streams and an English stream and river. *J. Anim. Ecol.*, **49**, 643–61.

Humphreys, W. F. (1979) Production and respiration in animal populations. *J. Anim. Ecol.*, **48**, 427–53.

Hunter, J. R. (1972) Swimming and feeding behavior of larval anchovy, *Engraulis mordax*. *Fishery Bull. Fish Wildl. Serv. U.S.*, **70**, 821–38.

Hunter, J. R. (1981) Feeding ecology and predation in marine fish larvae, in, *Marine Fish Larvae, Morphology, Ecology and Relation to Fisheries* (ed. R. Lasker), Univ. Washington Press, Seattle, pp. 32–77.

Hunter, J. R. and Kimbrell, C. (1981) Some aspects of the life history of laboratory reared Pacific mackerel larvae (*Scomber japonicus*), in. *The Early Life History of Fish: Recent Studies* (eds R. Lasker and K. Sherman), *Rapp. P.-v. Réun. Cons. perm. int. Explor. Mer*, **178**, 344.

Hurley, D. A. and Brannon, E. L. (1969) Effect of feeding before and after yolk absorption on the growth of sockeye salmon. *Int. Pac. Salmon Fish. Comm. Progr. Rep. No. 21*, 19 pp.

Hussain, N., Akatsu, S. and El-Zahr, C. (1981) Spawning, egg and early larval development, and growth of *Acanthopagrus cuvieri* (Sparidae). *Aquaculture*, **22**, 125–36.

Ignateva, G. M. (1979) *Early Embryogenesis of Fishes and Amphibians*, Nauka, Moscow.

Ilina, I. D. (1986) Development of digestive enzyme systems in carp fed artificial diets 'Equizo' and RK-S at high and low temperatures. *Sb. nauch. tr. VNII prud. ryb. kh-va*, **49**, 66–74. (In Russian)

Ilina, I. D. and Tureckij, V. I. (1986) Special features in the digestion of carp larvae. *Sb. nauch. tr. VNII prud. ryb. kh-va*, **49**, 45–54. (In Russian)

Imam, A. E. and Habashy, A. Ph. (1972) Artificial feeding of carp fry. *Bull. Inst. Oceanogr. Fish.*, **2**, 263–74.

Islam, M. A., Nose, Y. and Yasuda, F. (1973) Egg characteristics and spawning season of rainbow trout. *Bull. Jap. Soc. scient. Fish.*, **39**, 741–51.

Ivanov, D. I. (1986), Influence of ecological factors on biological properties of carp larvae receiving the starter Equizo. *Sb. nauch. trudov nauchno-proizv. obedin. po promyshl. i teplovod. rybov. GNII ozern. i rechn. ryb. kh-va*, **246**, 17–23. (in Russian)

Ivlev, V. S. (1939a) Transformation of energy by aquatic animals. Coefficient of energy consumption by *Tubifex* (Oligochaeta). *Int. Revue ges. Hydrobiol. Hydrog.* **38**, 449–58.

Ivlev, V. S. (1939b) Energy budget in growing larva of *Silurus glanis*. *Dokl. Akad. Nauk SSSR*, **25**, 87–9. (In Russian)

Ivlev, V. S. (1939c) Energy balance in carp. *Zool. Zh. Mosk.*, **18**, 303–18. (In Russian)

Ivleva, I. V. (1981) *Environmental Temperature and Rate of Energy Metabolism in Aquatic Animals*, Naukova Dumka, Kiev. (In Russian)

Iwai, T. and Hughes, G. M. (1977) Preliminary morphometric study on gill development in black sea bream (*Acanthopagrus schlegeli*). *Bull. Jap. Soc. scient. Fish.*, **43**, 929–34.

Iwata, K. (1986) Feeding, digestive and metabolic characters in the phytoplanktivorous cyprinids, *Carassius auratus cuvieri* and *Hypophthalmichthys molitrix*, with special reference to their developmental changes. *Bull. Fac. Educ. Wakayama Univ. Nat. Sci.*, **35**, 23–45.

Iwata, K., Arita, T. and Iino, K. (1983) Seasonal changes in energy reserves of *Scopimera globosa* (Crustacea: Ocypodidae), with special reference to its wandering behavior. *Zool. Mag. Tokyo*, **92**, 303–16.

Jacobsen, O. J. (1982) A review of food and feeding habits in coregonid fishes. *Pol. Arch. Hydrobiol.*, **29**, 179–200.

Jezierska, B. (1974) The effect of various types of food on the growth and chemical composition of the body of perch (*Perca fluviatilis* L.) in laboratory conditions. *Pol. Arch. Hydrobiol.*, **21**, 467–79.

Jobling, M. (1983) A short review and critique of methodology used in fish growth and nutrition studies. *J. Fish Biol.*, **23**, 685–703.

Jobling, M. (1985) Growth, in *Fish Energetics: New Perspectives* (eds P. Tytler and P. Calow), Croom Helm, London, pp. 213–30.

Jobling, M. (1986a) Mythical model of gastric emptying and implications for food consumption studies. *Env. Biol. Fishes*, **16**, 35–50.

Jobling, M. (1986b) Gastrointestinal overload – a problem with formulated feeds? *Aquaculture*, **51**, 257–63.

Jones, A. (1972) Studies on egg development and larval rearing of turbot *Scophthalmus maximus* L. and brill, *Scophthalmus rhombus* L. in the laboratory. *J. mar. biol. Ass. U.K.*, **52**, 965–86.

Jungwirth, M. and Winkler, H. (1984) The temperature dependence of embryonic development of grayling (*Thymallus thymallus*), Danube salmon (*Hucho hucho*), arctic char (*Salvelinus alpinus*) and brown trout (*Salmo trutta fario*). *Aquaculture*, **38**, 315–27.

Kainz, E. and Gollmann, H.P. (1980) Versuche zur Anfütterung der Brut von

Karpfen (*Cyprinus carpio* L.) und Graskarpfen (*Ctenopharyngodon idella* Val.) mit Trockenfutter. Öst. Fisch., **33**(4), 65–73.

Kaitaranta, J. K. and Ackman, R. G. (1981) Total lipid and lipid classes of fish roe. *Comp. Biochem. Physiol.*, **69B**, 725–29.

Kamler, E. (1970) The main parameters regulating the level of energy expenditure in aquatic animals. *Pol. Arch. Hydrobiol.*, **17**, 201–16.

Kamler, E. (1972a) Bioenergetical aspects of the influence of 2,4-D-Na on the early developmental stages in carp (*Cyprinus carpio* L.). *Pol. Arch. Hydrobiol.*, **19**, 451–74.

Kamler, E. (1972b) Respiration of carp in relation to body size and temperature. *Pol. Arch. Hydrobiol.*, **19**, 325–31.

Kamler, E. (1976) Variability of respiration and body composition during early developmental stages of carp. *Pol. Arch. Hydrobiol.*, **23**, 431–85.

Kamler, E. (1987) Comparison of energy contents in spawn of sea trout (*Salmo trutta* L.) from Rega, Słupia, Wieprza and Vistula Rivers. *Pol. Arch. Hydrobiol.*, **34**, 245–54.

Kamler, E. and Kato, T. (1983) Efficiency of yolk utilization by *Salmo gairdneri* in relation to incubation temperature and egg size. *Pol. Arch. Hydrobiol.*, **30**, 271–306.

Kamler, E. and Malczewski, B. (1982) Quality of carp eggs obtained by induced breeding. *Pol. Arch. Hydrobiol.*, **29**, 599–606.

Kamler, E. and Mandecki, W. (1978) Ecological bioenergetics of *Physa acuta* (Gastropoda) in heated waters. *Pol. Arch. Hydrobiol.*, **25**, 833–68.

Kamler, E., Lewkowicz, M., Lewkowicz, S., Uchmański, J. and Urban–Jezierska, E. (1986) Gravimetric techniques for measuring consumption of live foods and artificial diets by fish larvae. *Aquaculture*, **54**, 109–22.

Kamler, E. and Żuromska, H. (1979) Quality of eggs and production of *Coregonus albula* (L.) in three Masurian lakes. *Pol. Arch. Hydrobiol.*, **26**, 595–623.

Kamler, E., Matlak, O. and Srokosz, K. (1974) Further observations on the effect of sodium salt of 2,4-D on early developmental stages of carp (*Cyprinus carpio* L.). *Pol. Arch. Hydrobiol.*, **21**, 481–502.

Kamler, E., Szlamińska, M., Przybył, A., Barska, B., Jakubas, M., Kuczyński, M., Brzuska, E. and Dgebuadze, Yu.Yu. (1989) Stabilność w wodzie starterów dla wylęgu karpia [Stability of carp starters in water]. *Gosp. Rybna*, **41**, (2) 6–7. (In Polish)

Kamler, E., Szlamińska, M., Przybył, A., Barska, B., Jakubas, M., Kuczyński, M. and Raciborski, K. (1990) Developmental response of carp, *Cyprinus carpio*, larvae fed different foods or starved. *Env. Biol. Fishes*, **29**, 303–13.

Kamler, E., Szlamińska, M., Raciborski, K., Barska, B., Jakubas, M., Kuczyński, M. and Przybył, A. (in press). Bioenergetic evaluation of four artificial diets for carp (*Cyprinus carpo* L.) larvae as compared with zooplankton and starvation. *J. Anim. Physiol. Anim. Nutr.*

Kamler, E., Urban–Jezierska, E., Stanny, L. A., Lewkowicz, M. and Lewkowicz, S. (1987) Survival, development, growth, metabolism and feeding of carp larvae receiving zooplankton or starters. *Pol. Arch. Hydrobiol.*, **34**, 503–41.

Kamler, E., Żuromska, H. and Nissinen, T. (1982) Bioenergetical evaluation of environmental and physiological factors determining egg quality and growth in *Coregonus albula* (L.). *Pol. Arch. Hydrobiol.*, **29**, 71–121.

Kankaala, P. and Wulff, F. (1981) Experimental studies on temperature-dependent embryonic and postembryonic developmental rates of *Bosmina longispina maritima* (Cladocera) in the Baltic. *Oikos,* **36**, 137–46.

Kato, T. (1975) The relation between the growth and reproductive characters of rainbow trout, *Salmo gairdneri. Bull. Freshwat. Fish. Res. Lab. Tokyo,* **25**, 83–115. (In Japanese, English summ.)

Kato, T. (1978) Relation of growth to maturity of age and egg characteristics in kokanee salmon (*Oncorhynchus nerka*). *Bull. Freshwat. Fish. Res. Lab. Tokyo,* **28**, 61–75. (In Japanese, English summ.)

Kato, T. (1979) Selective breeding of rainbow trout with regard to reproduction characteristics. *Proc. 7th Japan–Soviet Joint Symp. Aquaculture, Tokyo, Sept, 1978,* 137–44.

Kato, T. (1980) Effect of water temperature on early developmental stages in salmonid fishes. Abstracts of 5th Conf. Salmon Culturists, June 18–19, 1980, Nagano Prefecture, Japan, 26–33.

Kato, T. and Kamler, E. (1983) Criteria for evaluation of fish egg quality, as exemplified for *Salmo gairdneri* (Rich.). *Bull. Natl. Res. Inst. Aquaculture,* **4**, 61–78.

Kausch, H. (1968) Der Einfluss der Spontanaktivität auf die Stoffwechselrate junger Karpfen (*Cyprinus carpio* L.) im Hunger und bei Fütterung. *Arch. Hydrobiol.,* **33**(Supp.), 263–330.

Kaushik, S. J. and Dąbrowski, K. (1983) Postprandial metabolic changes in larval and juvenile carp (*Cyprinus carpio*). *Reprod. Nutr. Dévelop.,* **23**(2A), 223–34.

Kaushik, S. J., Dąbrowski, K. and Luquet, P. (1982) Patterns of nitrogen excretion and oxygen consumption during ontogenesis of common carp (*Cyprinus carpio*). *Can. J. Fish. aquat. Sci.,* **39**, 1095–105.

Kawajiri, M. (1927a) On the optimum temperature of water for hatching the eggs of rainbow trout (*Salmo irideus* Gibbons). *J. imp. Fish. Inst., Tokyo,* **23**(3), 59–65.

Kawajiri, M. (1927b) The optimum temperature of water for the hatching of the eggs of trout, *Oncorhynchus masou* (Walbaum). *J. imp. Fish. Inst., Tokyo,* **23**(2), 14–21.

Kawajiri, M. (1927c) The influence of variation of temperature of water on the development of fish eggs. *J. imp. Fish. Inst., Tokyo,* **23**(3), 65–73.

Kawajiri, M. (1927d) On the preservation of the egg and sperm of *Oncorhynchus masou* (Walbaum). *J. imp. Fish. Inst., Tokyo,* **23**(2), 11–13.

Kazakov, R. V., Melnikova, M. N. and Ilenkova, S. A. (1981) Comparison of egg weight and colour in Atlantic salmon from the Neva River and Narova River. *Proc. 11th Session of the Sci. Council: Biologicheskie resursy Belogo Morya i vnutriennikh vodoemov Evropejskogo Severa,* Ministry of Fisheries and Acad. Sci. USSR, Petrozavodsk, pp. 108–9. (In Russian)

Kendall, R. L. (1988) Taxonomic changes in North American trout names. *Trans. Am. Fish. Soc.,* **117**, 321.

Kernehan, R. J., Headrick, M. R. and Smith, R. E. (1981) Early life history of striped bass in the Chesapeake and Delaware Canal and vicinity. *Trans. Am. Fish. Soc.,* **110**, 137–50.

Khmeleva, N. N. (1988) *Reproduction in Crustaceans,* Nauka i Tekhnika, Minsk. (In Russian)

Khmeleva, N. N. and Baichorov, V. M. (1987) Comparative estimate of reproductive cycles in two relict, ecologically different mysid species. *Pol. Arch. Hydrobiol.,* **34**, 321–9.

Khmeleva, N. N. and Golubev, A. P. (1984) *Production of Crustaceans*, Nauka i Tekhnika, Minsk. (In Russian)

Khmeleva, N. N., Klekowski, R. Z. and Nagorskaya, L. L. (1981) Respiration and yolk utilization efficiency during embryogenesis of *Streptocephalus torvicornis* (Waga) (Crustacea, Anostraca). *Pol. Arch. Hydrobiol.*, **28**, 43–53.

Kim, E. D. (1974a) Changes in the amino acid content in carp mature sexual products in relation to age and year of collection, in *Raznokachestvennost Rannego Ontogeneza u Ryb* (ed. V. I. Vladimirov), Naukova Dumka, Kiev, pp. 65–93. (In Russian)

Kim, E. D. (1974b) Changes in the cholesterol and phospholipid content in carp mature sexual products in relation to year of collection and age, in *Raznokachestvennost Rannego Ontogeneza u Ryb* (ed. V. I. Vladimirov), Naukova Dumka, Kiev, pp. 114–26. (In Russian)

Kim, E. D. (1981) Changes of amino acid and lipid contents in oocytes of some cyprinids during maturation, in, *Raznokachestvennost Ontogeneza u Ryb* (ed. V. D. Romanenko), Naukova Dumka, Kiev, pp. 61–84. (In Russian)

Kim, E. D. and Zhukinskij, V. N. (1978) Amino acid variability in bream spawn during maturation and overmaturation. *Dop. Akad. Nauk Ukrainskoj RSR, Ser. B.* **3**, 272–77. (In Ukrainian, English summ.)

Kimata, M. (1982) Changes of chemical composition during early development of egg and larva in the half beak *Hemiramphus sajori* (Temminck et Schlegel). *Bull. Jap. Soc. scient. Fish.*, **48**, 1663–71. (In Japanese, English summ.)

Kitajima, C., Fujita, S., Tsukashima, Y. and Arakawa, T. (1980) Rearing, development and change in feeding habit of larval and juvenile rabbitfish, *Siganus fuscescens. Bull. Nagasaki Pref. Inst. Fish.*, **6**, 61–70. (In Japanese, English summ.)

Kitajima, C., Fukusho, K., Iwamoto, H. and Yamamoto, H. (1976) Amount of rotifer, *Brachionus plicatilis*, consumed by red sea bream larvae, *Pagrus major. Bull. Nagasaki Pref. Inst. Fish.*, **2**, 105–12.

Kjørsvik, E. and Lønning, S. (1983) Effects of egg quality on normal fertilization and early development of the cod, *Gadus morhua* L. *J. Fish Biol.*, **23**, 1–12.

Kjørsvik, E., Mangor-Jensen, A. and Holmefjord, I. (1990) Egg quality in fishes. Adv. Mar. Biol., **26**, 71–113.

Kjørsvik, E., Stene, A. and Lønning, S. (1984) Morphological, physiological and genetical studies of egg quality in cod (*Gadus morhua* L.). *Flødevigen Rapportser.*, **1**, 67–86.

Kleiber, M. (1961) *The Fire of Life. An introduction to Animal Energetics*, Wiley, NY. 454 pp.

Klekowski, R. Z. (1970) Bioenergetic budgets and their application for estimation of production efficiency. *Pol. Arch. Hydrobiol.*, **17**, 55–80.

Klekowski, R. Z. and Duncan, A. (1975a) Physiological approach to ecological energetics, in, *Methods for Ecological Bioenergetics* (IBP Handbook No. 24) (eds W. Grodzinski, R. Z. Klekowski and A. Duncan), Blackwell Sci. Publ., Oxford, pp. 15–64.

Klekowski, R. Z. and Duncan, A. (1975b) Feeding and nutrition. 7A. Review of methods for identification of food and for measurement of consumption and assimilation rates, in *Methods for Ecological Bioenergetics* (IBP Handbook No. 24) (eds W. Grodziński, R. Z. Klekowski and A. Duncan) Blackwell Sci. Publ., Oxford, pp. 227–57.

Klekowski, R. Z. and Fischer, Z. (1975) Review of studies on ecological bioenergetics of aquatic animals. *Pol. Arch. Hydrobiol.*, **22**, 345–73.

Klekowski, R. Z. and Sazhina, L. (1985) Respiratory metabolism of some pelagic copepods from the equatorial countercurrent of the Indian Ocean. *Pol. Arch. Hydrobiol.*, **32**, 507–43.

Klekowski, R. Z., Fischer, E., Fischer, Z., Ivanova, M. B., Prus, T., Shushkina, E. A., Stachurska, T., Stępień, Z. and Żyromska–Rudzka, H. (1972) Energy budgets and energy transformation efficiencies of several animal species of different feeding types, in, *Productivity Problems in Freshwaters* (eds Z. Kajak and A. Hillbricht–Ilkowska), PWN, Warszawa-Kraków, pp. 749–63.

Klekowski, R. Z., Prus, T. and Żyromska–Rudzka, H. (1967) Elements of energy budget of *Tribolium castaneum* (Hbst.) in its developmental cycle, in *Secondary Productivity of Terrestrial Ecosystems* (ed. K. Petrusewicz), PWN Warszawa, pp. 859–79.

Klyashtorin, L. B. (1982) *Aquatic Respiration and Oxygen Requirements of Fishes*, Legkaya i pishch. promyshl, Moscow. (In Russian)

Knight, A. E. (1963) The embryonic and larval development of the rainbow trout. *Trans. Am. Fish. Soc.*, **92**, 344–55.

Knights, B. (1985) Energetics and fish farming, in, *Fish Energetics: New Perspectives* (eds P. Tytler and P. Calow) Croom Helm, London, pp. 309–40.

Knutsen, G. M. and Tilseth, S. (1985) Growth, development, and feeding success of Atlantic cod larvae *Gadus morhua* related to egg size. *Trans. Am. Fish. Soc.*, **114**, 507–11.

Koch, F. and Wieser, W. (1983) Partitioning of energy in fish: can reduction of swimming activity compensate for the cost of production? *J. exp. Biol.*, **107**, 141–46.

Köck, G. and Hofer, R. (1989) The effect of natural and artificial diets upon tryptic activity of roach (*Rutilus rutilus*) and whitefish (*Coregonus sp.*) larvae. *Pol. Arch. Hydrobiol.*, **36**, 443–53.

Kokurewicz, B. (1969a) O roli gruczołów wylęgu u ryb kostnoszkieletowych [On the role of hatching glands in teleostean fishes]. *Przegl. Zool.*, **13**, 181–7. (In Polish, English summ.)

Kokurewicz, B. (1969b) The influence of temperature on the embryonic development of the perches: *Perca fluviatilis* L. and *Lucioperca lucioperca* (L.). *Zoologica Pol.*, **19**, 47–67.

Kokurewicz, B. (1970) The effect of temperature on embryonic development of *Tinca tinca* (L.) and *Rutilus rutilus* (L.). *Zoologica Pol.*, **20**, 317–37.

Kokurewicz, B. (1971) Warunki termiczne a rozwój i rozród niektórych gatunków [Temperature conditions, development and reproduction of some fish species]. *Wrocław, Żabieniec, Samodzielna Pracownia Upowszechniania Postępu IRS. No. 47*, 18 pp. (In Polish)

Kokurewicz, B. (1981) Effect of different thermal regimes on reproductive cycles of tench (*Tinca tinca* L.). Part VII. Embryonal development of progeny. *Pol. Arch. Hydrobiol.*, **28**, 243–56.

Kokurewicz, B., Kowalewski, M. and Witkowski, A. (1980) Influence of constant and variable temperatures on the embryonic development of European grayling, *Thymallus thymallus* (L.). *Zoologica Pol.*, **27**, 335–62.

König, J. and Grossfeld, J. (1913) Fishrogen als Nahrungsmittel für den Menschen. *Biochem. Z.*, **54**, 351–95.

Konovalov, Yu. D. (1979) Viability of carp embryos and larvae as a function of the content of total protein and sulfhydryl groups in the gastrulation stage. *Gidrobiol. Zhurnal*, 15(6) (Transl. 1980 Scripta Publishing Co. ISSN0018-8166/79/0060–0065, pp. 65–8).

Konovalov, Yu. D. (1984) *Proteins and their Reactive Groups in the Early Ontogenesis of Fishes*, Naukova Dumka, Kiev. (In Russian)

Konstantinov, A. S. and Zdanovich, V. V. (1985) Influence of temperature fluctuations on growth and physiological state of young carps (*Cyprinus carpio* L.). *Dokl. Akad. Nauk SSSR*, 282, 760–4. (In Russian)

Korwin–Kossakowski, M. (1986) Etap krytyczny w rozwoju larwalnym karpia w zakwaszonej wodzie [Critical period for carp larval development in acidic water]. Abstracts, 13th Congr. Polish Hydrobiol. Soc., Szczecin, 16–19 September 1986, Pol. Tow. Hydrobiol. and Akademia Rolnicza, Szczecin, pp. 99–100. (In Polish)

Korwin–Kossakowski, M. (1988) Larval development of carp, *Cyprinus carpio* L. in acidic water. *J. Fish Biol.*, 32, 17–26.

Korwin–Kossakowski, M. and Jezierska, B. (1984) The influence of thermal conditions on postembryonic development of some species of Coregonidae and Cyprinidae. *Zoologica Pol.*, 31, 43–56.

Korzhuev, P. A., Nikolskaya, I. S. and Radzinskaya, L. I. (1960) Respiration of eggs of Acipenseridae during incubation. *Vop. Ikhtiol.*, 14, 113–18. (In Russian)

Kosiorek, D. (1979) Changes in chemical composition and energy content of *Tubifex tubifex* (Müll.), Oligochaeta, in life cycle. *Pol. Arch. Hydrobiol.*, 26, 73–89.

Kouřil, J. and Hamáčková, J. (1989) Research on fish larvae conducted at the Fisheries Institute at Vodňany, Czechoslovakia, in 1980–1988. *Pol. Arch. Hydrobiol.*, 36, 495–501.

Kowalska, A. (1959) O wpływie temperatury na rozwój embrionalny pstrąga potokowego (*Salmo trutta m. fario* L.) [On the influence of temperature on the embryonic development of the brook trout (*Salmo trutta m. fario* L.)]. *Przegl. zool.*, 3, 253–9. (In Polish, English summ.)

Koyama, H. and Kira, T. (1956) Intraspecific competition among higher plants. VIII. Frequency distribution of individual plant weight as affected by the interaction between plants. *J. Biol., Osaka City Univ.*, 7, 73–94.

Krasnoper, E. V. (1985) Relationship between food digestibility and assimilation coefficients in fish. *Vop. Ikhtiol.*, 25, 692–4. (In Russian)

Krishnamurthy, C., Purushothama, M. N., Narasimha Rao, Ch., Katre Shakuntala and Reddy, S. (1984) Efficiency of food conversion in the guppy, *Poecilia reticulata* in relation to sex differentiation. *Pol Arch. Hydrobiol.*, 31, 403–16.

Krivobok, M. N. (1953) Food utilization by sazan in the Azov–Dolgij Fish Farm. *Trudy VNIRO*, 24, 102–16. (In Russian)

Krogh, A. (1916) *The Respiratory Exchange of Animals and Man*, Longman's, London.

Krogh, A. (1919) The rate of diffusion of gases through animal tissues with some remarks on the coefficient of invasion. *J. Physiol., Lond.*, 52, 391.

Kryzhanovskij, S. G. (1940) On the significance of the size of the yolk sac surface in teleostean eggs for organogenesis. *Zool. Zh.*, 19, 456–70. (In Russian)

Kryzhanovskij, S. G. (1949) Eco-morphological principles of development in carps, loaches and catfishes. *Trudy Inst. Morf. Zhivot.*, 1, 5–332. (In Russian)

Kryzhanovskij, S. G. (1956) Data on the development of clupeids. *Trudy Inst. Morf. Zhivot.*, 17, 1–254. (In Russian)

Kryzhanovskij, S. G. (1960) Meaning of oil contents in the eggs of fishes. *Zool. Zh.*, **39**, 111–23. (In Russian)

Kudrinskaya, O. I. (1969) Metabolism intensity in larva of *Lucioperca lucioperca* L., *Perca fluviatilis* L., *Abramis brama* L. and *Rutilus rutilus* L. *Gidrobiol. Zh.* **5**(4), 117–21. (In Russian, English summ.)

Kudrinskaya, O. I., Sherstyuk, V. V., Kovalchuk, T. V. and Voronchuk, L. V. (1976) Energy flow through communities of fish larvae. *Tezisy dokladov, 23 Sezd Vsesoyuz. Gidrobiol. Obshch., Riga, 11–15 May 1976*, Vol. 2, Zinatne, Riga, pp. 135–7. (In Russian)

Kulikova, N. I. (1973) Changes in the content and qualitative composition of water-soluble proteins in oocytes and mature eggs from different spawn portions in *Gobius melanostomus* Pallas, in, *Ekologicheskaya Fiziologiya Ryb* (Tezisy dokl. vsesoyuzn. konf. po ekologisheskoj fiziologii ryb, Moskva, 1973) (ed. N. S. Stroganov), Minist. Rybnogo Khoz. SSSR, Moscow, pp. 103–4. (In Russian)

Kuznetsov, V. A. (1973) Fecundity and spawn quality in some economically important fish species inhabiting a dam reservoir, in, *Ekologicheskaya Fiziologiya Ryb* (Tezisy dokl. vseosoyuzn. konf. po ekologicheskoj fiziologii ryb, Moskva, 1973) (ed. N. S. Stroganov), Minist. Rybnogo Khoz., SSSR, Moscow pp. 106–7. (In Russian)

Lange, N. O. (1960) Developmental etaps of *Rutilus rutilus* in different ecological conditions. *Trudy Inst. Morf. Zhivot.*, **28**, 5–40. (In Russian)

Lange, N. O., Dmitrieva, E. N., Smirnova, E. N. and Peñaz, M. (1974) Methods of studying the morphological and ecological peculiarities of fish development during the embryonic, larval and juvenile periods, in, *Tipovye Metodiki Issledovaniya Produktivnosti Vidov Ryb v Predelakh ikh Arealov* (ed. R. S. Volskis), Mintis, Vilnius, pp. 56–71. (In Russian)

Lapin, V. I. and Matsuk, V. E. (1979) Yolk utilization and changes in the biochemical composition of eggs in the process of embryonic development of *Eleginus navaga* (Pallas). *Vop. Ikhtiol.*, **19**, 341–6. (In Russian)

Lasker, R. (1962) Efficiency and rate of yolk utilization by developing embryos and larvae of the Pacific sardine *Sardinops caerulea* (Girard). *J. Fish. Res. Bd Can.*, **19**, 867–75.

Lasker, R. and Theilacker, G. H. (1962) Oxygen consumption and osmoregulation by single Pacific sardine eggs and larvae (*Sardinops caerulea* Girard.). *J. Cons. perm. int. Explor. Mer*, **27**, 25–33.

Lasker, R., Feder, H. M., Theilacker, G. H. and May, R. C. (1970) Feeding, growth and survival of *Engraulis mordax* larvae reared in the laboratory. *Mar. Biol.*, **5**, 345–53.

Last, J. M. (1980) The food of twenty species of fish larvae in the west-central North Sea. *Fish. Res. Tech. Rep., MAFF Direct. Fish. Res., Lowestoft*, No. 60, 44 pp.

Lauff, M. and Hofer, R. (1984) Proteolytic enzymes in fish development and the importance of dietary enzymes. *Aquaculture*, **37**, 335–46.

Laurence, G. C. (1969) The energy expenditure of largemouth bass, *Micropterus salmoides*, during yolk absorption. *Trans. Am. Fish. Soc.*, **98.**, 398–405.

Laurence, G. C. (1973) Influence of temperature on energy utilization of embryonic and prolarval tautog, *Tautoga onitis*. *J. Fish. Res. Bd Can.*, **30**, 435–42.

Laurence, G. C. (1978) Comparative growth, respiration and delayed feeding abilities of larval cod (*Gadus morhua*) and haddock (*Melanogrammus aeglefinus*) as influenced by temperature during laboratory studies. *Mar. Biol.*, **50**, 1–7.

Leary, R. F., Allendorf, F. W. and Knudsen, K. L. (1985) Inheritance of meristic variation and the evolution of developmental stability in rainbow trout. *Evolution*, **39**, 308–14.

Lebedeva, O. A. (1981) Ecological tolerance of early developmental stages of coregonid fishes, in, *Tezisy dokl, vtorogo vsesoyuzn. soveshch. po biologii i biotekhnike razvedeniya sigovykh ryb, Petrozavodsk, 1981* (ed. Yu. S. Reshetnikov), Minist. Rybnogo Khoz. SSSR, Akademiya Nauk SSSR, Petrozavodsk, pp. 13–16. (In Russian)

Leiner, M. (1932) Die Entwicklungsdauer der Eier des dreistachelingen Stichlings in ihrer Abhängigkeit von der Temperatur. *Z. vergl. Physiol.*, **16**, 590–605.

Leitritz, E. and Lewis, R. C. (1976) Trout and salmon culture. *State of California, Dept Fish Game, Fish Bull.*, No. 164. 1–197.

Lewander, K., Dave, G., Johansson, M., Larsson, A. and Lidman, V. (1974) Metabolic and hematological studies on the yellow and silver phases of the European eel, *Anguilla anguilla* L. I. Carbohydrate, lipid, protein and inorganic ion metabolism. *Comp. Biochem. Physiol.*, **47B**, 571–81.

Lind, E. A. and Turunen, J. (1968) Talvikutuisen muikun levinneisyysalue hahmottumassa [Über das Laichen im Winter und die Verbreitung der winterlaichenden Kleinen Maräne, *Coregonus albula* L.]. *Eripainos Suomen Kalastuslehti*, **3**, 106–10. (In Finnish, German summ.)

Lindroth, A. (1942) Sauerstoffverbrauch der Fische. II. Verschiedene Entwicklungs– und Alterstadien vom Lachs und Hecht. *Z. vergl. Physiol.*, **29**, 583–94.

Lindroth, A. (1946) Zur Biologie der Befruchtung und Entwicklung beim Hecht. *Rep. Inst. Freshwat. Res. Drottningholm*, **24**, 1–173.

Lindsey, C. C. and Arnason, A. N. (1981) A model for responses of vertebral numbers in fish to environmental influences during development. *Can. J. Fish. aquat. Sci.*, **38**, 334–47.

Lirski, A. (1985) Reakcja karpia, amura białego, tołpygi białej i pstrej na temperaturę w pierwszych tygodniach życia [Reaction of *Cyprinus carpio, Ctenopharyngodon idella, Hypophthalmichthys molitrix* and *Aristichthys nobilis* on temperature during first weeks of life], PhD thesis, Inland Fisheries Institute, Żabieniec. (101 pp.) (In Polish)

Littak, A., Woźniewski, M. and Opuszyński, K. (1980) Wczesny podchów wylęgu karpia w kontrolowanych warunkach środowiskowych [Early rearing of carp larvae in controlled conditions]. *Inst. Ryb. Śródlądowego, Zakład Upowszechniania Postępu, Olsztyn*, No. 129, 32 pp. (In Polish)

Lizenko, E. I., Sidorov, V. S. and Potapova, O. I. (1973) On the content of lipids in the gonads of cisco, *Coregonus albula* L. from Lake Uros. *Vop. Ikhtiol.*, **13**, 303–12. (In Russian)

Loewe, H. and Eckmann, R. (1988) The ontogeny of the alimentary tract of coregonid larvae: normal development. *J. Fish Biol.*, **33**, 841–50.

Love, R. M. (1957) The biochemical composition of fish, in *The Physiology of Fishes*. Vol. 1. *Metabolism* (ed. M. E. Brown), Academic Press, NY, pp. 401–18.

Love, R. M. (1970) *The Chemical Biology of Fishes*, Academic Press, London.

Lowry, O. H., Farr, A. L., Randall, R. J. and Rosenbrough, N. J. (1951) Protein measurement with the Folin phenol reagent. *J. biol. Chem.*, **193**, 265–75.

Lukowicz, M. v. (1979) Experiences with krill (*Euphausia superba* Dana) in the diet for young carp (*Cyprinus carpio* L.), in, *Proc. World Symp. Finfish Nutrition Fishfeed*

Technol., Hamburg, 20–23 June 1978 (eds J. E. Halver and K. Tiews), Vol. II, Heenemann Verlagsgesellschaft, Berlin, pp. 293–302.

Lukowicz, M. v. and Rutkowski, F. (1976) Anfütterung von Karpfenbrut. *Fischer und Teichwirt*, **27**(6), 68–9.

Lyagina, T. N. (1975) Egg weight as related to the characteristics of female roach *Rutilus rutilus* (L.) under conditions of different food supply. *Vop. Ikhtiol.*, **15**, 652–63. (In Russian)

Łuczyński, M. (1987) Uwarunkowania termiczne wczesnego rozwoju osobniczego Coregoninae [Temperature conditions of the early development of Coregoninae]. *Acta Acad. Agricult. Techn. Olst., Protectio Aquarum et Piscatoria*, **14D**, 1–56. (In Polish)

Łuczyński, M. and Dettlaff, J. (1985) Precocious hatching in Coregoninae embryos by electric stimulation. *Z. angew. Ichthyol.*, **1**, 157–64.

Łuczyński, M. and Kirklewska, A. (1984) Dependence of *Coregonus albula* embryogenesis rate on the incubation temperature. *Aquaculture*, **42**, 43–55.

Łuczyński, M. and Kolman, R. (1987) Hatching of *Coregonus albula* and *C. lavaretus* embryos at different stages of development. *Env. Biol. Fishes*, **19**, 309–15.

Łuczyński, M., Dembiński, W. and Chybowski, L. (1986) Controlling rate of egg development in vendace (*Coregonus albula* L.) to increase larval growth rate in stocked lakes. *Aquaculture*, **51**, 195–205.

MacCrimmon, H. R. and Kwain, W. (1969) Influence of light on early development and meristic characters in the rainbow trout, *Salmo gairdneri* Richardson. *Can. J. Zool.*, **47**, 631–7.

MacCrimmon, H. R. and Twongo, T. K. (1980) Ontogeny of feeding behaviour in hatchery-reared rainbow trout, *Salmo gairdneri* Richardson. *Can. J. Zool.*, **58**, 20–6.

MacGregor, J. S. (1957) Fecundity of the Pacific sardine (*Sardinops caerulea*). *Fishery Bull. Fish Wildl. Serv. U.S.*, **57**, 427–49.

Macháček, J., Kouřil, J., Hamáčková, J. and Přikryl, I. (1986) The growth and food utilization of the early fry of carp (*Cyprinus carpio* L.) in relation to water temperature. *Pr. VÚRH Vodňany*, **15**, 42–51.

Maciolek, J.A. (1962) Limnological organic analyses by quantitative dichromate oxidation. *Research Rep. Fish Wildl. Serv. US*, **60**, 1–61.

Mahon, R., Balon, E. and Noakes, D. L. G. (1979) Distribution, community structure and production of fishes in the upper Speed River, Ontario: a preimpoundment study. *Env. Biol. Fishes*, **4**, 219–44.

Majkowski, J. and Hearn, W. S. (1984) Comparison of three methods for estimating the food intake of a fish. *Can. J. Fish. aquat. Sci.*, **41**, 212–15.

Majkowski, J. and Waiwood, K. G. (1981) A procedure for evaluating the food biomass consumed by a fish population. *Can. J. Fish. aquat. Sci.*, **38**, 1199–208.

Malyarevskaya, A. Ya. and Birger, T. I. (1965) Biochemical composition of spawn, eggs and larvae of *Rutilus rutilus heckeli* and *Abramis brama*, in, *Vliyanie Kachestva Proizvoditelej na Potomstvo u Ryb* (ed. V. I. Vladimirov), Naukova Dumka, Kiev, pp. 5–34. (In Russian)

Mann, R. H. K. and Mills, C. A. (1985) Variations in the sizes of gonads, eggs and larvae of the dace, *Leuciscus leuciscus*. *Env. Biol. Fishes*, **13**, 277–87.

Marciak, Z. (1962) Sezonowe zmiany w odżywianiu się i wzroście sielawy (*Coregonus albula* L.) z jeziora Pluszne. [Seasonal variation in feeding habits and growth of

Coregonus albula L. in Pluszne Lake]. *Roczn. Nauk roln., Ser. B*, **81**, 335–57. (In Polish, English summ.)

Mark, W., Hofer, R. and Wieser, W. (1987) Diet spectra and resource partitioning in the larvae and juveniles of three species and six cohorts of cyprinids from a subalpine lake. *Oecologia (Berlin)*, **71**, 388–96.

Marr, D. H. A. (1966) Influence of temperature on the efficiency of growth of salmonid embryos. *Nature*, **212**, 957–59.

Marr, J. C. (1956) The 'critical period' in the early life history of marine fishes. *J. Cons. perm. int. Explor. Mer*, **21**, 160–70.

Marsh, E. (1984) Egg size variation in central Texas populations of *Etheostoma spectabile* (Pisces: Percidae). *Copeia*, **1984**, 291–301.

Marsh, E. (1986) Effects of egg size on offspring fitness and maternal fecundity in the orangethroat darter, *Etheostoma spectabile* (Pisces: Percidae). *Copeia*, **1986**, 18–30.

Martyshev, F. G., Anisimova, J. M. and Privezentsev, Yu. A. (1967) *Age Selection in Carp Culture*, Kolos, Moscow. (In Russian)

Matlak, J. and Matlak, O. (1976) The natural food of carp fry (*Cyprinus carpio* L.). *Acta Hydrobiol., Kraków*, **18**, 203–28.

Matlak, O. (1966) Wzrost narybku dwóch rodzin karpia hodowanych w Gołyszu [Zuwachs der Karpfenbrut von zwei Karpfenfamilien aus Gołysz]. *Acta Hydrobiol, Kraków*, 8(Supp. 1), 255–91. (In Polish, German summ.)

Matlak, O. (1972) Sodium salt (2, 4-D) related to embryo and larval development of carp. *Pol. Arch. Hydrobiol.*, **19**, 437–50.

May, R. C. (1971) Effects of delayed initial feeding on larvae of the grunion, *Leuresthes tenuis* (Ayres). *Fish. Bull. nat. mar. Fish. Serv. U.S.*, **69**, 411–25.

McFadden, J. T., Cooper, E. L. and Andersen, J. K. (1965) Some effects of environment on egg production in brown trout (*Salmo trutta*). *Limnol. Oceanogr.*, **10**, 88–95.

McMullen, D. M. and Middaugh, D. P. (1985) The effect of temperature and food density on survival and growth of *Menidia peninsulae* larvae (Pisces: Atherinidae). *Estuaries*, **8**, 39–47.

McNeill, S. and Lawton, J. H. (1970) Annual production and respiration in animal populations. *Nature*, **225**, 472–4.

Mejen, V. A. (1940) Causes of egg variability in bony fishes. *Dokl. Akad. Nauk SSSR*, **28**, 654–6. (In Russian)

Melnichuk, G. L. (1969) Metabolism intensity in fry of food-fish from the Kremenchug Reservoir. *Gidrobiol. Zh.*, **5**(5), 66–71. (In Russian, English summ.)

Mills, C. A. (1982) Factors affecting the survival of dace, *Leuciscus leuciscus* (L.), in the early post-hatching period. *J. Fish Biol.*, **20**, 645–55.

Mills, C. A. and Eloranta, A. (1985) Reproductive strategies in the stone loach, *Noemacheilus barbatulus*. *Oikos*, **44**, 341–9.

Milman, L. S. and Yurowitzky, Yu. G. (1973) *Regulation of Glycolysis in the Early Development of Fish Embryos* (*Monographs in Developmental Biology*, Vol. 6), S. Karger, Basel.

Mironova, N. V. (1977) Energy expenditure for egg production in young *Tilapia mossambica* Peters and effect of maintenance conditions on spawning intensity. *Vop. Ikhtiol.*, **17**, 708–14. (In Russian)

Miura, T., Suzuki, N., Nagoshi, M. and Yamamura, K. (1976) The rate of production and food consumption of the biwamasu, *Oncorhynchus rhodeus*, population in lake Biwa. *Res. Popul. Ecol.*, **17**, 135–54.

Moliński, M., Penczak, T. and Bukowska–Madej, A. (1978) Materials for the ecology of dace, *Leuciscus leuciscus* (L.) from a polluted river in the region of the barbel (the River Pilica). 2. Dry weight, ash and content of some metals. *Acta Hydrobiol.*, Kraków **20**, 87–96.

Morawska, B. (1964) Płodność certy (*Vimba vimba* L.) systemu rzeki Wisły [Fertility of *Vimba vimba* L. in the Vistula River system]. *Roczn. Nauk. roln.*, *Ser. B*, **84**, 315–27. (In Polish, English summ.)

Morawska, B. (1967) Płodność troci (*Salmo trutta* L.) z rzeki Wisły [Sea-trout (*Salmo trutta* L.) fecundity]. *Roczn. Nauk roln.*, *Ser. H*, **90**, 249–65. (In Polish, English summ.)

Morawska, B. (1984) The effect of water temperature elevation on incipient and cumulative fecundity of batch-spawning tench, *Tinca tinca* (L.) *Aquaculture*, **42**, 273–88.

Morawska, B. (1986) Wpływ adaptacji termicznych lina (*Tinca tinca* L.) na jego reprodukcję [Effect of temperature adaptations on reproduction in tench (*Tinca tinca* L.)]. *Abstracts. 13th Congr. Polish Hydrobiol. Soc., Szczecin, 16–19 September 1986*. Pol. Tow. Hydrobiol. and Akademia Rolnicza, Szczecin, p. 134. (In Polish).

Morgan, R. P. II and Rasin, V. J., jun. (1982) Influence of temperature and salinity on development of white perch eggs. *Trans. Am. Fish. Soc.*, **111**, 396–8.

Morgan, R. P. II, Rasin, V. J., jun. and Copp, R. L. (1981) Temperature and salinity effects on development of striped bass eggs and larvae. *Trans. Am. Fish. Soc.*, **110**, 95–9.

Moroz, I. E. and Luzhin, B. P. (1973) Metabolism in carp embryonic development, in, *Ekologicheskaya Fiziologiya Ryb (Tezisy Dokl. Vsesoyuzn. Konf. po Ekologicheskoj Fiziologii Ryb, Moskva, 1973)* (ed. N. S. Stroganov), Minist. Rybnogo Khoz. SSSR, Moscow, pp. 203–5. (In Russian)

Moroz, I. E., Vovk, F. I. and Pashkin, L. M. (1973) Metabolic changes in sturgeon muscles, liver and gonad during migration, wintering and spawning in the dam reservoir of the Volgograd Electric Power Station, in *Ekologicheskaya Fiziologiya Ryb (Tezisy Dokl. Vsesoyuzn. Konf. po Ekologicheskoj Fiziologii Ryb, Moskva, 1973)* (ed. N. S. Stroganov) Minist. Rybnogo Khoz. SSSR, Moscow, pp. 109–11. (In Russian)

Mortensen, E. (1977) Population, survival, growth and production of trout *Salmo trutta* in a small Danish stream. *Oikos*, **28**, 9–15.

Mortensen, E. (1985) Population and energy dynamics of trout *Salmo trutta* in a small Danish stream. *J. Anim. Ecol.*, **54**, 869–82.

Mukhina, R. J. (1963) Changes of chemical composition of feeds after immersion in water. *Trudy VNII prud. ryb. kh-va*, **12**, 99–101. (In Russian)

Nakano, E. (1953) Respiration during maturation and fertilization of fish eggs. *Embryologia*, **2**, 21–31.

Nedyalkov, G. F. (1981) Embryo anomalies and mortality as a manifestation of egg variability in *Ctenopharyngodon idella*, in, *Raznokachestvennost Ontogeneza Ryb* (ed. V. D. Romanenko), Naukova Dumka, Kiev, pp. 37–49. (In Russian)

Needham, J. (1931) *Chemical Embryology*, Cambridge University Press, London.

Needham, J. (1950) *Biochemistry and Embryogenesis*, University Press, Cambridge.

Nejfakh, A. A. (1960) X-ray inactivation of cellular nuclei as a method for studying their function in the increase of respiration of fish embryos. *Biokhimiya*, **25**, 658–68. (In Russian)

Nelson, J. S. (1976) *Fishes of the World*, Wiley, NY.

Nielsen, L. A., Sheehan, R. J. and Orth, D. J. (1986) Impacts of navigation of riverine fish production in the United States. *Pol. Arch. Hydrobiol.*, **33**, 277–94.

Nikolskij, G. V. (1974) *Theory of Fish Population Dynamics*, Pishchevaya Promyshlennost, Moscow. (In Russian)

Nishino, M. (1980) Geographical variations in body size, brood size and egg size of a freshwater shrimp, *Palaemon paucidens* De Haan, with some discussion on brood habit. *Jap. J. Limnol.*, **41**(4), 185–202.

Nishino, M. (1981) Brood habits of two subspecies of a freshwater shrimp, *Paratya compressa* (Decapoda, Atyidae) and their geographical variations. *Jap. J. Limnol.*, **42**, 201–19.

Nishiyama, T. (1970) Caloric values of ovaries of sockeye salmon at last stage of . marine life. *Bull. Jap. Soc. Scient. Fish.*, **36**, 1095–100.

Ogino, C. and Yasuda, S. (1962) Changes in inorganic constituents of developing rainbow trout eggs. *Bull. Japn. Soc. Scient. Fish.*, **28**, 788–91. (In Japanese, English summ.)

Oikawa, S. and Itazawa, Y. (1985) Gill and body surface areas of the carp in relation to body mass, with special reference to the metabolism–size relationship. *J. exp. Biol.*, **117**, 1–14.

Okoniewska, G., Opuszyńska, W., Opuszyński, K., Wolnicki, J. and Myszkowski, L. (1986) Podchów wylęgu ryb karpiowatych w basenach przy użyciu pokarmu naturalnego i paszy [Tank rearing of cyprinid fish larvae fed natural food and formulated diets]. *Abstracts, 13th Congr. Polish Hydrobiol. Soc., Szczecin, 16–19 September 1986*. Pol. Tow. Hydrobiol. and Akademia Rolnicza, Szczecin, pp. 148–149. (In Polish)

Olifan, V. I. (1949) Cyclic character of fish growth in early stages of their postembryonic development. *Dokl. Akad. Nauk SSSR*, **66**, 1211–14. (In Russian)

Oosthuizen, E. and Daan, N. (1974) Egg fecundity and maturity of North Sea cod, *Gadus morhua*. *Neth. J. Sea Res.*, **8**, 378–97.

Opuszyński, K. (1983) *Podstawy Biologii Ryb* [*Fundamentals of Fish Biology*]. Państwowe Wyd. Rolnicze i Leśne, Warsaw. (In Polish)

Opuszyński, K., Myszkowski, L., Okoniewska, G., Opuszyńska, W., Szlamińska, M., Wolnicki, J. and Woźniewski, M. (1989) Rearing of common carp, grass carp, silver carp and bighead carp larvae using zooplankton and/or different feeds. *Pol. Arch. Hydrobiol.*, **36**, 217–30.

Orton, J. H. (1929) Reproduction and death in invertebrates and fishes. *Nature (Lond.)*, **123**, 14–15.

Osse, J. W. M. (1989) A functional explanation for a sequence of developmental events in the carp. The absence of gills in early larvae. *Acta morph. neerl. – scand.* **27**, 111–18.

Osse, J. W. M. and Drost, M. R. (1989) Hydrodynamics and mechanics of fish larvae. *Pol. Arch. Hydrobiol.*, **36**, 455–65.

Osse, J. M. W., Drost, M. R., Vial, J. and Aldunate, R. (1986) Growth and feeding in larval carp, *Cyprinus carpio* (L.). *Annls. Mus. r. Afr. cent. Sci. zool.*, **251**, 67–70.

Ostroumova, I. N., Tureckij, V. I., Ivanov, D. I. and Dementeva, M. A. (1980) High quality starter for carp larvae in warm waters. *Ryb. Khoz.*, **6**, 41–4. (In Russian)

Ozernyuk, N. D. (1970) Respiration intensity and ATP content in the course of oogenesis in *Misgurnus fossilis* L. *Dokl. Akad. Nauk SSSR*, **192**, 242–5. (In Russian)

Ozernyuk, N. D. (1985) *Energy Transformation in Fish Early Ontogenesis*, Nauka, Moscow. (In Russian)

Ozernyuk, N. D. and Lelyanova, V. G. (1985) Peculiarities of energy metabolism in the early ontogenesis of fishes and amphibians. *Zh. Obshch. Biol.*, **46**, 778–85. (In Russian)

Ozernyuk, N. D. and Lelyanova, V. G. (1987) Factors determining changes in the relative respiration rate during early ontogenesis of rainbow trout. *Dokl. Akad. Nauk SSSR*, **292**, 1510–12. (In Russian)

Ozernyuk, N. D. and Zotin, A. I. (1983) Changes in the respiratory rate during embryonic development of *Salmo salar* from the Neva River. *Ontogenez*, **14**, 539–42. (In Russian)

Paine, R. T. (1965) Natural history, limiting factors and energetics of the opisthobranch *Navanax inermis. Ecology*, **46**, 603–19.

Pandian, T. J. (1969) Yolk utilization in the gastropod, *Crepidula fornicata. Mar. Biol.*, **3**, 117–21.

Pandian, T. J. (1970) Ecophysiological studies on the developing eggs and embryos of the European lobster, *Homarus gammarus. Mar. Biol.*, **5**, 154–67.

Pandian, T. J. and Vivekanandan, E. (1985) Energetics of feeding and digestion, in, *Fish Energetics: New Perspectives* (eds P. Tytler and P. Calow), Croom Helm, London, pp. 99–124.

Panov, D. A. and Sorokin, Yu. K. (1967) Determination of the threshold concentration of food for fish larvae with the help of radioactive carbon. *Vop. Ikhtiol.*, **7**, 129–41. (In Russian)

Patalas, K. (1960a) Mieszanie wody jako czynnik określający intensywność krążenia materii w różnych morfologicznie jeziorach okolic Węgorzewa [Mixing of waters as the factor defining intensity of food materials circulation in morphologically different lakes of Węgorzewo District]. *Roczn. Nauk. roln.*, *Ser. B*, **77**, 223–42. (In Polish, English summ.)

Patalas, K. (1960b) Punktowa ocena pierwotnej produktywności jezior okolic Węgorzewa [The method of classification of primary productivity of a lake by point system, applied to the lakes of Węgorzewo District]. *Roczn. Nauk roln.*, *Ser. B*, **77**, 300–26. (In Polish, English summ.)

Pavlova, E. V. (1987) *Locomotion and Metabolism of Marine Planktonic Organisms*, Naukova Dumka, Kiev. (In Russian)

Pchelovodova, D. V. (1976) Correlation between eggs fat content and some biological indices in the rainbow trout females. *Izv. Gos. Nauchno-Issled. Inst. Ozern. Rechn. Rybn. Khoz.*, **113**, 12–15. (In Russian, English summ.)

Pechen, G. A. and Kuznetsova, A. P. (1966) Consumption and utilization of food by *Daphnia pulex* (De Geer). *Dokl. Akad. Nauk. BSSR*, **10**, 344–7. (In Russian)

Pedersen, B. H., Nilssen, E. M. and Hjelmeland, K. (1987) Variations in the content of trypsin and trypsinogen in larval herring (*Clupea harengus*) digesting copepod nauplii. *Mar. Biol.*, **94**, 171–81.

Peňáz, M. (1974) Influence of water temperature on incubation and hatching in *Chondrostoma nasus* (L.). *Zool. Entomol. Listy*, **23**, 53–60.

Peňáz, M., Prokeš, M., Kouřil, J. and Hamáčková, J. (1983) Early development of the carp. *Cyprinus carpio. Acta Sci. nat. Acad. Sci. Bohemoslov. (Brno)*, **17**(2), 1–39.

Penczak, T. (1985) Trophic ecology and fecundity of *Fundulus heteroclitus* in Chezzetcook Inlet, Nova Scotia. *Mar. Biol.*, **89**, 235–43.

Penczak, T., Galicka, W., Moliński, M., Kusto, E. and Zalewski, M. (1982) The enrichment of a mesotrophic lake by carbon, phosphorus and nitrogen from the cage aquaculture of rainbow trout. *Salmo gairdneri. J. appl. Ecol.*, **19**, 371–93.

Penczak, T., Moliński, M., Kusto, E., Ichniowska, B. and Zalewski, M. (1977) The ecology of roach, *Rutilus rutilus* (L.) in the barbel region of the polluted Pilica River. III. Lipids, protein, total nitrogen and caloricity. *Ekol. pol.*, **25**, 75–88.

Penczak, T., Moliński, M., Kusto, E., Palusiak, K., Panusz, H. and Zalewski, M. (1976) The ecology of roach, *Rutilus rutilus* (L.) in the barbel region of the polluted Pilica River. II. Dry weight, ash and contents of some elements. *Ekol. pol.*, **24**, 623–38.

Penczak, T., Moliński, M. and Ośka, M. (1978) Materials for the ecology of the dace, *Leuciscus leuciscus* (L.) from a polluted river in the region of barbel (the Pilica River). 3. Lipids, protein, total nitrogen and caloric value. *Acta Hydrobiol.* Kraków, **20**, 97–108.

Peslyak, Ya, K. (1967) Effect of spawners quality on quality of salmon fry cultured in fish farms, in, *Obmen Veshchestv i Biokhimiya Ryb* (ed. G. S. Karzinkin), Nauka, Moscow, pp. 73–5. (In Russian)

Peters, R. H. (1983) *The Ecological Implications of Body Size*, University Press, Cambridge.

Peterson, R. H., Daye, P. G. and Metcalfe, J. L. (1980) Inhibition of Atlantic salmon (*Salmo salar*) hatching at low pH. *Can. J. Fish. aquat. Sci.*, **37**, 770–4.

Petrusewicz, K. (1967) Concepts in studies on the secondary productivity of terrestrial ecosystems, in, *Secondary Productivity of Terrestrial Ecosystems* (ed. K. Petrusewicz), Państwowe Wyd. Naukowe, Warszawa–Kraków, pp. 17–49.

Philips, F. S. (1940) Oxygen consumption and its inhibition in the development of *Fundulus* and various pelagic fish eggs. *Biol. Bull. mar. biol. Lab.*, Woods Hole, **78**, 256–74.

Pianka, E. R. (1974) *Evolutionary Ecology*, Harper & Row, NY.

Pianka, E. R. and Parker, W. S. (1975) Age-specific reproductive tactics. *Am. Nat.*, **109**, 453–64.

Pilarska, J. (1977) Eco-physiological studies on *Brachionus rubens* Ehrbg (Rotatoria). II. Production and respiration. *Pol. Arch. Hydrobiol.*, **24**, 329–41.

Pliszka, F. (1953) Rozród i rozwój certy (*Vimba vimba* L.) [Reproduction and development of *Vimba vimba* L.] *Pol. Arch. Hydrobiol.*, **1**, 137–63. (In Polish, English summ.)

Pope, J. A., Mills, D. H. and Shearer, W. M. (1961) The fecundity of the Atlantic salmon (*Salmo salar* Linn.). *Freshwat. Salm. Fish. Res.*, **26**, 1–12.

Potapova, O. I. (1978) *The 'Ripus' Vendace, Coregonus albula L.*, Nauka, Leningrad. (In Russian)

Price, E. E., Stauffer, J.R., jun. and Swift, M. C. (1985) Effect of temperature on growth of juvenile *Oreochromis mossambicus* and *Sarotherodon melanotheron. Env. Biol. Fishes.*, **13**, 149–52.

Priede, I. G. (1985) Metabolic scope in fishes, in, *Fish Energetics: New Perspectives* (eds P. Tytler and P. Calow), Croom Helm, London, pp. 33–64.

Privolnev, T. I. (1938) Wachstum und atmung der Lachs–Embryonen (*Salmo salar*). *Arkh. Anat. Gist. Embryol.*, **18**, 165–77. (In Russian, German summ.)

Privolnev, T. I. (1953) Critical periods in the development and their importance in fish acclimatization. *Izv. GosNIORKH*, **32**, 238–48. (In Russian)

Prosser, C. L. (1961) Oxygen: respiration and metabolism, in *Comparative Animal Physiology*, 2nd edn (eds C. L. Prosser and F. A.. Brown), W. B. Saunders, Philadelphia, pp. 153–197.

Prus, T. (1970) Caloric value of animals as an element of bioenergetical investigations. *Pol. Arch. Hydrobiol.*, **17**, 183–99.

Prus, T. (1975) Measurement of calorific value using Phillipson microbomb calorimeter, in, *Methods for Ecological Bioenergetics* (IBP Handbook No. 24) (eds W. Grodzinski, R. Z. Klekowski, and A. Duncan), Blackwell Sci. Publ., Oxford, pp. 149–160.

Quantz, G. (1985) Use of endogenous energy sources by larval turbot, *Scophthalmus maximus*. *Trans. Am. Fish. Soc.*, **114**, 558–63.

Raciborski, K. (1987) Energy and protein transformation in sea trout (*Salmo trutta* L.) larvae during transition from yolk to external food. *Pol. Arch. Hydrobiol.*, **34**, 437–502.

Radziej, J. (1965) Studium limnologiczne pod kątem zróżnicowania sielawy (*Coregonus albula* L.) jeziora Narie i jeziora Łańskiego [Limnological study of Narie and Łańskie lakes from the point of view of vendace (*Coregonus albula* L.) differentiation] PhD thesis, Wyższa Szkoła Rolnicza, Olsztyn. (83 pp.) (In Polish)

Radziej, J. (1973) Wpływ środowiska na wolno rosnącą sielawę wsiedloną z jeziora Narie do jeziora Wierzbiczany [The effect of site upon slowly growing *Coregonus albula* (Linnaeus 1758) transferred from the Lake Narie to the Lake Wierzbiczany]. *Roczn. Nauk roln.*, Ser. H, **95**, 129–46. (In Polish, English summ.)

Rahn, C. H., Sand, D. M. and Schlenk, H. (1977) Metabolism of oleic, linoleic and linolenic acids in gourami (*Trichogaster cosby*) fry and mature females. *Comp. Biochem. Physiol.*, **58B**, 17–20.

Rakusa–Suszczewski, S. (1972) Respiration of the antarctic fish eggs (*Trematomus borchgrevinki* Boul.). *Pol. Arch. Hydrobiol.*, **19**, 399–401.

Rask, M. (1983) The effect of low pH on perch, *Perca fluviatilis* L. I. Effect of low pH on the development of eggs of perch. *Annls zool. fenn.*, **20**, 73–6.

Redfield, G. W. (1981) Nutrition and degeneration of eggs in a limnetic daphnid. *Verh. int. Verein. Limnol.*, **21**, 1550–4.

Renard, P., Billard, R. and Christen, R. (1985) Changes in the chorion during the cortical reaction in carp (*Cyprinus carpio*) eggs. *Posters summary, Symp. Aquaculture Carp and Related Species*, Evry, 2–5 Sept. 1985, INRA, ITAVI, p. 21.

Reznichenko, P. N., Solovev, L. G. and Gulidov, M. V. (1967) Polarographic modelling of oxygen access to fish embryo respiratory surfaces, in *Obmen Veshchestv i Biokhimiya Ryb* (ed. G. S. Karzinkin), Nauka, Moscow, pp. 148–155. (In Russian)

Reznick, D. (1985) Costs of reproduction: an evaluation of the empirical evidence. *Oikos*, **44**, 257–67.

Richman, S. (1958) The transformation of energy by *Daphnia pulex*. *Ecol. Monogr.*, **28**, 273–91.

Ricker, W. E. (1968) *Methods for Assessment of Fish Production in Freshwaters* (IBP Handbook No. 3) Blackwell Sci. Publ., Oxford.

Ricker, W. E. (1973) Linear regression in fishery research. *J. Fish. Res. Bd Can.*, **30**, 409–34.

Ricker, W. E. (1979) Growth rates and models, in *Fish Physiology*, Vol. VIII (eds W. S. Hoar, D. J. Randall and J. R. Brett), Academic Press, NY, pp. 677–743.

Ridelman, J. M., Hardy, R. W. and Brannon, E. L. (1984) The effect of short-term starvation on ovarian development and egg viability in rainbow trout (*Salmo gairdneri*). *Aquaculture*, **37**, 133–40.

Rogers, B. A. and Westin, D. (1981) Laboratory studies on effect of temperature and delayed initial feeding on development of striped bass larvae. *Trans Am. Fish. Soc.*, **110**, 100–10.

Romanenko, V. D. (1981) *Variability of Fish Ontogenesis*, Naukova Dumka, Kiev. (In Russian).

Rombough, P. J. (1985) Initial egg weight, time to maximum alevin wet weight, and optimal ponding times for chinook salmon (*Oncorhynchus tshawytscha*). *Can. J. Fish. aquat. Sci.*, **42**, 287–91.

Rösch, R. (1989) Beginning of food intake and subsequent growth of larvae of *Coregonus lavaretus* L. *Pol. Arch. Hydrobiol.*, **36**, 475–84.

Rösch, R. and Applebaum, S. (1985) Experiments on the suitability of dry food for larvae of *Coregonus lavaretus* L. *Aquaculture*, **48**, 291–302.

Rösch, R. and Dąbrowski, K. (1986) Tests of artificial food for larvae of *Coregonus lavaretus* from Lake Constance. *Arch. Hydrobiol. Beih Ergebn. Limnol.*, **22**, 273–82.

Rösch, R. and Segner, H. (1990) Development of dry food for larvae of *Coregonus lavaretus* L. I. Growth, food digestion and fat absorption. *Aquaculture*, **91**, 101–15.

Roshchin, V. E. and Mazelev, K. L. (1978) Respiration and energy utilization efficiency in embryonic development of *Asellus aquaticus* L. (Crustacea). *Vestsi Akad. Nauk BSSR*, **5**, 112–14. (In Russian, English summ.)

Rothbard, S. (1981) Induced reproduction in cultivated cyprinids. The common carp and the group of Chinese carps: 1. The technique of induction, spawning and hatching. *Bamidgeh*, **33**, 103–21.

Ryzhkov, L. P. (1966) Development of *Salmo ischchan*. *Trudy Karelskogo Otdeleniya. GosNIORKH*, **4**, 100–14. (In Russian)

Ryzhkov, L. P. (1973) Water changes in tissues of larvae juveniles of salmonid fishes, in, *Ekologicheskaya Fiziologiya Ryb* (*Tezisy Dokl. Vsesoyuzn. Konf. po Ekologicheskoj Fiziologii Ryb. Moskva, 1973*) (ed. N. S. Stroganov), Minist. Rybnogo Khoz. SSSR, Moscow, pp. 234–36. (In Russian)

Ryzhkov, L. P. (1976) *Morpho-physiological Peculiarities and Transformation of Matter and Energy in Early Development of Freshwater Salmonid Fishes*, Kareliya Petrozavodsk. (In Russian)

Ryzhkov, L. P. (1979) Influence of temperature on morphology and physiology of salmonid embryos, larvae and alevins. *Pol. Arch. Hydrobiol.* **26**, 397–425.

Ryzhkov, L. P. (1981) Characteristics of changes of energy budget in coregonid fishes, in, *Tezisy Dokladov Vtorogo Vsesoyuznogo Soveshchaniya po Biologii i Biotekhnike Razvedeniya Sigovykh Ryb, Petrozavodsk, October 1981* (ed. Yu, S. Reshetnikov), Min. Rybnogo Rkoz. SSSR, Akad. Nauk SSSR, Petrozavodsk, pp. 224–6. (In Russian)

Ryzhkov, L. P., Polina, A. V. and Korenev, O. N. (1982) Ecological and physiological properties of *Oncorhynchus kisutch* and *Salvelinus alpinus* cultured in controlled conditions, in, *Vosproizvodstvo Zapasov Lososevykh Ryb vo Vnutriennykh Vodoemakh*

Karelskoj ASSR (ed. L.P. Ryzhkov), Min. Ryb. Khoz. SSSR, Murmansk, pp. 29–51. (In Russian)

Sarig, S. (1966) Synopsis of biological data on common carp, *Cyprinus carpio*. (L.) (Near East and Europe) *F.A.O. Fish. Synopses*, **31.**2, 35 pp.

Satia, B. P., Donaldson, L. R., Smith, L. S. and Nightingale, J. N. (1974) Composition of ovarian fluid and eggs of the University of Washington strain of rainbow trout (*Salmo gairdneri*). *J. Fish. Res. Bd Can.*, **31**, 1796–9.

Savostyanova, G. G. and Nikandrov, V. Ya. (1976) Dependence of some biometrical indices of eggs on age of the rainbow trout females. *Izv. Gos. Nauchno-issled. Inst. Ozernogo i Rechnogo Rybn. Khoz*, **113**, 3–7 (In Russian, English summ.)

Schiemer, F., Duncan, A. and Klekowski, R. Z. (1980) A bioenergetic study of a benthic nematode, *Plectus palustris* de Man 1880, throughout its life cycle. *Oecologia (Berl.)*, **44**, 205–12.

Schiemer, F., Kockeis, H. and Wanzenböck, J. (1989) Foraging in Cyprinidae during early development. *Pol. Arch. Hydrobiol.* **36**, 467–74.

Schmidt-Nielsen, K. (1972) Locomotion: energy cost of swimming, flying and running. *Science*, **177**, 222–8.

Schneider, J. C. (1985) Yellow perch maturity and fecundity as a function of age and growth. *Mich. Dep. Nat. Resour. Fish. Res. Rep.*, No. 1915, 22 pp.

Scott, D. P. (1962) Effect of food quantity on fecundity of rainbow trout, *Salmo gairdneri*. *J. Fish. Res. Bd Can.*, **19**, 715–30.

Segner, H., Rösch, R., Schmidt, H. and Poeppinghausen, K. J. von (1988) Studies on the suitability of commerical dry diets for rearing of larval *Coregonus lavaretus* from Lake Constance. *Aquat. Living Resour.*, **1**, 231–8.

Segner, H., Rösch, R., Schmidt, H. and Poeppinghausen, K. J. von (1989) Digestive enzymes in larval *Coregonus lavaretus* L. *J. Fish Biol.*, **35**, 249–63.

Semenov, K. I., Konovalov, Yu. D., Nesen, Eh. N., Babitskaya, L. F. and Nagirnyj, S. N. (1974) Water and total protein contents during embryogenesis of carp different offspring and their viability, in *Raznokachestvennost Rannego Ontogeneza u Ryb* (ed. V. I. Vladimirov), Naukova Dumka, Kiev, pp. 139–69. (In Russian)

Shamardina, I. P. (1954) Changes in the respiratory rate of fishes in the course of their development. *Dokl. Akad. Nauk SSSR*, **98**, 689–92 (In Russian)

Shatunovskij, M. I. (1963) Lipid and water contents in muscles and gonads and their relation to gonad maturation in *Pleuronectes flesus*. *Vop. Ikhtiol.*, **3**, 652–67. (In Russian)

Shatunovskij, M. I. (1980) *Ecological Regularities of Metabolism in Marine Fishes*, Nauka, Moscow. (In Russian)

Shatunovskij, M. I. (ed.) (1985) *Ecology and Economical Potential of the Fishes of the Monogolian People's Republic*, Nauka, Moscow. (In Russian)

Sherman, K. and Lasker, R. (1981) Symposium on the early life history of fish. Introduction and background. *Rapp. P.-v. Réun. Cons. perm. int. Explor. Mer*, **178**, III–VI.

Shimizu, M. and Yamada, J. (1980) Ultrastructural aspects of yolk absorption in the vitelline syncytium of the embryonic rockfish, *Sebastes schlegeli*. *Jap. J. Ichthyol.*, **27**, 56–63.

Shimma, Y., Kato, T. and Fukuda, Y. (1978) Effects of feeding methanol-grown yeast on growth and maturity of yearling rainbow trout and egg hatchability. *Bull. Freshwat. Fish. Res. Lab. Tokyo*, **28**, 21–8. (In Japanese, English summ.)

Shireman, J. V. and Smith, C. R. (1983) Synopsis of biological data on the grass carp, *Ctenopharyngodon idella* (Cuvier and Valenciennes, 1844). *F.A.O. Fish. Synopses*, 135, 86 pp.

Shireman, J. V., Opuszyński, K. and Okoniewska, G. (1988) Food and growth of hybrid bass fry (*Morone saxatilis* × *Morone chrysops*) under intensive culture conditions. *Pol. Arch. Hydrobiol*, 35, 109–17.

Shulman, G. E. (1972) *Physiological and Biochemical Properties of Annual Cycles in Fishes*, Pishchevaya Promyshlennost, Moscow. (In Russian)

Skácelová, O. and Matěna, J. (1981) Přirozena potrava pludku kapra (*Cyprinus carpio* L.) v prvnich dnech života [The natural food of the fry of carp (*Cyprinus carpio* L.) during first days of life]. *Bull. VURH Vodň*, 17(4), 20–6 (In Czech, English summ.)

Skadovskij, S. N. and Morosova, T. E. (1936) Sauerstoffverbrauch von *Acipenser stellatus* Eiern auf verschiedenen Entwicklungsstadien. *Uchenye Zap. Mosk. Gosud. Univ.*, 9, 128–31. (In Russian, German summ.)

Smirnov, A. I., Kamyshnaya, M. S. and Kalashnikova, Z. M. (1968) Size, biochemical factors and calorie content of mature eggs of some fishes belonging to genera *Oncorhynchus* and *Salmo* Vop. Ikhtiol., 8, 653–61. (In Russian).

Smith, L. L. and Chernoff, B. (1981) Breeding populations of cyprinodontoid fishes in a thermal stream. *Copeia*, 3, 701–2.

Smith, S. (1947) Studies in the development of the rainbow trout (*Salmo irideus*). I. The heat production and nitrogenous excretion. *J. exp. Biol*, 23, 357–73.

Smith, S. (1957) Early development and hatching, in *The Physiology of Fishes*. Vol. 1 – *Metabolism* (ed. M. E. Brown), Academic Press, NY, pp. 323–59.

Smith, S. (1958) Yolk utilization in fishes, in, *Embryonic Nutrition* (ed. D. Rudnick), Univ. of Chicago Press, Chicago, pp. 33–53.

Snedecor, G. W. (1956) *Statistical Methods*, Iowa State Univ. Press, Annes, Iowa.

Snedecor, G. W. and Cochran, W. G. (1967) *Statistical Methods*, Iowa State Univ. Press, Ames, Iowa.

Soin, S. G. (1968) *Adaptative Properties of Fish Development*. MGU Press, Moscow. (In Russian)

Sokal, R. R. and Rohlf, F. J. (1969) *Biometry*, W. H. Freeman, San Francisco.

Springate, J. R. C. and Bromage, N. R. (1985) Effects of egg size on early growth and survival in rainbow trout (*Salmo gairdneri* Richardson). *Aquaculture*, 47, 163–72.

Springate, J. R. C., Bromage, N. R. and Cumaranatunga, P. R. T. (1985) The effect of different ration on fecundity and egg quality in the rainbow trout (*Salmo gairdneri*), in *Nutrition and Feeding in Fish* (eds C. B. Cowey, A. M. Mackie and J. G. Bell), Academic Press, London, pp. 371–93.

Springate, J. R. C., Bromage, N. R., Elliott, J. A. K. and Hudson, D. L. (1984) The timing of ovulation and stripping and their effects on the rates of fertilization and survival to eying, hatch and swim-up in the rainbow trout (*Salmo gairdneri* R.). *Aquaculture*, 43, 313–22.

Staples, D. J. and Nomura, M. (1976) Influence of body size and food ration on the energy budget of rainbow trout *Salmo gairdneri* Richardson. *J. Fish Biol.*, 9, 29–43.

Stearns, S. C. (1976) Life history tactics: a review of ideas. *Q. Rev. Biol.* 51, 3–47.

Stearns, S. C. (1977) The evolution of life-history traits: a critique of the theory and a review of the data. *Ann. Rev. Ecol. Syst.*, 8, 145–71.

Stepanova, G. M. and Tyutyunik, S. N. (1973) Biochemical characteristics of muscles and gonads of grass carp spawners bred in Moldavia, in, *Ekologicheskaya Fiziologiya*

Ryb (*Tezisy Dokl. Vsesoyuzn. Konf. po Ekologicheskoj Fiziologii Ryb, Moskva, 1973*) (ed. N. S. Stroganov), Minist. Rybnogo Khoz. SSSR, Moscow, pp. 149–51. (In Russian)

Streit, B. (1975) Experimentelle Untersuchungen zum Stoffhaushalt von *Ancylus fluviatilis* (Gastropoda – Basommatophora). 1. Ingestion, Asimilation, Wachstum und Eiablage. *Arch. Hydrobiol.* **47**(Supp. 4), 458–514.

Striganova, B. R. (1980) *Feeding of Soil Saprophages.* Nauka, Moscow (In Russian)

Stroband, H. W. J. and Dąbrowski, K. R. (1981) Morphological and physiological aspects of the digestive system and feeding in fresh-water fish larvae, in, *Nutrition des Poissons* (ed. M. Fontaine), Editions du CNRS, Paris, pp. 355–76.

Stroganov, N. S. (1956) *Physiological Adaptation of Fish to Environmental Temperature,* Izdat. Akad. Nauk SSSR, Moscow. (In Russian).

Stroganov, N. S. (1962) *Ecological Physiology of Fishes,* Izdat. Moskovskogo Universiteta, Moscow. (In Russian).

Suyama, M. (1958a) Changes in chemical composition during the maturation of rainbow trout eggs. *Bull. Jap. Soc. scient. Fish.,* **24**, 656–9. (In Japanese, English summ.)

Suyama, M. (1958b) Changes in the amino acid composition of protein during the development of rainbow trout eggs. *Bull. Jap. Soc. scient. Fish.,* **23**, 789–92.

Suyama, M. and Ogino, C. (1958) Changes in chemical composition during development of rainbow trout eggs. *Bull. Jap. Soc. scient. Fish.,* **23**, 785–8.

Suzuki, R. (1974) Influence of environmental condition for rearing adult on percent fertilization and hatching in the loach, cyprinid fish. *The Aquiculture,* **22**, 72–7. (In Japanese, English summ.)

Suzuki, R. (1975) Duration of development capacity of eggs after ovulation in the loach, cyprinid fish. *The Aquiculture,* **23**, 93–9. (In Japanese, English summ.)

Suzuki, R. (1976) Number of ovarian eggs and spawned eggs and their size composition in the loach, cyprinid fish. *Bull. Jap. Soc. scient. Fish.,* **42**, 961–7. (In Japanese, English summ.)

Suzuki, R. (1983) Multiple spawning of the cyprinid loach, *Misgurnus anguillicaudatus. Aquaculture,* **31**, 233–43.

Svärdson, G. (1949) Natural selection and egg number in fish. *Rep. Inst. Freshwat. Res. Drottningholm,* **29**, 115–22.

Swift, D. (1965) Effect of temperature on mortality and rate of development of the eggs of pike and perch. *Nature,* **206**, 528.

Szczerbowski, J. (1966) Relation between the body length of salmon trouts females (*Salmo trutta m. lacustris* L.) and the diameter of its eggs. *Zool. pol.,* **16**, 195–201.

Szlamińska, M. (1984) Studium tempa wzrostu larw karpia (*Cyprinus carpio* L.) żywionych zmodyfikowanymi paszami. Celuloza jako balast w starterach karpiowych [Growth rate of carp (*Cyprinus carpio* L.) larvae fed the modified dry feeds diets. Cellulose as a ballast in carp starters]. PhD thesis, Inland Fisheries Institute, Żabieniec. (58 pp.) (In Polish).

Szlamińska, M. (1987a) Survival, growth and size structure of carp. (*Cyprinus carpio* L.) larvae populations fed dry feed Ewos Larvstart C-10 or zooplankton or else starved. *Pol. Arch. Hydrobiol.,* **34**, 331–46.

Szlamińska, M. (1987c), Intestine evacuation rate in cyprinid larvae fed on two kinds of dry feeds at 24°C. *Acta Ichthyol. et Piscatoria,* **17**, 35–41.

Szlamińska, M. and Przybył, A. (1986) Feeding of carp (*Cyprinus carpio* L.) larvae with an artificial dry food, living zooplankton and mixed food. *Aquaculture*, **54**, 77–82.

Szlamińska, M. (1988) Przegląd badań nad podchowem ryb karpiowatych w warunkach sztucznych [A review of studies on cyprinid fish rearing in artificial conditions]. *Roczn. Nauk roln.*, *Ser. H*, 101 **4**, 85–109 (In Polish, English summ.)

Szlamińska, M., Kamler, E., Bergot, P., Barska, B., Jakubas, M., Kuczyński, M. and Raciborski, K. (1989) The course of relationship between weight and length during development of carp (*Cyprinus carpio* L.) larvae. *Pol. Arch. Hydrobiol.*, **36**, 149–60.

Szubińska, B. (1961) Further observations on the morphology of the yolk in Teleostei. *Acta biol. cracov.*, *Ser. Zool.*, **4**, 1–19.

Szubińska-Kilarska, B. (1959) The morphology of the yolk in certain Salmonidae. *Acta biol. cracov.*, *Ser. Zool.*, **2**, 97–111.

Talbot, C. (1985) Laboratory methods in fish feeding and nutritional studies, in *Fish Energetics: New Perspectives* (eds P. Tytler and P. Calow), Croom Helm, London, pp. 125–154.

Talbot, C. and Higgins, P. J. (1983) A radiographic method for feeding studies on fish using metallic iron powder as a marker. *J. Fish Biol.*, **23**, 211–20.

Taniguchi, A.K. (1981) Survival and growth of larval spotted seatrout (*Cynoscion nebulosus*) in relation to temperature, prey abundance and stocking densities. *Rapp. P.-v. Réun. Cons. perm. int. Explor. Mer*, **178**, 507–8.

Tashiro, F., Tachikawa, W., Kamata, T., Tamura, E., Aoe, H. and Yabe, K. (1974) *Rainbow Trout* (*Fish Culture Course*, No. 10), Midori Shobo, Tokyo.

Tatarko, K.I. (1965) Effect of temperature on the embryonic development of pond carp. *Gidrobiol. Zh.*, **1**(1), 62–6. (In Russian, English summ.)

Tatarko, K.I. (1966) Effect of temperature on the early stages of postembryonic development of pond carp. *Gidrobiol. Zh.*, **2**(3), 53–9. (In Russian, English summ.)

Tatarko, K.I. (1970) Sensitivity of pond carp to high temperature at early stages of post-embryonic development. *Gidrobiol. Zh.*, **6**(2), 102–5. (In Russian, English summ.)

Terner, C. (1979) Metabolism and energy conversion during early development, in *Fish Physiology*, Vol. 8 (eds W. S. Hoar, D. J. Randall and J. R. Brett) Academic Press, NY, pp. 261–78.

Thorpe, J. E. (1990) Variations in life-history strategy in salmonids. *Pol. Arch. Hydrobiol.*, **37**, 3–12.

Thorpe, J. E., Miles, M. S. and Keay, D. S. (1984) Developmental rate, fecundity and egg size in Atlantic salmon, *Salmo salar* L. *Aquaculture*, **43**, 289–305.

Toetz, D. W. (1966) The change from endogenous to exogenous sources of energy in bluegill sunfish larvae. *Invest. Indiana Lakes and Streams*, **7**, 115–46.

Tomita, M., Iwahashi, M. and Suzuki, R. (1980) Number of spawned eggs and ovarian eggs and egg diameter and per cent eyed eggs with reference to the size of female carp. *Bull. Jap. Soc. scient. Fish.*, **46**, 1077–81. (In Japanese, English summ.)

Townsend, C. R. and Winfield, I. J. (1985) The application of the optimal foraging theory to feeding behaviour in fish, in, *Fish Energetics: New Perspectives* (eds P. Tytler and P. Calow), Croom Helm, London, pp. 67–98.

Townshend, T. J. and Wootton, R. J. (1984) Effects of food supply on the reproduction of the convict cichlid, *Cichlasoma nigrofasciatum*. *J. Fish Biol.*, **24**, 91–104.

Trifonova, A. N. (1949) Critical periods in embryonic development. *Usp. sovrem. Biol.*, **28**, 154–8. (In Russian)

Trojan, P. (1985) *Bioklimatologia Ekologiczna [Ecological Bioclimatology]*, Państwowe Wyd. Naukowe, Warsaw. (In Polish)

Trzebiatowski, R. and Domagała, J. (1986) Skład chemiczny ikry troci rzeki Regi w okresie tarła [Chemical composition of eggs of sea trout from Rega River during spawning]. *Abstracts, 13th Congr. Polish Hydrobiol. Soc., Szczecin, 16–19 September 1986*. Pol. Tow. Hydrobiol. and Akademia Rolnicza, Szczecin, pp. 216–7. (In Polish)

Tytler, P. and Calow, P. (eds) (1985) *Fish Energetics: New Perspectives*, Croom Helm, London.

Urban, E. (1982) Energy balance and nitrogen changes of grass carp (*Ctenopharyngodon idella* Val.) larvae and fry under different food conditions. *Ekol. pol.*, **30**, 327–91.

Urban, E. (1984) Energy balance and nitrogen changes of grass carp (*Ctenopharyngodon idella* Val.) larvae and fry under different conditions. II. Assimilation and balance of nitrogen compounds. *Ekol. pol.*, **32**, 341–82.

Urban-Jezierska, E., Kamler, E. and Barska, B. (1984) Dry matter and nitrogen compounds leached from six carp starters. *Pol. Arch. Hydrobiol.*, **31**, 119–34.

Urban-Jezierska, E., Kamler, E., Jezierski, H., Barska-Klyta, B., Lewkowicz, M. and Lewkowicz, S. (1989) Evaluation of natural foods and starters for carp larvae from the angle of nitrogen assimilation. *Pol. Arch. Hydrobiol.*, **36**, 231–62.

Van der Wind, J. J. (1979) Feeds and feeding in fry and fingerling culture. *EIFAC Tech. Pap.* No. 35 (Supp. 1), 59–72.

Vasileva, L. P. and Averyanova, V. V. (1981) Fecundity and biochemical composition of *Coregonus* eggs from Topo–Pyaozerskoe dam reservoir, in, *Biologicheskie Resursy Belogo Morya i Vnutrennykh Vodoemov Evropejskogo Severa* (ed. L. P. Ryzkhov), Minist. Rybnogo Khoz. SSSR, Akademiya Nauk SSSR, Petrozavodsk, pp. 103–4. (In Russian)

Vasnetsov, V. V. (1953) Developmental stages of bony fishes, in, *Ocherki po Obshchim Voprosam Ikhtiologii*, (ed. E. N. Pavlovskij) Akademiya Nauk Press, Moscow, pp. 207–17. (In Russian)

Vasnetsov, V. V., Eremeeva, E. F., Lange, N. O., Dmitrieva, E. N. and Braginskaya, R. Ya. (1957) Developmental stages of commercial fishes of Volga and Don – *Abramis brama, Cyprinus carpio, Rutilus rutilus caspicus, Rutilus rutilus heckeli* and *Lucioperca lucioperca. Trudy Inst. Morf. Zhivot.*, **16**, 7–76 (In Russian)

Velsen, F. P. J. (1987) Temperature and incubation in Pacific salmon and rainbow trout: compilation of data on median hatching time, mortality and embryonic staging. *Can. Data Rep. Fish. aquat. Sci.* No. 625, 58 pp.

Vernier, J. -M. (1969) Table chronologique du dévelopment embryonnaire de la Truite Arc-en-ciel, *Salmo gairdneri* Rich. 1836. *Annls Embryol. Morphogen.*, **2**, 495–520.

Vetter, R. D., Hodson, R. E. and Arnold, C. (1983) Energy metabolism in a rapidly developing marine fish egg, the red drum (*Sciaenops ocellata*). *Can. J. Fish. aquat. Sci.*, **40**, 627–34.

Viljanen, M. (1988) Relations between egg and larval abundance, spawning stock and recruitment in vendace (*Coregonus albula* L.). *Finn. Fish. Res.*, **9**, 271–89.

Vinogradov, A. K. (1973) Biochemical and energetical properties of spawn of some Black Sea fishes, in, *Ekologicheskaya Fiziologiya Ryb (Tezisy Dokl. Vsesoyuzn. Konf. po Ekologicheskoj Fiziologii Ryb, Moskva, 1973)* (ed. N. S. Stroganov), Minist. Rybnogo Khoz. SSSR, Moscow, pp. 147–49. (In Russian)

Vladimirov, V. I. (1965a) *Influence of Spawners Quality on Fish Offspring*, Naukova Dumka, Kiev. (In Russian)

Vladimirov, V. I. (1965b) Influence of feeding state of fish females on their offspring quality during early life, in *Vliyanie Kachestva Proizvoditelej na Potomstwo u Ryb* (ed. V. I. Vladimirov), Naukova Dumka, Kiev, pp. 35–93. (In Russian)

Vladimirov, V. I. (1970) Difference of ontogeny quality as one of factors in dynamics of fish school quantity. *Gidrobiol. zh.*, **6**(2), 14–27. (In Russian, English summ.)

Vladimirov, V. I. (1973) Relation of carp offspring quality on age of females, on amino acid content in eggs and addition of amino acids to water at the beginning of development, in *Ekologicheskaya Fiziologiya Ryb (Tezisy Dokl. Vsesoyuzn. Konf. po Ekologicheskoj Fiziologii Ryb, Moskva, 1973)* (ed. N. S. Stroganov), Minist. Rybnogo Khoz. SSSR, Moscow, pp. 93–5. (In Russian).

Vladimirov, V. I. (1974a) Size variability and survival of fish early developmental stages, in *Raznokachestvennost Rannego Ontogeneza u Ryb* (ed. V. I. Vladimirov), Naukova Dumka, Kiev, pp. 227–54. (In Russian)

Vladimirov, V. I. (1974b) *Variability of Fish Early Ontogenesis*, Naukova Dumka, Kiev. (In Russian)

Vladimirov, V. I., Semenov, K. I. and Zhukinskij, V. N. (1965) Quality of spawners and viability of offspring at the early life stages in some fish species, in, *Teoreticheskie Osnovy Rybovodstva*. Nauka, Moscow, pp. 19–32. (In Russian)

Vovk, P. S. (1974) Reactions of grass carp embryos and larvae on temperature, in, *Raznokachestvennost Rannego Ontogeneza u Ryb* (ed. V. I. Vladimirov), Naukova Dumka, Kiev, pp. 191–226. (In Russian)

Wallace, R. A. and Selman, K. (1981) Cellular and dynamic aspects of oocyte growth in teleosts. *Am. Zool.*, **21**, 325–43.

Walzer, C. and Schönenberger, N. (1979) Ultrastructure and cytochemistry study of the yolk syncytial layer in the alevin of trout (*Salmo fario trutta* L.) after hatching. *Cell Tissue Res.*, **196**, 59–73.

Ware, D. M. (1975) Relation between egg size, growth and natural mortality of larval fish. *J. Fish. Res. Bd Can.*, **32**, 2503–12.

Ware, D. M. (1982) Power and evolutionary fitness of teleosts. *Can. J. Fish. aquat. Sci.*, **39**, 3–13.

Watanabe, T., Arakawa, T., Kitajima, C., Fukusho, K. and Fujita, S. (1978) Proximate and mineral compositions of living feeds used in seed production of fish. *Bull. Jap. Soc. Scient. Fish.* **44**, 979–84.

Watanabe, T., Kitajima, C. and Fujita, S. (1983) Nutritional values of live organisms used in Japan for mass propagation of fish: a review, *Aquaculture*, **34**, 115–43.

Welch, H. E. (1968) Relationships between assimilation efficiencies and growth efficiencies for aquatic consumers. *Ecology*, **49**, 755–9.

Wickett, W. P. (1954) The oxygen supply to salmon eggs in spawning beds. *J. Fish Res. Bd Can.*, **11**, 933–53.

Wieser, W. and Forstner, H. (1986) Effects of temperature and size on the routine rate of oxygen consumption and on the relative scope for activity in larval cyprinids. *J. comp. Physiol.*, **156B**, 791–6.

Wieser, W., Forstner, H., Medgyesy, N. and Hinterleitner, S. (1988a) To switch or not to switch: partitioning of energy between growth and activity in larval cyprinids (Cyprinidae: Teleostei). *Funct. Ecol.*, **2**, 499–507.

Wieser, W., Forstner, H., Schiemer, F. and Mark, W. (1988b) Growth rates and growth efficiencies in larvae and juveniles of *Rutilus rutilus* and other cyprinid species: effects of temperature and food in the laboratory and in the field. *Can. J. Fish. aquat. Sci.*, **45**, 943–50.

Wilbur, H. M. and Collins, J. P. (1973) Ecological aspects of amphibian metamorphosis. *Science*, **182**, 1305–14.

Wilkońska, H. and Żuromska, H. (1982) Effect of environmental factors and egg quality on the mortality of spawn in *Coregonus albula* (L.) and *Coregonus lavaretus* (L.). *Pol. Arch. Hydrobiol.*, **29**, 123–57.

Wilkońska, H. and Żuromska, H. (1988) Effect of environment on *Coregonus albula* (L.) spawners and influence of their sexual products on the numbers and quality of offspring. *Finn. Fish. Res.*, **9**, 81–8.

Winberg, G. G. (1956) Rate of metabolism and food requirements of fishes. Belorussian State Univ. Minsk; *Fish. Res. Bd Can. Transl. Ser.*, No. 194 (1960), pp. 1–253.

Winberg, G. G. (ed.) (1971) *Methods for the Estimation of Production of Aquatic Animals*, Academic Press, London.

Winberg, G. G. (1983) Van't Hoff temperature coefficient and Arrhenius equation in biology. *Zh. Obshch. Biol.* **44**, 31–42 (In Russian)

Winberg, G. G. (1987) Effect of temperature on the rate of ontogenetic development, in, *Produktsionno-gidrobiologicheskie Issledovaniya Vodnykh Ekosistem* (ed. A. F. Alimov) *Trudy Zool. Inst. A. N. SSSR*, **165**, 5–34. (In Russian)

Winberg, G. G. and Khartova, L. E. (1953) Metabolic rate in carp fry. *Dokl. Akad. Nauk SSSR*, **89**, 1119–22. (In Russian)

Winberg, G. G. and Collaborators (eds) (1971) *Symbols, Units and Conversion Factors in Studies of Fresh Water Productivity*, IBP Central Office, London, 23 pp.

Winnicki, A. (1960) Waterless environment as a factor causing a prolongation of embryonic development of the brook trout (*Salmo fontinalis* (Mitchill.)). *Zool. pol.*, **9**, 131–9.

Winnicki, A. (1967) Embryonic development and growth of *Salmo trutta* L. and *Salmo gairdneri* Rich. in conditions unfavourable to respiration. *Zool. pol.*, **17**, 45–58.

Winnicki, A. (1968) Respiration of the embryos of *Salmo trutta* L. and *Salmo gairdneri* Rich. in media differing in gaseous diffusion rate. *Pol. Arch. Hydrobiol.*, **5**, 23–38.

Woodward, I. O. and White, R. W. (1981) Effects of temperature and food on the fecundity and egg development rates of *Boeckella symmetrica* Sars (Copepoda: Calanoida). *Aust. J. mar. Freshwat. Res.*, **32**, 997–1002.

Wootton, R. J. (1977) Effect of food limitation during the breeding season on the size, body components and egg production of female sticklebacks (*Gasterosteus aculeatus*). *J. Anim. Ecol.*, **46**, 823–34.

Wootton, R. J. (1979) Energy costs of egg production and environmental determinants of fecundity in teleost fishes. *Symp. zool. Soc. Lond.*, **44**, 133–59.

Wootton, R. J. (1982) Environmental factors in fish reproduction, in *Reproductive Physiology of Fish* (eds C. J. J. Richter and H. J. Th. Goos), Pudoc, Wageningen, pp. 210–19.

Wootton, R. J. (1984) Introduction: strategies and tactics in fish reproduction, in, *Fish Reproduction: Strategies and Tactics* (eds G. W. Potts and R. J. Wootton), Academic Press, London, pp. 1–12.

Wootton, R. J. (1985) Energetics of reproduction, in, *Fish Energetics: New Perspectives* (eds P. Tytler and P. Calow), Croom Helm, London, pp. 231–54.

Wootton, R. J. (1986) Problems in the estimation of food consumption and fecundity in fish production studies. *Pol. Arch. Hydrobiol.*, **33**, 263–76.

Wootton, R. J., Allen, J. R. M. and Cole, S. J. (1980) Energetics of the annual reproductive cycle in female sticklebacks, *Gasterosteus aculeatus* L. *J. Fish Biol.*, **17**, 387–94.

Wourms, J. P. and Evans, D. (1974) The embryonic development of the black prickleback, *Xiphister atropurpureus*, a Pacific coast blenninoid fish. *Can J. Zool.*, **52**, 879–87.

Wyatt, T. (1972) Some effects of food density on the growth and behaviour of plaice larvae. *Mar. Biol.*, **14**, 210–16.

Yamashita, Y. and Aoyama, T. (1985) Hatching time, yolk sac absorption, onset of feeding, and early growth of the Japanese sand eel *Ammodytes personatus*. *Bull. Jap. Soc. scient. Fish.*, **51**, 1777–80.

Yarzhombek, A. A. (ed.) (1986) *Fish Physiology Tables*, Agropromizdat, Moscow. (In Russian)

Zaika, V. E. (1972) *Specific Production of Aquatic Invertebrates*, Naukova Dumka, Kiev.

Zalewski, M. (1986) Regulacja zespołów ryb w rzekach przez kontinuum czynników abiotycznych i biotycznych [Regulation of fish communities in rivers by a continuum of abiotic and biotic factors]. Acta Univ. Lodziensis 1986, Dr hab. thesis, 7–86. (In Polish)

Zalewski, M. and Naiman, R. J. (1985) The regulation of riverine fish communities by a continuum of abiotic–biotic factors, in, *Habitat Modification and Freshwater Fisheries* (ed. J. S. Alabaster), Butterworth, London, pp. 3–9.

Zalicheva, I. N. (1981) Embryonic development of *Salmo trutta labrax* in different temperatures, in, *Biologicheskie Resursy Belogo Morya i Vnutrennykh Vodoemov Evropejskogo Severa* (ed. L. P. Ryzhkov), Minist. Rybnogo Khoz. SSSR, Akademiya Nauk SSSR, Petrozavodsk, pp. 107–8. (In Russian)

Zawisza, J. and Backiel, T. (1970) Gonad development, fecundity and egg survival in *Coregonus albula*, in, *Biology of Coregonid Fishes* (eds C. C. Lindsey and C. S. Woods), Univ. Manitoba Press, Winnipeg, pp. 363–97.

Zdor, V. I., Yarzhombek, A. A. and Mikheev. V. P. (1978) Effect of collagen on formation of amino acid composition in *Cyprinus carpio* and *Acipenser ruthenus*. *Sb. nauch. trudov VNIIPRKHa*, **21**, 156–70. (In Russian)

Zeuthen, E. (1955) Comparative physiology (respiration). *A. Rev. Physiol.*, **17**, 459–82.

Zeuthen, E. (1970) Rate of living as related to body size in organisms. *Pol. Arch. Hydrobiol.*, **17**, 21–30.

Zhukinskij, V. N. (1965) Dependence of reproductive product quality and of embryo viability on spawner age in *Rutilus rutilus heckeli*, in, *Vliyanie Kachestva Proizvoditelej na Potomstwo u Ryb* (ed. V. I. Vladimirov), Naukova Dumka, Kiev, pp. 94–122. (In Russian)

Zhukinskij, V. N. (1981) Preface, in, *Raznokachestvennost Ontogeneza u Ryb* (ed. V. D. Romanenko), Naukova Dumka, Kiev, pp. 3–6. (In Russian)

Zhukinskij, V. N. (1986) *Influence of Abiotic Factors on Variability and Viability of Fishes During Early Onthogenesis*, Agropromizdat, Moscow. (In Russian)

Zhukinskij, V. N. and Gosh, R. I. (1970) Viability of the embryos depending on the respiration intensity of ovulated spawn in *Rutilus rutilus heckeli* (Nordmann) and *Abramis brama* L. of different age. *Gidrobiol. Zh.*, **6**(4), 60–70. (In Russian, English summ.)

Zhukinskij, V. N. and Gosh, R. I. (1974) Viability of *Rutilus rutilus heckeli* and *Abramis brama* embryos as related to metabolic rate of ovulated eggs and sperm produced by spawners of different age, in, *Raznokachestvennost Rannego Ontogenea u Ryb* (ed. V. I. Vladimirov), Naukova Dumka, Kiev, pp. 7–64. (In Russian)

Zhukinskij, V. N. and Gosh, R. I. (1988) Biological quality of ovulated eggs in commercial fishes: criteria and method of express-test. *Vop. Ikhtiol.*, **28**, 773–81. (In Russian)

Zhukinskij, V. N. and Nedyalkov, G. F. (1980) Endogenous differentiation of early ontogenesis quality as one of the factors in dynamics of fish reproduction (exemplified by *Ctenopharyngodon idella* (Val.) and *Cyprinus carpio* L.). *Gidrobiol. Zh.* **16** (2), 57–71. (In Russian, English summ.)

Zhukinskij, V. N., Gosh, R. I., Konovalov, Yu. D., Kim, E. D. and Kovtun, E. I. (1981) Physiological and biochemical aspects of overripeness and resorption of ripe ova in *Rutilus rutilus heckeli* and *Abramis brama*, in, *Raznokachestvennost Ontogeneza u ryb* (ed. V. D. Romanenko), Naukova Dumka, Kiev, pp. 85–126. (In Russian)

Zotin, A. I. (1954) Mechanism of production of perivitelline fluid by eggs of salmonid fishes. *Dokl. Akad. Nauk SSSR*, **96**, 421–4. (In Russian)

Żuromska, H. (1982) Egg mortality and its causes in *Coregonus albula* (L.) and *Coregonus lavaretus* (L.) in two Masurian lakes. *Pol. Arch. Hydrobiol.*, **29**, 29–69.

Żuromska, H. and Markowska, J. (1984) The effect of sexual products quality on offspring survival and quality in tench (*Tinca tinca* L.). *Pol. Arch. Hydrobiol.*, **31**, 287–313.

Species Index

(Non-fish species are marked with an *)

Subject Index